Environmental Chemistry of Aerosols

Environmental Chemistry of Aerosols

Edited by

Ian Colbeck
Director
Centre for Environment and Society
University of Essex
UK

Blackwell
Publishing

Blackwell Publishing editorial offices:
Blackwell Publishing Ltd, 9600 Garsington Road, Oxford OX4 2DQ, UK
Tel: +44 (0)1865 776868
Blackwell Publishing Professional, 2121 State Avenue, Ames, Iowa 50014-8300, USA
Tel: +1 515 292 0140
Blackwell Publishing Asia Pty Ltd, 550 Swanston Street, Carlton, Victoria 3053, Australia
Tel: +61 (0)3 8359 1011

First published 2008 by Blackwell Publishing Ltd

ISBN: 978-1-4051-3919-9

Library of Congress Cataloging-in-Publication Data
Environmental chemistry of aerosols / edited by Ian Colbeck. — 1st ed.
p. cm.
Includes bibliographical references and index.
ISBN-13: 978-1-4051-3919-9 (hardback : alk. paper)
ISBN-10: 1-4051-3919-6 (hardback : alk. paper) 1. Aerosols. 2. Aerosols—Environmental aspects. 3. Atmospheric chemistry. 4. Air—Pollution—Environmental aspects. 5. Environmental chemistry. I. Colbeck, I. (Ian).

QC882.42.E58 2008
551.51′13—dc22

2007042414

A catalogue record for this title is available from the British Library

Set in 10/12 pt Minion by Aptara Inc., New Delhi, India
Printed and bound in Singapore by Fabulous Printers Pte Ltd

The publisher's policy is to use permanent paper from mills that operate a sustainable forestry policy, and which has been manufactured from pulp processed using acid-free and elementary chlorine-free practices. Furthermore, the publisher ensures that the text paper and cover board used have met acceptable environmental accreditation standards.

For further information on Blackwell Publishing, visit our website:
www.blackwellpublishing.com

Contents

List of Contributors

Professor Urs Baltensperger	Laboratory of Atmospheric Chemistry, Paul Scherrer Institute, Villigen, Switzerland
Dr Charles Clement	Physics and Astronomy Department at University College, London, UK
Professor Ian Colbeck	Centre for Environment and Society, University of Essex, Colchester, UK
Dr Markus Furger	Laboratory of Atmospheric Chemistry, Paul Scherrer Institute, Villigen, Switzerland
Dr Irena Grgić	National Institute of Chemistry, Ljubljana, Slovenia
Professor Ari Laaksonen	Department of Physics, University of Kuopio, Kuopio, Finland
Dr Mihalis Lazardis	Department of Environmental Engineering, Technical University of Crete, Polytechneioupolis, Chania, Greece
Dr Kari Lehtinen	Department of Physics, University of Kuopio, Kuopio and Finnish Meteorological Institute, Kuopio Unit, Finland
Dr Rob MacKenzie	Department of Environmental Science, Lancaster University, Lancaster, UK
Dr David Topping	School of Earth, Atmospheric and Environmental Science, The University of Manchester, Manchester, UK
Professor Jay Turner	Environmental Engineering Program, Washington University, St Louis, Missouri, USA

Preface

Aerosol particles are ubiquitous in the earth's atmosphere and are central to many environmental issues such as climate change, stratospheric ozone depletion and air quality. In urban environments, aerosol particles can affect human health through their inhalation. A recent assessment of health damages from exposure to the high levels of particulates in 126 cities worldwide reveals that these damages may amount to near 130 000 premature deaths each year.

Atmospheric aerosols originate from either naturally occurring processes or anthropogenic activity. Major natural aerosol sources include volcanic emissions, sea spray and mineral dust emissions, while anthropogenic sources include emissions from industry and combustion processes. Understanding the ways in which aerosols behave, evolve and exert these effects requires knowledge of their formation and removal mechanisms, transport processes, as well as their physical and chemical characteristics. The chemical composition of aerosol particles varies dramatically in different localities. In urban areas, anthropogenic particles contribute more compared to rural areas, whereas in coastal areas, sea-salt particles contribute more. Aerosols present pathways for reactions, transport and deposition that would not occur in the gas phase alone. It is recognised that current knowledge of heterogeneous or multiphase chemistry, in which atmospheric aerosol particles participate, is noticeably less well understood than homogeneous gas-phase atmospheric chemistry. There have been significant developments recently due to advanced capabilities of instruments designed specifically to investigate fundamental chemical and physical processes critical to atmospheric chemistry. For example particle beam mass spectrometers being used in studies of particle formation in combustion systems, such as diesel engines, and for analysis of ambient atmospheric particles.

Motivated by climate change and adverse health effects of traffic-related air pollution, aerosol research has increasingly intensified over the past couple of decades. Numerous books have considered the physics of aerosols but few have addressed the chemistry of atmospheric aerosols. Recent scientific advances now offer an improved understanding of the mechanisms and factors controlling the chemistry of atmospheric aerosols.

This book intends to bring together the state-of-the-art knowledge of environmental aerosol chemistry with each chapter written by a leader in the field. It has not been possible to cover all the topics that might find a place in a comprehensive coverage of atmospheric aerosols. The book presents a coherent structure on which to base the study of environmental chemistry of aerosols. It is intended to be used as a resource, identifying the themes and relevant concepts required.

Chapter 1 summarises key physical and chemical properties of atmospheric aerosols, while Chapters 2 and 3 describe the key processes in the formation of new atmospheric particles and their subsequent growth to larger sizes. Recent studies have shown that atmospheric aerosols are a complex mixture and that the organic fraction can comprise hundreds of species, representing a wide variety of compound classes and origins. This is discussed in Chapter 4, while Chapter 5 considers the role of metals in atmospheric aerosol chemistry. Chapter 6 provides an overview of aerosol–water vapour interactions and equilibrium thermodynamics. The final two chapters offer examples of environmental aerosols. Chapter 7 gives a general sense of the behaviour of the stratospheric aerosol layer and its impact on environmental chemistry. Finally, Chapter 8 describes aerosol in remote locations and highlights the need for harmonised procedures worldwide.

I thank Stephan Nyeki, Ian Ford, Jonathan Barrett, Helmuth Horvath, Lucy Carpenter and Simon Clegg for their constructive comments on the various draft chapters. The staff of Wiley-Blackwell, especially Sarahjayne Sierra, have been supportive throughout.

Ian Colbeck

Chapter 1

Physical and Chemical Properties of Atmospheric Aerosols

Jay Turner and Ian Colbeck

1.1 Introduction

Particles in the atmosphere are distinguished by their size, shape and composition. They can be directly emitted from sources and can be formed in the atmosphere by chemical reactions and physical processes. Once particles are formed, their properties can be modulated in space and time by atmospheric physical and chemical processes, such as condensation, evaporation and coagulation. Eventually, the particles are removed from the atmosphere by wet or dry deposition, with such removal occurring minutes to weeks after their release or formation, and after travelling metres to thousands of kilometres. This chapter summarises key physical and chemical properties of atmospheric aerosols. These properties can be described at various levels of aggregation (bottom-up, from the single-particle perspective) or discretisation (top-down, from the aerosol perspective). An accounting of each particle by size and composition would provide the finest resolution. In contrast, the bulk composition over all particles would be the coarsest resolution. Furthermore, given the dynamic nature of atmospheric aerosols, the degree of time integration is important.

Particles in the atmosphere are directly emitted by sources (primary emissions) or formed in the atmosphere by physicochemical processes (secondary formation) [1]. Combustion and other high-temperature processes are largely responsible for primary emissions of fine-mode particles, while mechanical processes such as grinding, entrainment of dust and soil and droplet formation by waves generate coarse-mode particles. Condensable gases can homogeneously nucleate to form new particles in the atmosphere if there is insufficient surface area for their uptake. Heterogeneous condensation is the uptake of vapour by pre-existing particles, with the reverse process of evaporation also important.

A summary of the atmospheric aerosol properties must necessarily reflect upon the methods used to arrive at the current state of knowledge. As tools to probe the atmosphere continue to advance, refinements will be made to the understanding of atmospheric aerosols. Starting with the dimensions of size, composition, space and time, observational data are collected from measurements that integrate across these dimensions. Most observational data have been collected at fixed locations, although more data are being obtained from satellite and other profiling instruments that provide spatial resolution [2, 3]. The spatial zone represented by monitoring at a fixed site will depend on the property of interest. It may be possible to infer

Table 1.1 Example of monitoring networks involving aerosol composition measurements in the US

Network	Start year	Number of sites	Measurements
Clean Air Status and Trends Network (CASTNet)	1990	87	Sampling schedule: filters exposed over 7 days to produce 7-day averages (Tuesday to Tuesday) Gaseous: SO_2, HNO_3, O_3 Particulate: SO_4^{2-}, NO_3^-, NH_4^+, other relevant ions
Interagency Monitoring of Protected Visual Environments (IMPROVE)	1987	Over 160	Sampling schedule: 24 h every third day Particulate: element and organic C, SO_4^{2-}, NO_3^-, Cl^-, elements between Na and Pb PM_{10} and $PM_{2.5}$ (mass) Selected sites: light scattering and/or extinction coefficient

the zone of representation from temporal patterns in the data, but in most cases the zone of representation is determined by monitoring at several sites over the spatial scale of interest, such as an urban area. This is tractable for relatively simple measurements but is not tractable for more advanced, research measurements.

Time resolution of the measurements also shapes the understanding of dynamical aspects of aerosol properties. Routine networks for particle mass and, more recently, composition typically collect samples over 24-h periods with frequency of every day to once every several days. Continuous monitoring of particle mass concentration is now common with data typically reported as hourly averages [4]. This is very important for real-time health advisory reporting and to bring the particle mass measurements closer to the characteristic timescales for variations in the weather, emission rates and atmospheric processes that drive variations in particle properties.

Routine measurements are integrated across size and composition. Routine networks typically sample particles in one size range ($PM_{2.5}$, PM_{10}: particulate matter (PM) below a specified size; e.g. 10 refers to 10 μm) and report the bulk composition of the collected particles [5]. The composition measurements are incomplete; some compounds are not measured (e.g. water) and some components are characterised by lumped parameters (e.g. carbonaceous matter) rather than a detailed accounting of each compound. Table 1.1 shows the parameters reported for two routine particle composition networks operated in the United States. In Europe speciation of ambient PM is mainly focused on determination of the secondary inorganic aerosol constituents, sulphate (SO_4^{2-}), nitrate (NO_3^-) and ammonium (NH_4^+) via EMEP (Co-Operative Programme for Monitoring and Evaluation of the Long-Range Transmission of Air Pollutants in Europe) stations [6]. The carbonaceous content of the aerosols are measured only at a very few stations, but has been extensively addressed during measurements campaign, such as the EMEP Elemental/Organic Carbon campaign [7].

Averaging takes place not only in data collection but also in data reduction and analysis. Sites in routine networks with 24-h sampling are typically compared as annual or seasonal averages. Data collection at higher time resolution is often expressed as diurnal profiles, including weekend versus weekday differences [8, 9]. Finally, there can be cataloguing of the frequency of certain types of episodes or events.

Capabilities have existed for some time to measure the size of individual particles [10]. Small particles are typically differentiated by their mobility in an electric field, while larger particles are differentiated by differences in their light scattering or aerodynamic behaviour. There are subtle issues involved in each of these measurements, which should be considered when interpreting the data. Instruments that size, count and determine the chemical composition of individual particles have been developed over the past 15 years. Various single-particle spectrometers now exist, which couple a size measurement to a composition measurement [11, 12]. There are limitations to the size ranges, composition measurements and counting statistics, and significant data aggregation is needed to make the analysis and interpretation tractable. However, these approaches do collect a spectrum for each particle and the analyst can decide how to best reduce the data to fit the project needs.

This chapter summarises the key physical and chemical properties of the ambient aerosol. Emphasis is placed on fine PM in urban areas, as viewed from observational data (rather than modelling). It is beyond the scope of this chapter to exhaustively summarise the myriad features that are observed on the various spatial and temporal scales. Instead, examples are used to introduce the key characteristics with the caveat that emissions, atmospheric dynamics, weather and other factors that influence aerosol chemical and physical properties can lead to significant differences between settings.

1.2 Ambient aerosol size distributions

Many physical properties of ambient aerosols can be estimated from their size distribution and chemical properties [13–15]. Examples include light scattering and absorption properties that are important in both visibility and climate change contexts. This section summarises key features of atmospheric aerosol size distributions.

1.2.1 Idealised size distributions of aerosol number and mass

An aerosol in which all particles are of the same size is termed monodisperse. Such an aerosol is rare and those generated in the laboratory for instrument calibration have a spread in particle diameter of a few percent. Typically, aerosols contain a wide range of sizes and are termed polydisperse. Because of the presence of a range of sizes the problem arises of how to describe the size distribution. In the atmosphere particle size is determined by the formation process and subsequent physical and chemical reactions.

In practice, the form of size distributions, although containing the same modes, can look very different according to whether the distribution is plotted as number distribution or mass distribution. The reason for this is that particles at the small end of the size distribution can be very abundant in number, but because mass depends upon the cube of diameter, such particles may contribute only a small amount of the total mass. Hence, a size distribution expressed as the number of particles per size fraction will give far more emphasis to the smaller particles than will occur in a distribution expressed by mass per size fraction.

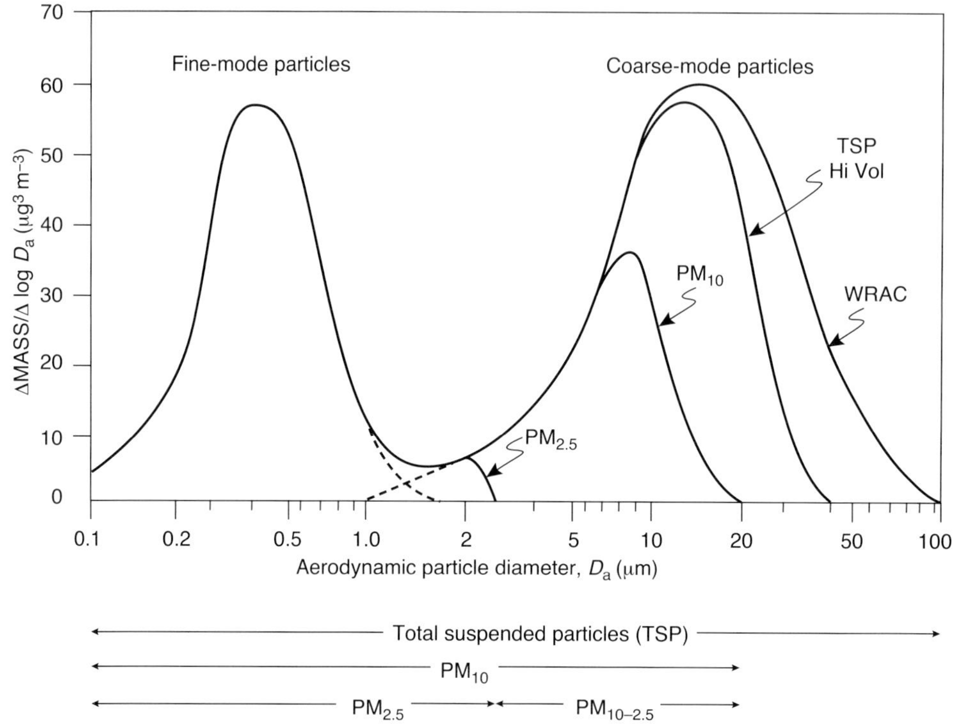

Figure 1.1 Aerosol mass distributions for an idealised ambient urban aerosol. Sampler cut points are also shown. (Reproduced from [16] with permission. Copyright Elsevier 2002.)

Figure 1.1 shows the distribution of aerosol mass as a function of particle size for an idealised urban aerosol. There are two modes – a fine mode and a coarse mode – with particles in these modes generally originating from different source types and having different atmospheric fates (to be discussed later in this chapter). The distribution of aerosol mass and other properties between these two modes can vary in both space and time. The fine-mode particle can be further subdivided into nuclei and accumulation modes. Figure 1.2 shows the aerosol volume distribution for an idealised traffic aerosol. Here there are three modes. The smallest one is the nucleation mode whereby particles have recently been emitted from processes involving condensation of hot vapours, or freshly formed within the atmosphere by gas to particle conversion. Such particles account for the preponderance of particles by number, but because of their small size account for little of the total mass of airborne particles. Due to their high number concentration they are subject to rapid coagulation and/or condensation of vapours and they enter the accumulation mode. Here they generally account for most of the aerosol surface area and a substantial part of the aerosol mass. Accumulation-mode particles have a typical atmospheric lifetime of around 1–2 weeks, as removal by precipitation scavenging or dry deposition is inefficient. These particles are therefore involved in long-range transport. The coarse mode relates to particles greater than 2 μm and which are mainly formed by mechanical attrition processes, and hence soil dust, sea spray and many industrial dusts fall within this mode.

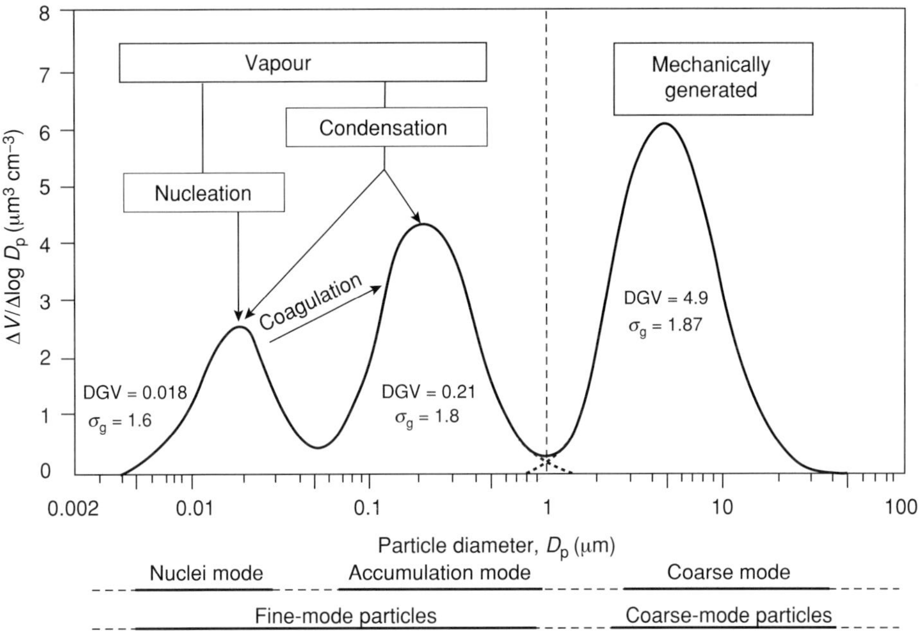

Figure 1.2 Aerosol volume distribution for traffic-dominated aerosol. Abbreviations: DGV, volume median diameter; σ_g, geometric standard deviation. (Reproduced from [16] with permission. Copyright Elsevier 2002.)

Several studies have demonstrated that size distributions of ambient aerosols can be represented by modes. A distinction must be made between aerosol characterisation based on size cuts and modes. For example, aerosol mass typically has a fine mode and a coarse mode with an overlap of these modes in the range of about 1–3 μm in diameter. For regulatory and indeed most research applications, however, size ranges are defined by the sampler characteristics. The extent to which $PM_{2.5}$, which itself is not a perfect size discriminator but rather depends on the sampler, represents fine-mode aerosol depends on the extent to which the coarse-mode particle extends below 2.5 μm in diameter. This is an important distinction that must be considered when describing the aerosol in terms of fine- and coarse-mode properties [17].

An examination of the evolution of PM air-quality standards in the U.S. demonstrates the relationships between commonly reported PM size cuts and the dominant aerosol mass modes. Starting in 1971, the Ambient Air Quality Standard (NAAQS) for PM was based on the mass concentration of total suspended particulate (TSP). In 1987, this indicator was replaced with PM_{10} because TSP levels were often driven by a few very large particles, while health effects generally correlated with smaller particles, indeed those particles in the respirable size range, which is generally characterised by PM_{10}. Finally, in the light of health effects data, United States Environmental Protection Agency revised the PM NAAQS in 1997 to include an indicator for the fine mode only – namely, a $PM_{2.5}$ standard. As shown in Figure 1.1, the samplers used to collect a given size fraction do not exhibit a sharp cut at the stated particle size nor do they rigorously represent a specific mode. For example, a PM_{10} sampler will collect a few particles as large as 20 μm and will not collect all particles in the size range 5–10 μm. For this reason, the

PM NAAQS are actually defined as the mass concentration for particles in a given size range as collected by a specific sampler design.

In summary, the following size ranges are defined:

• TSP	Total suspended particulate matter (up to ~35 μm in diameter)
• PM_{10}	Particulate matter smaller than 10 μm in diameter
• $PM_{2.5}$	Particulate matter smaller than 2.5 μm in diameter
• Fine PM	Particles smaller than about 1–3 μm in diameter; we shall operationally define as $PM_{2.5}$
• Coarse PM	Particles larger than 1–3 μm in diameter; we shall operationally define as PM_{10} minus $PM_{2.5}$ but it is actually the PM_{10} contribution to the coarse PM mode
• Ultrafine PM	Particles smaller than about 100 nm
• Nuclei mode	Particles in the range 5–100 nm
• Accumulation mode	Particles in the range 100 nm–2 μm

Additional size ranges may be reported with names such as 'inhalable' or 'respirable'.

Figure 1.3 shows typical distributions for aerosol number, surface area and volume as a function of particle size for ambient air. These three curves represent the same aerosol; we are merely considering different properties of this ensemble of particles as a function of particle size. The aerosol number concentration, N, is unimodal, with a peak around 0.01 μm. The aerosol volume concentration per unit volume of air, V, is bimodal, with a minimum in the range 1–3 μm. The idealised aerosol mass distribution of Figure 1.2 at least qualitatively resembles the volume distribution of Figure 1.3, with some differences arising from differences in particle density as a function of size because the composition is also changing as a function of size.

1.2.2 Size distribution measurements

The distribution of aerosol properties as function of size has significant implications to behaviour of the ensemble of particles. There are various approaches to characterising such behaviour. For a given property of interest, it might be possible to directly measure that property for size-selected aerosols. Alternatively, if the functional relationship between particle size and the property of interest is known, then a measurement of the number size distribution can be used to determine the ensemble behaviour. For this reason, this section focuses on number size distributions of atmospheric aerosol, although other size distributions (especially mass) are discussed.

Time-integrated sampling has been used to determine aerosol size distributions for a number of properties. In some cases, overlapping size ranges are sampled and the size distribution is determined by difference. For example, $PM_{2.5}$ and PM_{10} mass are measured at numerous sites with the coarse aerosol contribution to PM_{10} determined by difference (i.e. $PM_{10.0-2.5}$). The differencing method can be simple, but the propagated sampling and analytical uncertainties can become quite large in some cases. Another approach is size-selective sampling, where the sampler sorts the particles into size bins. For the example of determining the coarse aerosol contribution to PM_{10}, a dichotomous sampler can be used to separate the aerosol into fine particulate matter (PM_f) and coarse particulate matter (PM_c) bins. A correction must be applied for fine particle intrusion into the coarse particle bin, and typically the rate of change

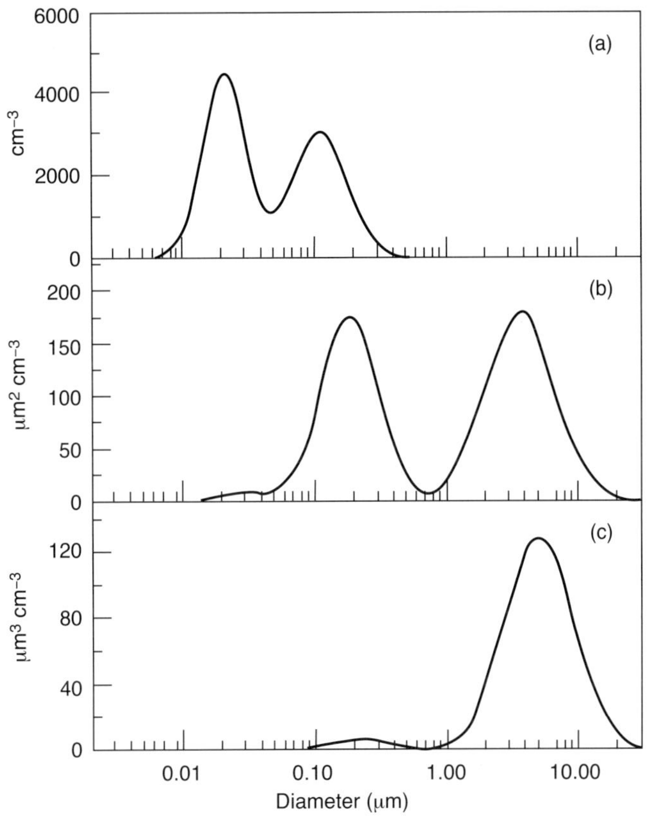

Figure 1.3 Number, surface and volume aerosol distributions.

of selection with size (the so-called sharpness of the size cut) is broader, which leads to some sampling overlap between the bins. In other cases, the goal is not to split the aerosol into two or more streams for collection but rather to sequentially select out particles from the sample stream. Cascade impactors are used to deposit aerosols of increasing smaller, discrete size ranges onto substrates for subsequent analysis [11]. Various impactor designs are available to meet a variety of measurement objectives, ranging from personal sampling for exposure studies to the collection of large sample volumes for toxicological studies.

Since the particle size of interest in aerosol behaviour ranges from molecular clusters of 0.001 μm to fog droplets and dust particles as large as 100 μm, a variation of 10^5 in size, one cannot use a single instrument to cover the full range. There is constant effort to improve aerosol instrumentation so that various physical or chemical properties can be determined over a wider size range, to develop new measurement techniques and to increase sensitivity. To characterise the chemical composition of the complex atmosphere, a complementary collection of different analytical methodologies is typically used. Several factors must be taken into consideration when instruments are being developed for these measurements. Ideally, the technique should be fast enough to track changes in concentration, composition, or both as they occur, sensitive enough to detect the species of interest and free of interferences from other species present in the same air mass.

Typically, instrumentation is split into two approaches: direct-reading measurements, and collection and analysis [12]. The former yields virtually instantaneous results and results in many measurements to be made, while the latter is generally less expensive and more time consuming, but provides both qualitative and quantitative measurements.

It is often assumed that the sample of particles collected for analysis is representative of those airborne particles from which the sample was taken. For gases this assumption is reasonable, provided buoyancy (and density) considerations are taken into account during the sampling, and the collected samples are kept under appropriate temperature conditions. However, particles do not behave the same way as gas molecules when dispersed in air; they deposit under gravity, impact on bends due to particle inertia, are deposited on internal surfaces by molecular and turbulent diffusion and are affected by thermal, electrostatic and acoustic forces. In addition, the efficiency with which particles are sampled is governed mainly by inertial and sedimentation forces, with other forces playing a role dependent upon the nature and size of the particles and the sampler, and upon the environmental conditions during which the samples were taken. In situ measurements are the least invasive whereby measurements, generally by optical techniques, are made without moving a sample from a flowing aerosol stream.

The instrument selected for a particular application will depend on several factors. In particular, one needs to know why does one want to sample PM and how the results will be used.

All of the above approaches, regardless of the level of time integration, provide information in discrete size bins. The final representation of the data might be in such bins, either in raw form (the parameter as measured for size-segregated sample) or after processing the data through an 'inversion algorithm' to account for the size segregation not being step functions. Another approach is to fit a continuous distribution to the data and report the results as properties of the continuous distribution, such as its 'moments'. A reconstruction of the size distribution requires knowing the transmission efficiency through the instrument as a function of size, as only a fraction of the particles are actually measured and there is size-selective bias in the penetration of particles into and through the analyser.

Single-particle analysers can be used to continuously measure the particle properties [18–20]. The advantage offered by single-particle mass spectrometers relative to pre-existing means of characterising airborne particles is that the single-particle mass spectrometer is able to generate information individually on the chemical composition of large numbers of particles as a function of their size rather than integrate that information across all of the particles collected over a period of hours, as happens with traditional bulk chemical measurements. It therefore provides direct information on the chemical associations of different elements within individual particles, whilst traditional bulk analysis would require uncertain inferences to be drawn regarding such chemical associations [21].

No attempt is made in this chapter to exhaustively examine the opportunities and challenges in constructing size distributions of atmospheric aerosols. The intent is merely to articulate that the measurement issues should be carefully considered in the interpretation of reported size distributions.

1.2.3 Dynamics of atmospheric aerosol size distributions

The aerosol size distribution measured at any instant will be a complex function of the emissions contributing to the aerosol burden in the air mass, atmospheric processing of the air mass including the formation of new particles and condensation onto existing particles, and removal processes through evaporation, scavenging and deposition.

Size distributions for primary emissions from sources with discrete emission points (e.g. stacks and tailpipes) are typically characterised by inserting a probe into the emission zone and carefully diluting the exhaust to [22] conditions. This approach provides a framework to collect samples from various sources – within a given source category and across numerous source categories – under a common set of conditions. While a standardised approach has many advantages, it does not necessarily capture the dynamics that occur proximate to the emission point which could substantially influence the size distribution that represents the contribution of this source to ambient burdens. A simplistic viewpoint is one of each emission source releasing particles that rapidly mix with 'background' air and simply add to the total particle burden. In many cases, however, the process is much more complex, with plume dynamics occurring downwind of the release point.

For example, Zhang *et al.* [23–25] developed a conceptual model for the evolution of particle number distributions from motor vehicle exhaust. Plume processing is divided into two stages – an 'tailpipe-to-roadway' stage and a 'roadway-to-ambient' stage. These stages have distinguishing characteristics that influence the particle size distribution observed at ambient conditions. The tailpipe-to-roadway stage features a steep temperature gradient with mixing driven by vehicle-induced turbulence and dilution on the order of 1000-fold in about 1 s. The roadway-to-ambient stage features relatively constant temperature with mixing driven by atmospheric shear and stability and dilution less than tenfold dilution in 3–10 min. Spatial scales for these two stages are 2–3 and 50–100 m, respectively. Plumes from individual vehicles are relatively isolated during the first stage but become co-mingled during the second stage. In the tailpipe-to-roadway stage, sulphuric-acid-induced nucleation provides fresh aerosol that rapidly grows by organic vapour condensation. In the tailpipe-to-roadway stage, particles can continue to grow by condensation but might also evaporate depending on the partial pressures on condensed organics in the diluted gas phase. The net effect is large numbers of particles smaller than 6 nm being emitted from the roadway, with these particles growing to above 10 nm within the first 30–90 m downwind. Thereafter, some of these particles will continue to grow into the accumulation mode, while others will shrink. Figure 1.4 shows measurements by Zhu *et al.* [26] that were simulated by Zhang *et al.* [24], using a sectional aerosol model to elucidate the governing dynamics. There are dramatic changes to the particle size distribution over the first 300 m downwind of the roadway. These dynamics have important implications to understanding exposures at dwellings proximate to roadways and also the particle size distributions used to represent motor vehicle emissions in Eulerian transport models where the finest resolution for temporal allocation of mobile source emissions is often the modelling grid cell (typically 4 × 4 km or larger). Similar issues are encountered with other emission sources such as point-source plumes (e.g. combustors) and even re-entrained dust. In the latter case, the size distribution used for Eulerian transport modelling and certain other applications needs to take into consideration near-field losses by deposition.

Atmospheric dynamics can modulate the ambient size distribution. Transformations occurring during transport can have profound effects on size distributions. Homogeneous nucleation events generate large number concentrations of ultrafine particles that can grow by vapour condensation or coagulation. Particles can evaporate, and they can be removed by wet and dry processes. The dynamics of size distributions are typically reported from a Eulerian perspective; the measurement location is fixed and the air parcels are probed as they advect past the site.

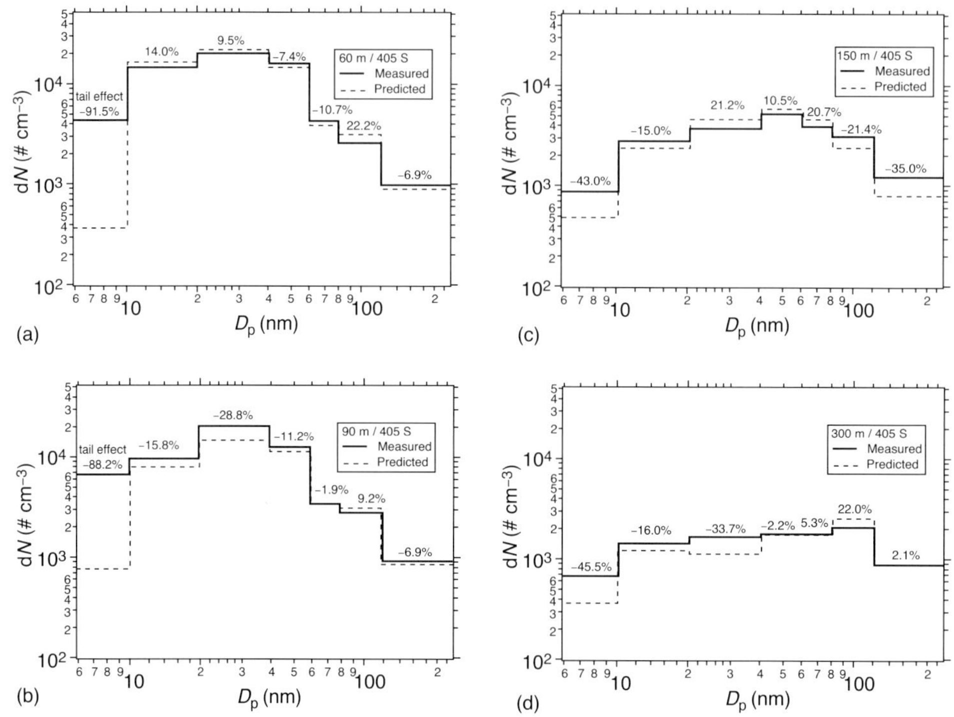

Figure 1.4 Number size distributions for ultrafine aerosols measured 30–300 m downwind of a major freeway. The number over each size bin indicates relative difference between measurement and predict; positive for overprediction and negative for underprediction. (Reproduced from [24] with permission. Copyright Elsevier 2004.)

Since real-time monitoring is relatively expensive and one of the main reasons for monitoring is to aid in the assessment of human exposure to pollution, numerous papers have been published on methodology to identify idea sites for pollution monitoring. A full understanding of the complex interactions behind the aerosol environmental effects requires a multidisciplinary approach, involving both experimental and theoretical methods [27–30].

Modern instrumentation allows quantitative measurements of aerosol formation and growth rates. An example of this is shown in Figure 1.5, where size distribution data over the range 3–680 nm have been collected at 5-min resolution, using a scanning mobility particle spectrometer. The measurements were carried out between July 2001 and June 2002 as part of the Pittsburgh Air Quality Study [31]. Analysis of the data indicates that nucleation occurred on 50% of the study days and regional-scale formation of ultrafine particles on 30% of the days. Figure 1.5 depicts the size distributions measured during 10 August and 11 August 2001. The evolution of the size distributions is shown in Figures 1.5a–1.5b, while Figure 1.5c indicates the integrated particle concentration. It is evident that on 10 August 2001 there was no detectable nucleation activity, while an intense nucleation event was observed around 9 am EST on 11 August, followed by rapid growth of the particles to a size around 100 nm [31]. Kulmala *et al.* [32] have reviewed recent observations of particle formation and growth. The measurements span a broad range of both geographical locations and ambient conditions. They conclude that

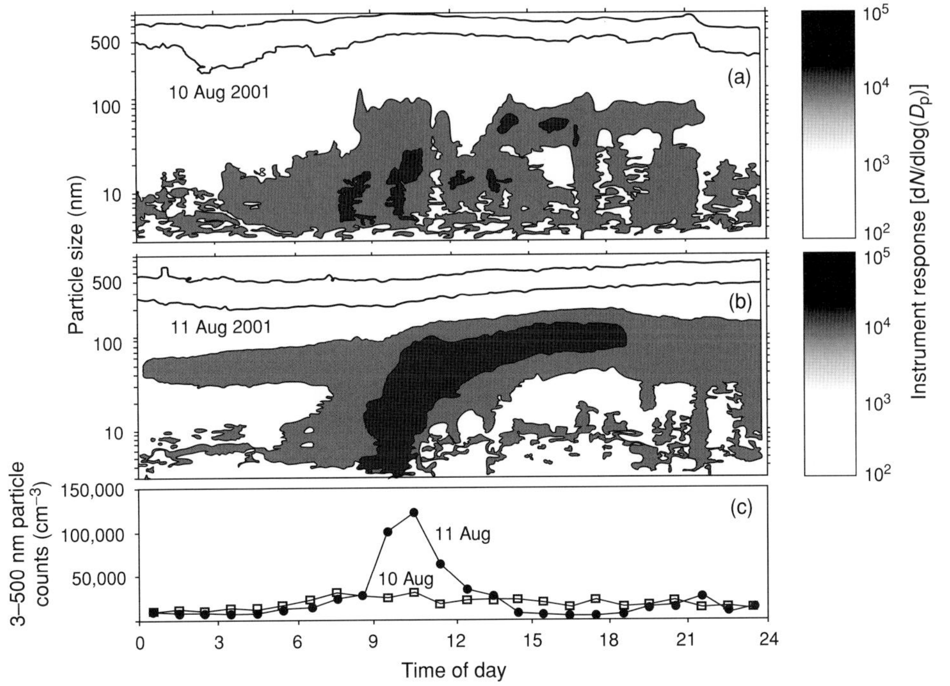

Figure 1.5 Evolution of particle size distributions and particle concentration on a day without nucleation (10 August) and a day with nucleation (11 August). The top two plots show instrument response over all size channels. The bottom plot shows the integrated particle concentration time series. Nucleation is apparent at 9 a.m. EST on 11 August. (Reproduced from [31] with permission. Copyright 2004. American Association for Aerosol Research, Mount Laurel, NJ.)

the formation rate of 3-nm particles is often in the range 0.01–10 $cm^{-3}\ s^{-1}$ in the boundary layer. However, in urban areas formation rates are often higher than this (up to 100 $cm^{-3}\ s^{-1}$), and rates as high as 10^4–10^5 $cm^{-3}\ s^{-1}$ have been observed in coastal areas and industrial plumes.

Removal processes such as wet and dry deposition also alter the aerosol size distribution. For example, Freiman *et al.* [33] measured particle size distributions at five sites in Haifa, Israel, including two pairs of sites that were closely matched in the characteristics that govern variability in particle matter properties over short spatial scales [34, 35]. Mass size distributions were measured over the range 0.23–10 μm and differences were observed between the paired sites. For all size ranges, the less vegetated sites exhibited higher mass concentrations than the more vegetated sites and these differences could not be explained by factors such as meteorology or local emissions. It was concluded that vegetative cover was responsible for differential removal rates of particles between the sites, leading to differences in the aerosol mass–size distributions.

1.2.4 Features of selected ambient aerosol size distributions

In light of the above factors, it is evident that aerosol size distributions can vary in space in time due to differences in emission sources and atmospheric processes. The construct of a 'representative' size distribution (e.g. temporally averaged for a fixed site) necessarily smoothes

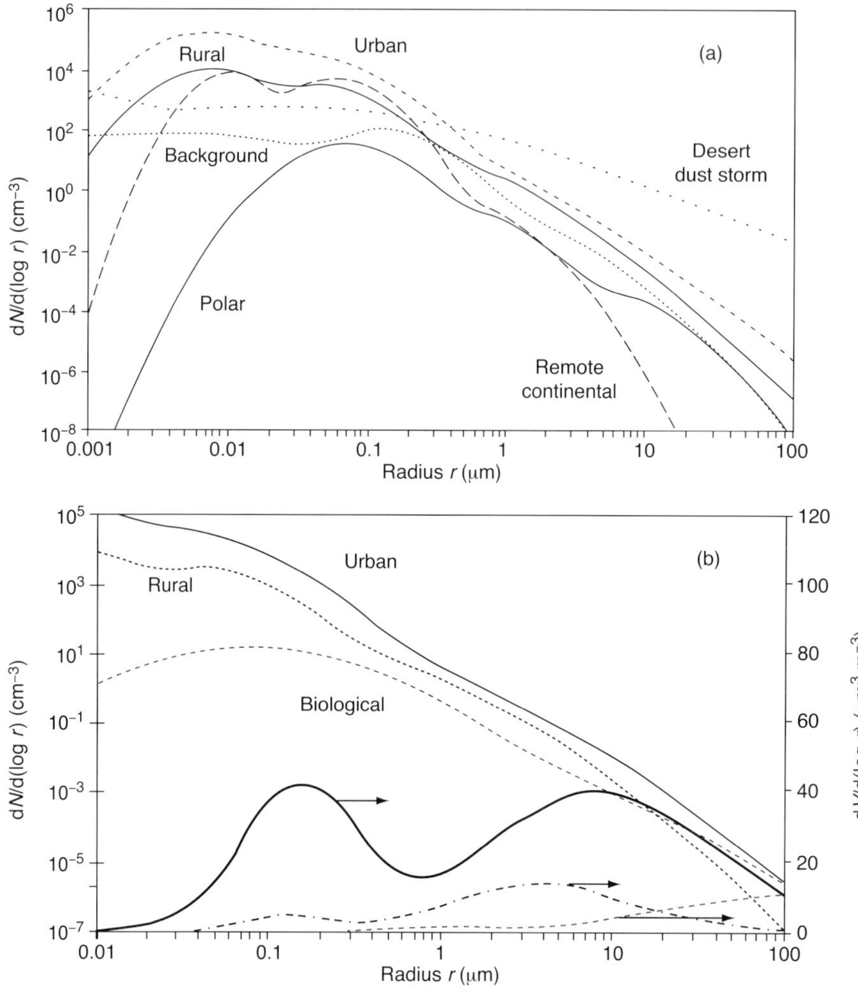

Figure 1.6 Model size distributions for atmospheric aerosols: (a) number size distributions for selected aerosol classes and (b) number and volume distributions for selected aerosol classes.

out many of the features that may be of interest but nonetheless has value. Jaenicke [36, 37] presented idealised number and volume distributions for several classes of tropospheric aerosols (Figure 1.6). The key features of each class are as follows:

- *Polar aerosols* are present near the ground in the Arctic and Antarctica. Number concentrations are very low with significant contributions from aged, globally transported aerosols.
- *Background aerosols* are measured in the free troposphere (FT) above the clouds, operationally above 5 km. When measured in remote regions they can represent conditions over large spatial domains, such as continents.
- *Maritime aerosols* are present over the oceans. They are a mixture of background aerosols and sea salt. The sea salt is predominantly coarse aerosol (by mass).

- *Remote continental aerosols* are measured at the surface at locations with small influences by humans.
- *Desert dust storm aerosols* originate from locations such as the Gobi and Saharan deserts. They can be transported around the globe.
- *Rural aerosols* are continental aerosols at sites with some degree of anthropogenic impacts.
- *Urban aerosols* are measured at the surface in population centres and are influenced by a variety of anthropogenic activities.
- *Biological aerosols* are particles derived from living organisms. They can be viable (alive) or non-viable (dead) in the atmosphere and include materials such as pollens, spores, bacteria and plant fragments.

As Figure 1.6 demonstrates, there are dramatic differences in these aerosol size distributions. For instance, polar aerosols reflect their aged nature as well as their low concentrations. For the urban environment, the maximum particle concentration is observed for small particle sizes and drops off significantly towards minimum values of just a few particles per cubic centimetre for particles in the 10-μm range. Urban aerosols are a mixture of primary particulate emissions from industries, transportation, power plants and natural sources. Secondary material from gas-to-particle conversion also contributes to the urban aerosols. Generally, the size distribution of urban aerosols is very variable. High concentrations of small particles are found near sources, and fall off rapidly with distance from these sources [1]. Desert aerosols exhibit the greatest concentration of coarse particles of any of the classes. In fact, aerosols resuspended from the world's deserts can be transported over considerable distances. Such particles contain essential minerals and can play a major role in fertilising nutrient-limited regions. Annual global dust emissions range from 1000 to 3000 Tg per year [38]. In Figure 1.6b it is clear that the mode diameter increases when going from a number through to a volume distribution, again emphasising the way in which submicron aerosols dominate the number distribution and supermicron aerosols the volume distribution.

Over the past 15 years there has been a dramatic increase in measurements of atmospheric aerosols in various settings. For example, long-term measurements have been conducted at the Jungfraujoch (JFJ) High Alpine Research Station in Switzerland. This site is predominantly in the free troposphere (FT) but is periodically in the planetary boundary layer (PBL), and is presumed representative of the continental aerosol over remote regions of Europe. Figure 1.7 shows seasonal average FT size distributions for the JFJ site compared to the polar, background and remote continental model distributions [13]. The JFJ measurements more closely resemble the model remote continental distribution than the model background distribution.

Desert dust storm aerosols are characterised by high concentrations of coarse particles. The sizes distributions will be attenuated during transport due to preferential removal of the largest particles by sedimentation. Figure 1.8 shows aerosol number distributions measured in Seoul, Korea [39], during April 1998. The baseline number distribution for this urban setting is high (open circles) but excursions were nonetheless observed during periods with moderate (closed circles) and heavy (closed triangles) dust periods from dust storms over the Gobi Desert in Western China and Mongolia.

Biological aerosol particles can be a significant component of the ambient aerosol. Figure 1.9 shows primary biological aerosol particle size distributions measured at a site with both urban and rural influences (Mainz, Germany), a remote continental site (Lake Baikal, Russia) and a marine site (on the Atlantic Ocean). Particle number concentrations ranged from 0.6 cm^{-3} at

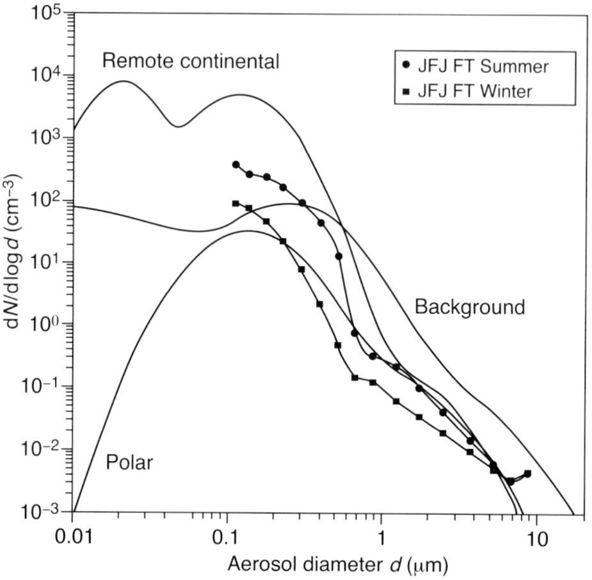

Figure 1.7 Seasonal average number size distributions measured for the free troposphere at Jungfraujoch, Switzerland, and model size distributions from remote continental, polar and background aerosol classes. (Reproduced from [13] by permission of American Geophysical Union. Copyright 1998 American Geophysical Union.)

the marine and remote continental sites to 3 cm^{-3} at the urban/rural site and, owing to the relatively large particle sizes, they account for a significant fraction of the total aerosol volume at all three locations [40–42].

Size distributions for urban aerosols have been measured in various settings around the globe [43–58]. Figure 1.10 shows mean size distributions for three sites in and around Houston, Texas, between June and October 2001 [59]. Mean $PM_{2.5}$ mass concentrations were 13.34 μg m^{-3} at HRM-3 (east Houston, industrial source region), 12.25 μg m^{-3} at Aldine (north-central Houston, urban receptor site) and 9.66 μg m^{-3} at Deer Park (east Houston, industrial/residential source region). The size distributions suggest that there are significant primary aerosols released near the HRM-3 site; these primary-mode (<0.1 μm diameter) particles have generally coagulated into larger particles by the time they reach Aldine. Analysis of the size distributions as a function of time of day indicates that the HRM-3 site remains dominated by the primary-mode particles throughout the day, while the Aldine site and the Deer Park site show a much stronger diurnal pattern, with morning and afternoon peaks in number concentrations of particles [59].

An example of the daily variation in the total particle number and mass concentration in developing countries is shown in Figure 1.11 [60]. For New Delhi mean PM_{10} concentration during the measurements was 360 μg m^{-3}, which is at the high end of the PM_{10} concentrations reported in other developing countries (Table 1.2). The long-term average PM_{10}levels in urban air over Europe and Northern America are much lower, being typically in the range 10–40 μg m^{-3} and practically always <100 μg m^{-3} [68–70]. The average total particle number

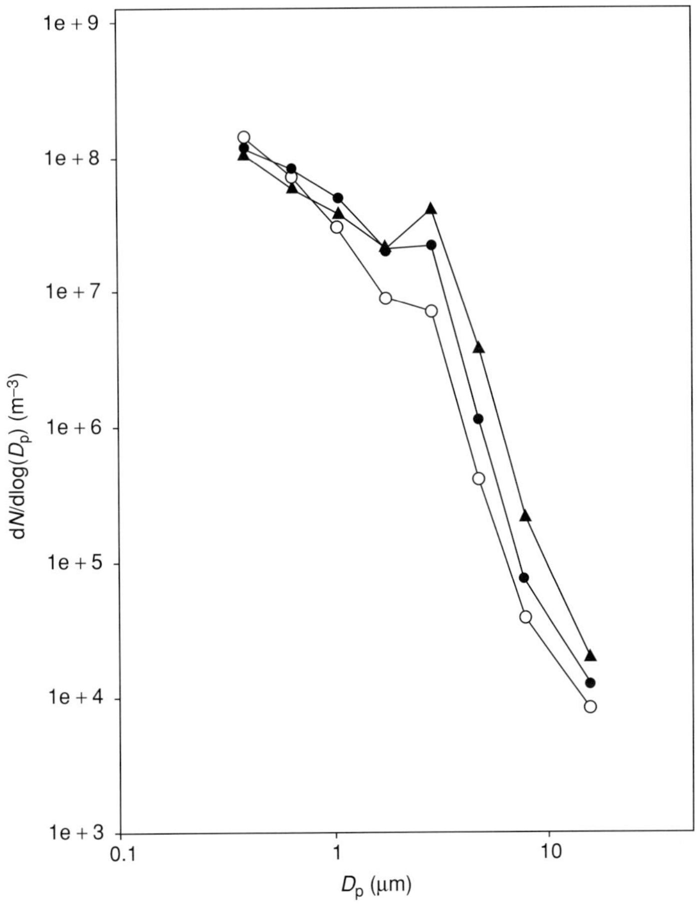

Figure 1.8 Aerosol number distributions for three conditions in Seoul, 1998: non-dust period during March–April (open circles), moderate dust event during 14–22 April (closed circles) and heavy dust event during 19–22 April (closed triangles). (Reproduced from [39] with permission. Copyright Elsevier 2001.)

concentration (3–800 nm) during our measurement period was 63 000 cm^{-3}, which is two to five times higher than the annual average particle number concentrations measured in various urban sites in Europe and the United States [55, 56, 71]. For Beijing, concentrations were just below those for New Delhi.

1.3 Major chemical components of ambient fine particulate matter

The composition of atmospheric aerosols is seldom simple, with their composition being determined by the sources of primary particles and for secondary particles by the processes forming the particles. Coarse particles ($\geq$2.5 μm) are generally formed by mechanical processes

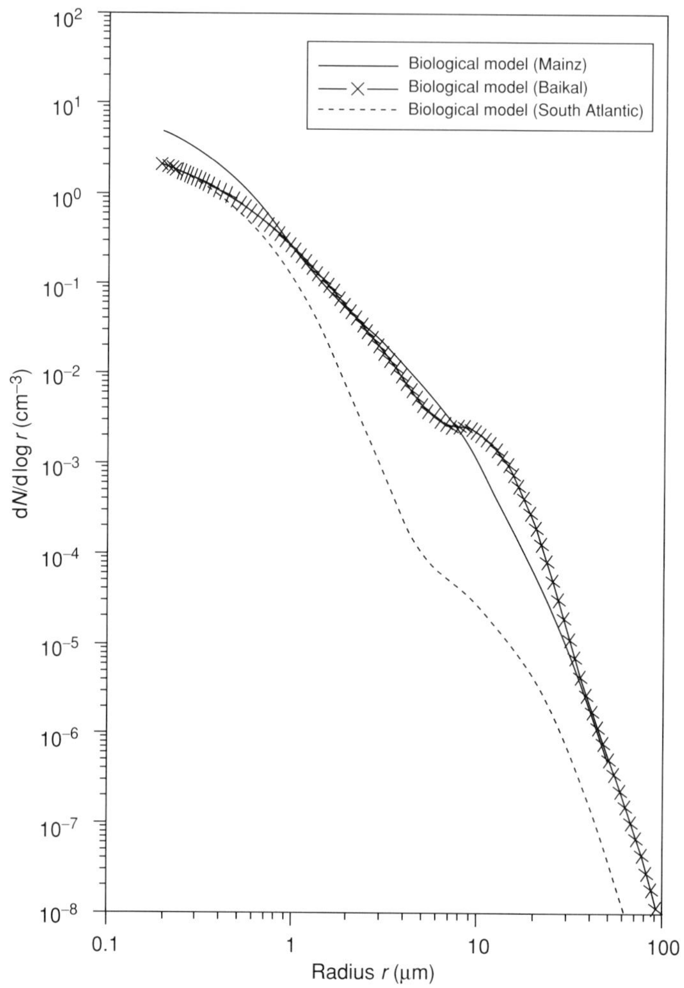

Figure 1.9 Primary biological aerosol model particle number distributions for three site classifications. (Reproduced from [40] with permission. Copyright Elsevier 2000.)

such as fragmentation and are rich in Ca, Fe, Si and other naturally occurring earth constituents, as well as sea salt, NaCl. Fine particles (<2.5 μm) are usually formed by combustion or gas-to-particle conversions and can be rich in C, sulphates, ammonium and nitrate ions, as well as trace toxic species (As, Cd, Cs, Sr, Zn and Se).

The concentration, composition and size distribution of atmospheric aerosol particles are temporally and spatially highly variable. Particles from specific sources can fall into characteristic size ranges and, therefore, particle composition can vary with particle size, reflecting this effect. For example, sulphate arising from the oxidation of SO_2 is typically present in fine particles, whereas silicon from the resuspension of soils and surface dusts is normally found in coarse particles. The composition of the individual particles can be fairly uniform (internally

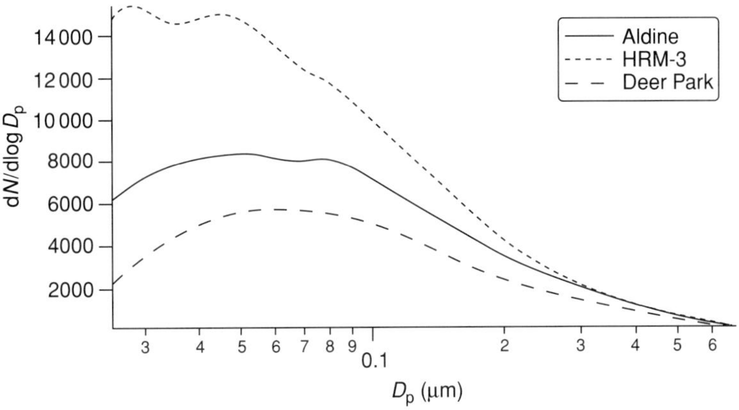

Figure 1.10 Mean size distributions for particulate matter at HRM-3 (east Houston, industrial source region), Aldine (north-central Houston, urban receptor site) and Deer Park 2 (east Houston, industrial/residential source region). Mean distributions are over several measurements taken during July to October 2001. (Reproduced from [59] with permission. Copyright 2004. American Association for Aerosol Research, Mount Laurel, NJ.)

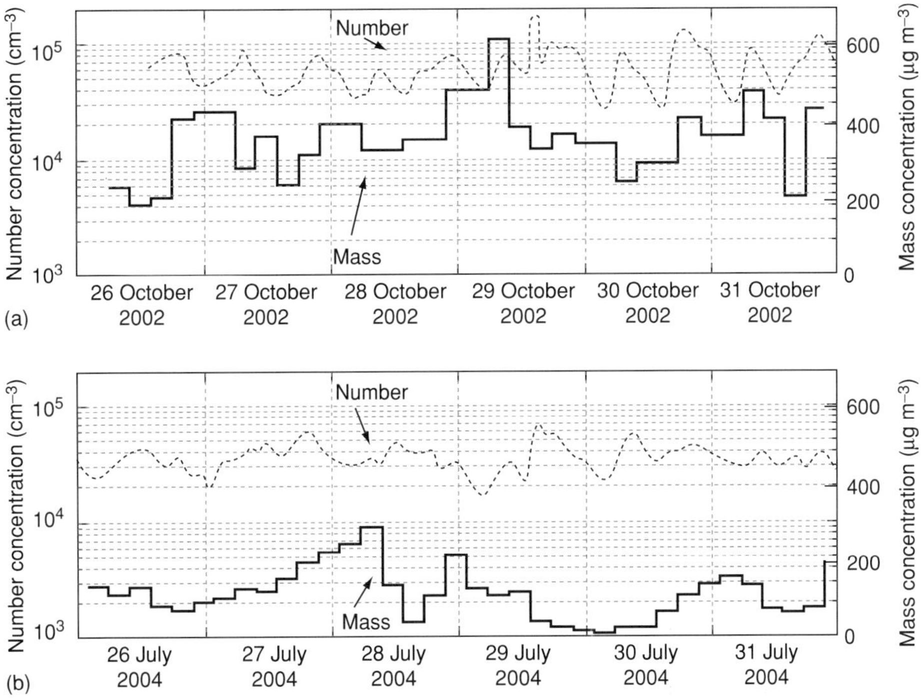

Figure 1.11 (a) Four-hour averages of number (3–800 nm) and mass concentrations (<10 μm) between 26 October and 2 November in New Delhi, India. Left axes are for number concentration (dashed line) and right axes for mass concentration (solid line). (b) Four-hour averages of total number (3–800 nm) and mass concentration (<10 μm) in Beijing, China, from 26–31 July 2004 (Monday to Saturday). (Reproduced from [60] with permission from Springer Science and Business Media.)

Table 1.2 Particle mass concentrations in some cities of the developing world

Country	City	Environment type	Population	Representativeness	Season/monsoon season	TSP ($\mu g\,m^{-3}$)	PM_{10} ($\mu g\,m^{-3}$)	$PM_{2.5}$ ($\mu g\,m^{-3}$)
India	Mumbai	Residential	10 million	Few days, 1996 [61]	Winter		197	
		Slum		Few days, 1996	Winter		127	
		Coastal urban		1 month, 1999 [62]	Winter		182	
				1 month, 1999	Winter		128	
	New Delhi	Urban	14 million	1 week, 2002 [60]	Autumn		360	
China	Beijing	Urban	11 million	1 year, 1999–2000 [63]	All	335	176	97
		Residential			All	350	185	102
	Beijing	Residential	11 million	1 week, 2004 [60]	Summer		120	
	Xian	Urban	3.66 million	1 year, 1996–1997 [64]	All	410	260	
					Autumn	350	224	
					Winter	1000	640	
					Spring	380	243	
					Summer	190	122	
Bangladesh	Dhaka	Urban	8.6 million	1.5 years	Low rain		227	88
					Medium rain		85	34
					High rain		30	17
					All		123	51
	40 km from Dhaka	Rural	8.6 million	1 year	Low rain		98	35
					Medium rain		66	24
					High rain		23	12
					All		56	21
Pakistan	Karachi	Urban	14 million	1 month [65]	Winter	660	420	
	Islamabad	Urban	7.8 million		Winter	680	430	
Mexico	Ciudad de Mexico	Urban	20 million	1 month, 1997 [66]	Winter		80	40
Botswana	Serove	Village	55 000	2 months, 1999 [67]	Winter		29	10

mixed aerosols) or very different from the ensemble composition (externally mixed aerosols), depending on the particle sources and atmospheric aging processes involved.

In general, the predominant chemical components of PM are sulphate, nitrate, ammonium, sea salt, mineral dust, organic compounds and black or elemental carbon, each of which typically contributes about 10–30% of the overall mass load. The predominance of these chemical components in PM_{10} and $PM_{2.5}$ and their size distribution are closely linked to the emitting source and the formation mechanisms of the particles.

- Sulphate – arises primarily as a secondary component from atmospheric oxidation of SO_2;
- Nitrate – typically present as NH_4NO_3, resulting from the neutralisation of HNO_3 vapour by NH_3, or as sodium nitrate ($NaNO_3$), due to displacement of hydrogen chloride from NaCl by HNO_3 vapour;
- Ammonium – usually present in the form of ammonium sulphate ($(NH_4)_2SO_4$) or NH_4NO_3;
- Sodium and chloride – from sea salt;
- Elemental carbon – formed during the high-temperature combustion of fossil and biomass fuels;
- Organic carbon – carbon in the form of organic compounds, could be either primary, resulting from automotive or industrial sources, or secondary, resulting from the oxidation of volatile organic compounds;
- Mineral components – mainly present in the coarse fraction and rich in elements such as aluminium, silicon, iron and calcium;
- Water may also be present within, for example $(NH_4)_2SO_4$, NH_4NO_3 and NaCl. The water-soluble components can take up water from the atmosphere at high relative humidity, thereby turning from crystalline solids into liquid droplets.

In addition to the above there are many minor chemical components present in airborne particles. Their detection is often a function of the sensitivity of the analytical procedure. Such components include the following:

- Trace metals – such as lead, cadmium, mercury, nickel, chromium and zinc;
- Trace organic compounds – although the total mass of organic compounds can comprise a significant part of the overall mass of particles, it is made up of a very large number of individual organic compounds, each of which is present at a very low concentration.

At different locations, times, meteorological conditions and particle size fractions, however, the relative abundance of different chemical components can vary by an order of magnitude or more [1, 72–74]. Figure 1.12 illustrates how particle number concentration, size distribution and chemical composition of fine particulate vary in urban and high Alpine air. The dotted lines indicate characteristic particle size modes, which can be attributed to different sources, sinks and aging processes of atmospheric particles: nucleation (Aitken), accumulation and coarse modes.

In the UK intensive research on the chemical characterisation of PM_{10} has been summarised by Air Quality Expert Group [4]. Data for four roadside sites and four urban background sites have been investigated and show that [75]:

- The marine contribution is estimated at around 2 $\mu g\ m^{-3}$ in PM_{10} (8 and 6% in urban background and roadside sites) and only 0.3 $\mu g\ m^{-3}$ in the $PM_{2.5}$ fraction (1–2% of $PM_{2.5}$

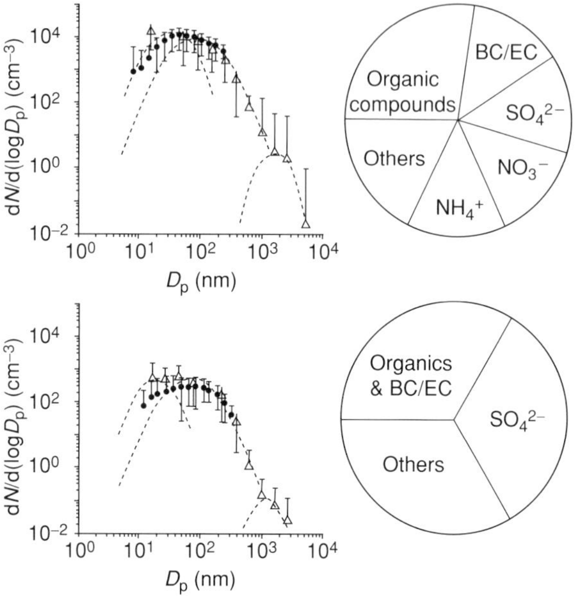

Figure 1.12 Characteristic examples of aerosol particle size distribution and chemical composition in urban (top) and high Alpine air (bottom). Graphs (left): number size distribution function d*N*/d(log d*p*) (symbols and error bars: arithmetic mean values and standard deviations, ~ ELPI, * SMPS, characteristic particle size modes). Pie charts (right): typical mass proportions of main components. (Reproduced from [74] with permission from Wiley-VCH Verlag.)

mass). These values are very similar to those reported for Berlin and Spain, and lower than the Dutch levels.

- Road dust levels reach 5.5 and 6.5 μg m^{-3} in PM_{10} (22 and 18% at urban background and roadside sites) and only 1.6 μg m^{-3} in the $PM_{2.5}$ fraction (11% of $PM_{2.5}$ mass). These values are relatively low when compared with data obtained from Spain, and similar to the data reported for the Netherlands.
- The levels of carbonaceous components reach 9.3 and 18.1 μg m^{-3} in PM_{10} (38 and 50% at urban background and roadside sites). These values are in the range of data reported for Berlin and Spain, and higher than the values reported for the Netherlands.
- Organic carbon (OC) predominates over elemental carbon (EC) at urban background sites (26% versus 12% in PM_{10}), whereas in roadside sites similar levels are determined for both type of carbonaceous in PM_{10}.
- The carbonaceous components are mostly present in $PM_{2.5}$. Thus, around 90 and 80% of the OC and EC mass, respectively, are present in fine fractions at both background and roadside sites. Around 50 and 65% of the $PM_{2.5}$ mass at urban background and roadside sites is made up by carbonaceous compounds.
- Levels of secondary inorganic are close to 8 μg m^{-3} in both urban background and roadside sites, accounting for 32 and 23% of the PM_{10} mass, respectively, in the range of levels found for the other European Union examples reported here.

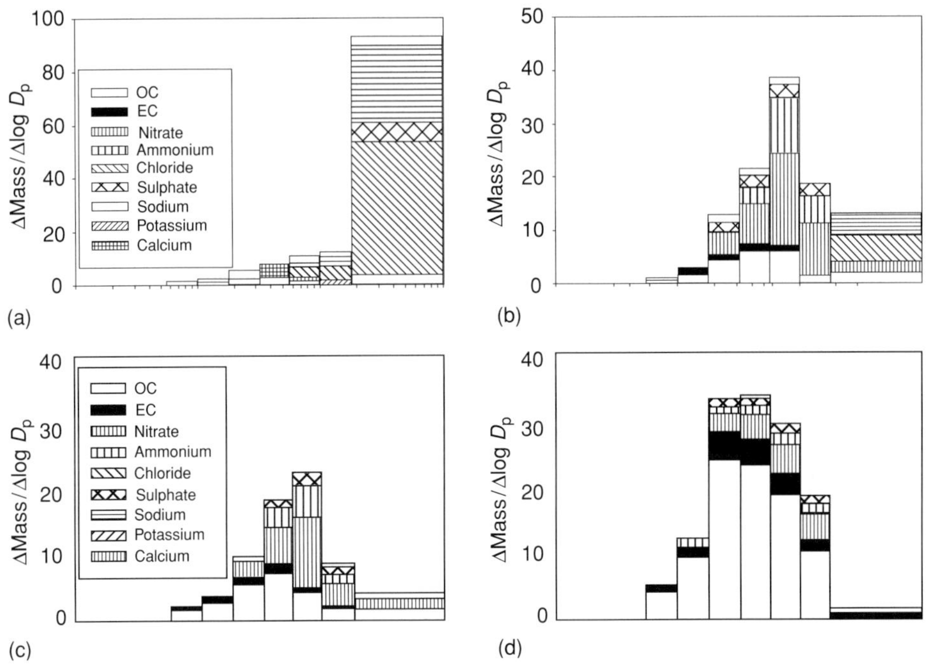

Figure 1.13 Size–composition distributions for winter aerosol mass: coastal site – Bodega Bay, CA, (a) onshore winds, (b) offshore winds; and inland site – Sacramento, CA, (c) daytime, (d) nighttime. (Reproduced from [76] with permission. Copyright 2006. American Association for Aerosol Research, Mount Laurel, NJ.)

- The ratio ammonium sulphate/ammonium nitrate in PM_{10} is also constantly 1:1 at both type of sites, and around 30% of nitrate in PM_{10} is present in the coarse mode ($PM_{2.5-10.0}$, probably as sodium nitrate), whereas this proportion is around 12% in the case of ammonium sulphate.

1.4 Aerosol composition as a function of size

Preceding sections focused on particle size distributions without regard to particle composition, and particle composition for fine and coarse PM without further size discrimination. In this section, we draw attention to joint size–composition distributions that can be particularly valuable in understanding the physical and chemical mechanisms that govern ambient aerosols. In the preceding section it is evident that specific compounds will occur in certain size ranges. Crustal minerals such as calcium, magnesium, aluminium and iron are typically found in the coarse fractions, whereas sulphate, ammonium, organic and elemental carbon and certain transition metals occur mainly in the fine fractions.

Herner *et al.* [76] examined size–composition distributions collected for winter conditions at five sites in (or near) the San Joaquin Valley of California as part of the California Regional Particulate Air Quality Study. Distributions for a coastal site were stratified into periods with onshore versus offshore winds, while distributions for inland sites were stratified into daytime versus nighttime periods. Examples of these average distributions are shown in Figure 1.13.

PM_{10} mass at the coastal site is dominated by coarse PM sea salt during periods with onshore winds (Figure 1.13a). Offshore winds coincide with much lower PM_{10} mass concentrations with more fine mass than coarse mass and a peak at 0.5–1.0 μm which is primarily ammonium nitrate (Figure 1.13b). Average daytime aerosol in Sacramento (Figure 1.13c), a large inland city, has a distribution qualitatively similar to the coastal offshore distribution with the most excess mass at the coastal site being coarse PM. The average Sacramento nighttime distribution (Figure 1.13d) is dramatically different; it is dominated by carbonaceous aerosol arising from primary emissions. The size–composition distributions were simulated using a box model to determine the relative roles of condensation/evaporation, hygroscopic growth, dry deposition, emissions and coagulation to arrive at a detailed conceptual model for ambient PM in the San Joaquin Valley.

An example of the size distribution for the various inorganic compounds is shown in Figure 1.14. Here the size distribution is presented for one urban and two rural sites in British Columbia, Canada [77]. The dominant inorganic chemical component shows a bimodal size distribution at all three sites, with the accumulation mode dominating the mass relative to the coarse-mode peak. SO_4^{2-} dominates the accumulation mode at all sites, with a median peak concentration at 0.3–0.55 μm. NO_3^- shows a prominent coarse mode at >1 μm that resembles that of Na^+, but also exhibits a secondary accumulation mode between 0.3 and 1.0 μm. It can be seen that sea salt aerosols, as represented by Na^+, are confined to the coarse mode and peak between 1 and 18 μm, depending on location and time. The chloride ion has a similar coarse-mode distribution [77].

Figure 1.15 summarises the size distributions of trace elements collected in the centre of Athens during the summer of 2001 [78]. It is evident that metallic elements in the Athens urban aerosol can be classified into two groups. The first group comprises Cd, V, Ni, Pb and Mn emitted from anthropogenic sources. The second group consists of Al, Fe, Cu and Cr, and is mostly present in the coarse fraction and has its origin in soil dust or mechanical abrasion processes. The second group holds by far the bulk of the aerosol elemental metal mass concentration [78].

1.5 Summary and conclusions

We often assume that air pollution is a modern phenomenon, and that it has become worse in recent times. However, since the dawn of history, mankind has been burning biological and fossil fuel to produce heat. The walls of caves, inhabited millennia ago, are covered with layers of soot and many of the lungs of mummified bodies from Palaeolithic times have a black tone [15, 79]. However, it is only relatively recently that it has become increasingly clear that aerosol particles are of major importance for atmospheric chemistry and physics, the hydrological cycle, climate and human health.

An important aspect in the study of PM is its complexity of the physicochemical characteristics, the multiple sources, morphology and its dynamics that is correlated with the size of the particles.

Atmospheric aerosols originate either from naturally occurring processes or from anthropogenic activity. Major natural aerosol sources include volcanic emissions, sea spray and mineral dust emissions, while anthropogenic sources include emissions from industry and combustion processes. Within both categories further distinction of so-called primary and secondary

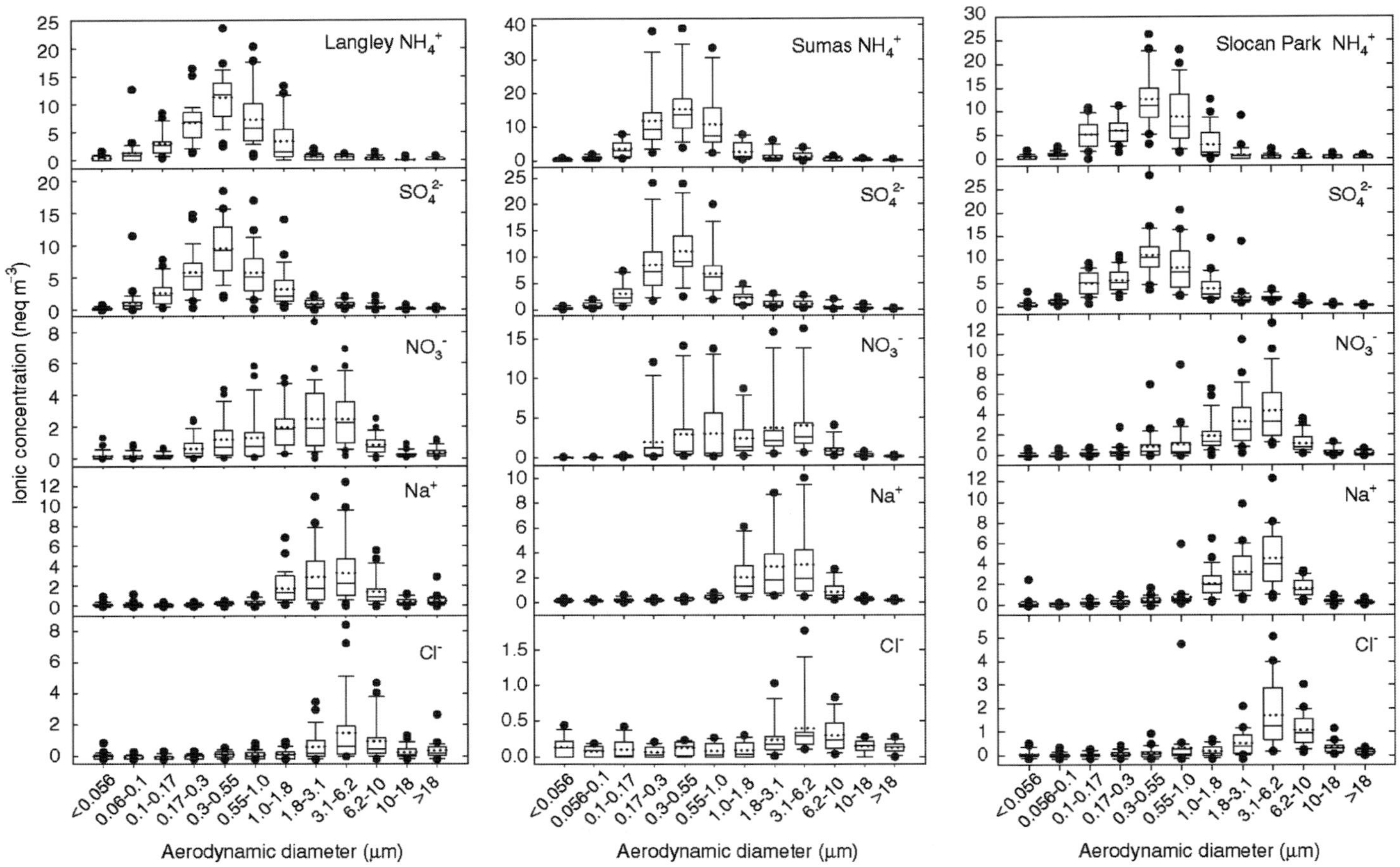

Figure 1.14 Box and whisker plots of ionic particle size distributions at the three sites. The dotted line and solid bar in each box represent the mean and median, respectively; the upper and lower parts of each box represent the 75th and 25th percentiles and the extensions represent the 90th and 10th percentiles; any points outside these limits are shown as large filled circles. (Reproduced from [77] with permission. Copyright Elsevier 2006.)

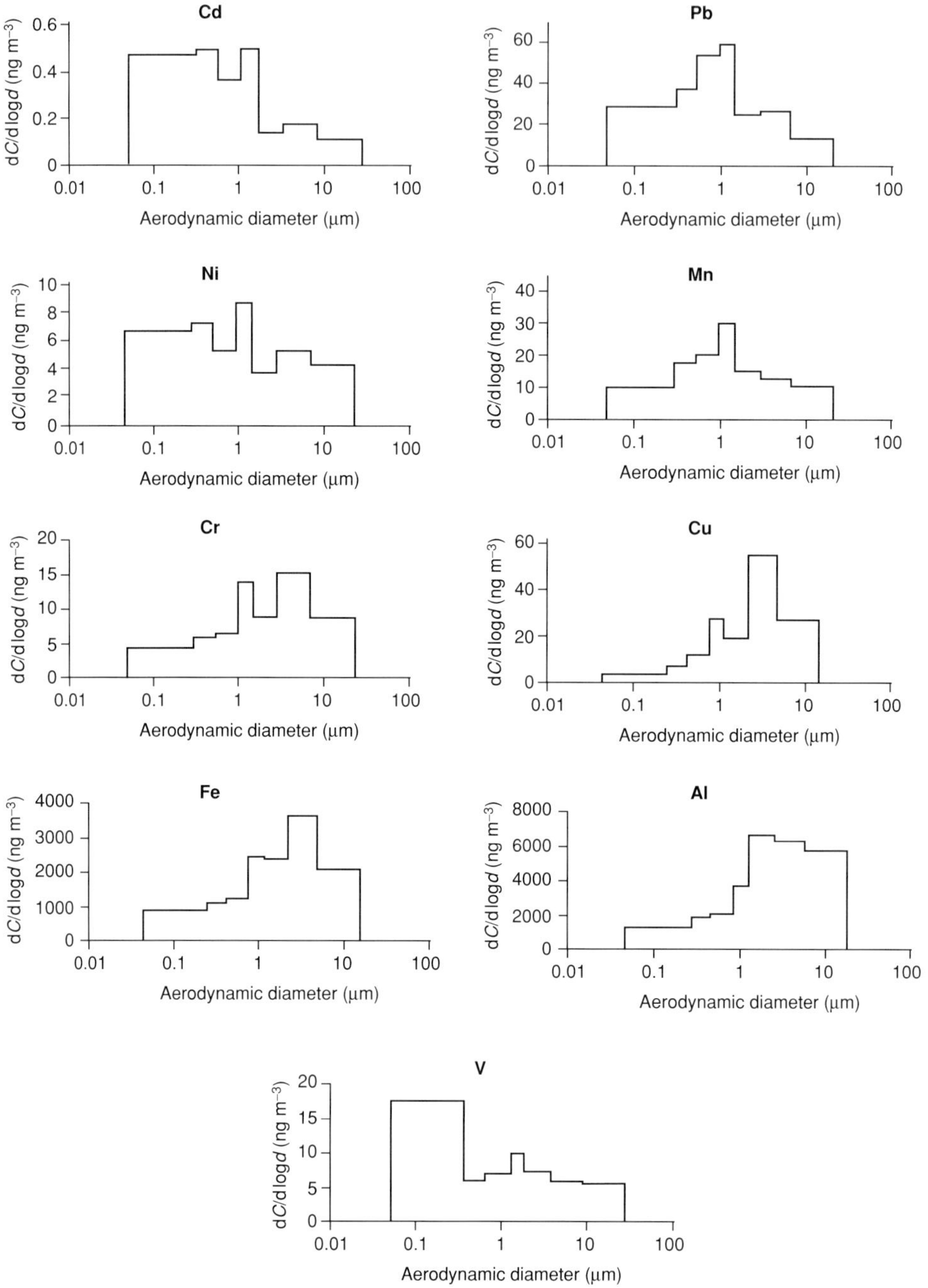

Figure 1.15 Mean size distribution of nine metals (Cd, Pb, V, Ni, Cr, Mn, Cu, Fe and Al). Values are geometric means calculated from 16 samples. (Reproduced from [78] with permission. Copyright Elsevier 2007.)

sources may be made. The direct emission of aerosols into the atmosphere constitute primary sources, while secondary sources arise from the gas-to-particle conversion of gaseous precursor compounds, such as nitric oxide and nitrogen dioxide, sulphur dioxide and hydrocarbons. Particles in the accumulation range arise typically from the condensation of low-volatility vapours and from coagulation of smaller particles in the nucleation mode with themselves or with the larger accumulation-mode particles. Particles tend to accumulate in this mode, as there is a minimum efficiency in sink processes.

Particles in the coarse mode are usually produced by weathering and wind-erosion processes. Chemically, their composition reflects their sources and hence inorganic compounds such as mineral dust and sea salt are found in addition to organic compounds such as biological and biogenic particles resulting from direct emission of hydrocarbons into the atmosphere.

Recent scientific advances now offer an improved understanding of the mechanisms and factors controlling the chemistry of atmospheric aerosols.

References

1 Seinfeld, J.H. and Pandis, S.N. *Atmospheric Chemistry and Physics*. John Wiley and Sons, New York: 1998.

2 Kapustin, V.N., Clarke, A.D., Shinozuka, Y., Howell, S., Brekhovskikh, V., Nakajima, T. and Higurashi, A. On the determination of a cloud condensation nuclei from satellite: challenges and possibilities. *Journal of Geophysical Research – Atmospheres* 2006; **111**: D04202.

3 Castanho, A.D.D., Martins, J.V., Hobbs, P.V., Artaxo, P., Remer, L., Yamasoe, M. and Colarco, P.R. Chemical characterization of aerosols on the east coast of the United States using aircraft and ground-based stations during the CLAMS experiment. *Journal of the Atmospheric Sciences* 2005; **62**: 934–946.

4 AQEG. *Particulate Matter in the United Kingdom*. Air Quality Expert Group, UK: 2005.

5 Demerjian, K.L. A review of national monitoring networks in North America. *Atmospheric Environment* 2000; **34**: 1861–1884.

6 Yttri, K.E. and Aas, W. *Transboundary Particulate Matter in Europe: Status Report 2006*. EMEP Report 4/2006. EMEP CCC & MSC-W, Norway: 2006; p. 137.

7 Yttri, K.E., Aas, W., Bjerke, A., Ceburnis, D., Dye, C., Emblico, L., Facchini, M.C., Forster, C., Hanssen, J.E., Hansson, H.C., Jennings, S.G., Maenhaut, W. and Putaud, J.P. Elemental and organic carbon in PM_{10}: a one year measurement campaign within the European Monitoring and Evaluation Programme EMEP. *Atmospheric Chemistry and Physics Discussions* 2007; **7**: 3859–3899.

8 Atkinson-Palombo, C.M., Miller, J.A. and Balling, R.C. Quantifying the ozone 'weekend effect' at various locations in Phoenix, Arizona. *Atmospheric Environment* 2006; **40**: 7644–7658.

9 Riga-Karandinos, A.N., Saitanis, C. and Arapis. G. Study of the weekday-weekend variation of air pollutants in a typical Mediterranean coastal town. *International Journal of Environment and Pollution* 2006; **27**: 300–312.

10 Baron, P.A. and Willeke, K., eds. *Aerosol Measurement*, 2nd edn. John Wiley and Sons, New York: 2001.

11 Baron, P.A. and Heitbrink, W.A., eds. An approach to performing aerosol measurements. In: P.A. Baron and K. Willeke, eds. *Aerosol Measurement*, 2nd edn. John Wiley and Sons, New York: 2001; pp. 117–139.

12 McMurry, P.H. A review of atmospheric aerosol measurements. *Atmospheric Environment* 2000; **34**: 1959–1999.

13 Nyeki, S., Li, F., Weingartner, E., Streit, N., Colbeck, I., Gaggeler, H.W. and Baltensperger, U. The background aerosol size distribution in the free troposphere: an analysis of the annual cycle at a high-alpine site. *Journal of Geophysical Research – Atmospheres* 1998; **103**: 31749–31761.

14 Nyeki, S., Baltensperger, U., Colbeck, I., Jost, D.T., Weingartner, E. and Gaggeler, H.W. The Jungfraujoch high-Alpine research station (3454m) as a background clean continental site for the measurement of aerosol parameters. *Journal of Geophysical Research – Atmospheres* 1998; **103**: 6097–6107.
15 John, W. Size distribution characteristics of aerosols. In: P.A. Baron and K. Willeke, eds. *Aerosol Measurement*, 2nd edn. John Wiley and Sons, New York: 2001.
16 Wilson, W.E., Chow, J.C., Claiborn, C., Wei, F.S., Engelbrecht, J. and Watson, J.G. Monitoring of particulate matter outdoors. *Chemosphere* 2002; **49**: 1009–1043.
17 Wilson, W.E. and Suh, H.H. Fine particles and coarse particles: concentration relationships relevant to epidemiologic studies. *Journal of the Air & Waste Management Association* 1997; **47**: 1238–1249.
18 Canagaratna, M.R., Jayne, J.T., Jimenez, J.L., Allan, J.D., Alfarra, M.R., Zhang, Q., Onasch, T.B., Drewnick, F., Coe, H., Middlebrook, A., Delia, A., Williams, L.R., Trimborn, A.M., Northway, M.J., DeCarlo, P.F., Kolb, C.E., Davidovits, P. and Worsnop, D.R. Chemical and microphysical characterization of ambient aerosols with the aerodyne aerosol mass spectrometer. *Mass Spectrometry Reviews* 2007; **26**: 185–222.
19 Dall'Osto, M. and Harrison, R.M. Chemical characterisation of single airborne particles in Athens (Greece) by ATOFMS. *Atmospheric Environment* 2006; **40**: 7614–7631.
20 Toner, S. M., Sodeman, D.A. and Prather, K.A. Single particle characterization of ultrafine and accumulation mode particles from heavy duty diesel vehicles using aerosol time-of-flight mass spectrometry. *Environmental Science & Technology* 2006; **40**: 3912–3921.
21 Wexler, A.S. and Johnston, M.V. Real-time single-particle analysis. In: P.A. Baron and K. Willeke, eds. *Aerosol Measurement*, 2nd edn. John Wiley and Sons, New York: 2001.
22 Lipsky, E.M. and Robinson, A.L. Effects of dilution on fine particle mass and partitioning of semivolatile organics in diesel exhaust and wood smoke. *Environmental Science & Technology* 2006; **40**: 155–162.
23 Zhang, K.M. and Wexler, A.S. Evolution of particle number distribution near roadways – Part I: analysis of aerosol dynamics and its implications for engine emission measurement. *Atmospheric Environment* 2004; **38**: 6643–6653.
24 Zhang, K.M., Wexler, A.S., Zhu, Y.F., Hinds, W.C. and Sioutas, C. Evolution of particle number distribution near roadways – Part II: the 'road-to-ambient' process. *Atmospheric Environment* 2004; **38**: 6655–6665.
25 Zhang, K.M., Wexler, A.S., Niemeier, D.A., Zhu, Y.F., Hinds, W.C. and Sioutas, C. Evolution of particle number distribution near roadways – Part III: traffic analysis and on-road size resolved particulate emission factors. *Atmospheric Environment* 2005; **39**: 4155–4166.
26 Zhu, Y.F., Hinds, W.C., Kim, S., Shen, S. and Sioutas, C. Study of ultrafine particles near a major highway with heavy-duty diesel traffic. *Atmospheric Environment* 2002; **36**: 4323–4335.
27 Martuzevicius, D., Luo, J.X., Reponen, T., Shukla, R., Kelley, A.L., St Clair, H. and Grinshpun, S.A. Evaluation and optimization of an urban $PM_{2.5}$ monitoring network. *Journal of Environmental Monitoring* 2005; **7**: 67–77.
28 Goswami, E., Larson, T., Lumley, T. and Liu, L.J.S. Spatial characteristics of fine particulate matter: identifying representative monitoring locations in Seattle, Washington. *Journal of the Air & Waste Management Association* 2002; **52**: 324–333.
29 Karaca, F., Nikov, A. and Alagha, O. NN-AirPol: a neural-networks-based method for air pollution evaluation and control. *International Journal of Environment and Pollution* 2006; **28**: 310–325.
30 Kukkonen, J., Partanen, L., Karppinen, A., Ruuskanen, J., Junninen, H., Kolehmainen, M., Niska, H., Dorling, S., Chatterton, T., Foxall, R. and Cawley, G. Extensive evaluation of neural network models for the prediction of NO_2 and PM10 concentrations, compared with a deterministic modelling system and measurements in central Helsinki. *Atmospheric Environment* 2003; **37**: 4539–4550.
31 Stanier, C.O., Khlystov, A.Y. and Pandis, S.N. Nucleation events during the Pittsburgh air quality study: description and relation to key meteorological, gas phase, and aerosol parameters. *Aerosol Science and Technology* 2004; **38**: 253–264.

32 Kulmala, M., Vehkamaki, H., Petajda, T., Dal Maso, M., Lauri, A., Kerminen, V.M., Birmili, W. and McMurry, P.H. Formation and growth rates of ultrafine atmospheric particles: a review of observations. *Journal of Aerosol Science* 2004; **35**: 143–176.

33 Freiman, M.T., Hirshel, N. and Broday, D.M. Urban-scale variability of ambient particulate matter attributes. *Atmospheric Environment* 2006; **40**: 5670–5684.

34 Pinto, J.P., Lefohn, A.S. and Shadwick, D.S. Spatial variability of PM2.5 in urban areas in the United States. *Journal of the Air & Waste Management Association* 2004; **54**: 440–449.

35 Wilson, J.G., Kingham, S., Pearce, J. and Sturman, A.P. A review of intraurban variations in particulate air pollution: implications for epidemiological research. *Atmospheric Environment* 2005; **39**: 6444–6462.

36 Jaenicke, R. Aerosol physics and chemistry. In *Landolt-Börnstein: Zahlenwerte und Funktionen aus Naturwissenschaften und Technik*. Springer, Berlin: 1988; pp. 391–457.

37 Jaenicke, R. Tropospheric aerosols. In: P.V. Hobbs, ed. *Aerosol Cloud Climate Interactions*. Academic Press, New York: 1993; pp. 1–31.

38 Cakmur, R.V., Miller, R.L., Perlwitz, J., Geogdzhayev, I.V., Ginoux, P., Koch, D., Kohfeld, K.E., Tegen, I. and Zender, C.S. Constraining the magnitude of the global dust cycle by minimizing the difference between a model and observations. *Journal of Geophysical Research – Atmospheres* 2006; **111**: D06207 02006.

39 Chun, Y., Kim, J., Choi, J.C., Boo, K.O., Oh, S.N. and Lee, M. Characteristic number size distribution of aerosol during Asian dust period in Korea. *Atmospheric Environment* 2001; **35**: 2715–2721.

40 Matthias-Maser, S., Obolkin, V., Khodzer, T. and Jaenicke, R. Seasonal variation of primary biological aerosol particles in the remote continental region of Lake Baikal/Siberia. *Atmospheric Environment* 2000; **34**: 3805–3811.

41 Matthias-Maser, S. and Jaenicke, R. Examination of atmospheric bioaerosol particles with radii greater than 0.2 Mu M. *Journal of Aerosol Science* 1994; **25**: 1605–1613.

42 Matthias-Maser, S. and Jaenicke, R. The size distribution of primary biological aerosol particles with radii >0.2 μm in an urban rural influenced region. *Atmospheric Research* 1995; **39**: 279–286.

43 Olivares, G., Johansson, C., Strom, J. and Hansson, H.C. The role of ambient temperature for particle number concentrations in a street canyon. *Atmospheric Environment* 2007; **41**: 2145–2155.

44 Duan, J.C., Bi, X.H., Tan, J.H., Sheng, G.Y. and Fu, J.M. Seasonal variation on size distribution and concentration of PAHs in Guangzhou city, China. *Chemosphere* 2007; **67**: 614–622.

45 Viana, M., Querol, X., Gotschi, T., Alastuey, A., Sunyer, J., Forsberg, B., Heinrich, J., Norback, D., Payo, F., Maldonado, J.A. and Kunzli, N. Source apportionment of ambient PM2.5 at five Spanish centres of the European Community Respiratory Health Survey (ECRHS II). *Atmospheric Environment* 2007; **41**: 1395–1406.

46 Flocas, H.A., Assimakopoulos, V.D. and Helmis, C.G. An experimental study of aerosol distribution over a Mediterranean urban area. *Science of the Total Environment* 2006; **367**: 872–887.

47 Minoura, H. and Takekawa, H. Observation of number concentrations of atmospheric aerosols and analysis of nanoparticle behavior at an urban background area in Japan. *Atmospheric Environment* 2005; **39**: 5806–5816.

48 Tuch, T.M., Wehner, B., Pitz, M., Cyrys, J., Heinrich, J., Kreyling, W.G., Wichmann, H.E. and Wiedensohler, A. Long-term measurements of size-segregated ambient aerosol in two German cities located 100km apart. *Atmospheric Environment* 2003; **37**: 4687–4700.

49 McMurry, P.H. and Woo, K.S. Size distributions of 3-100-nm urban Atlanta aerosols: measurement and observations. *Journal of Aerosol Medicine – Deposition, Clearance, and Effects in the Lung* 2002; **15**: 169–178.

50 Wiedensohler, A., Wehner, B. and Birmili, W. Aerosol number concentrations and size distributions at mountain-rural, urban-influenced rural, and urban-background sites in Germany. *Journal of Aerosol Medicine – Deposition, Clearance, and Effects in the Lung* 2002; **15**: 237–243.

51 Hering, S.V., Kreisberg, N.M., Stolzenburg, M.R. and Lewis, G.S. Comparison of particle size distributions at urban and agricultural sites in California's San Joaquin Valley. *Aerosol Science and Technology* 2007; **41**: 86–96.
52 Querol, X., Alastuey, A., Ruiz, C.R., Artinano, B., Hansson, H.C., Harrison, R.M., Buringh, E., ten Brink, H.M., Lutz, M., Bruckmann, P., Straehl, P. and Schneider, J. Speciation and origin of PM10 and PM2.5 in selected European cities. *Atmospheric Environment* 2004; **38**: 6547–6555.
53 Shi, J.P., Evans, D.E., Khan, A.A. and Harrison, R.M. Sources and concentration of nanoparticles (<10 nm diameter) in the urban atmosphere. *Atmospheric Environment* 2001; **35**: 1193–1202.
54 Harrison, R.M., Shi, J.P., Xi, S.H., Khan, A., Mark, D., Kinnersley, R. and Yin, J.X. Measurement of number, mass and size distribution of particles in the atmosphere. *Philosophical Transactions of the Royal Society of London. Series A – Mathematical Physical and Engineering Sciences* 2000; **358**: 2567–2579.
55 Harrison, R.M. and Jones, A.M. Multisite study of particle number concentrations in urban air. *Environmental Science & Technology* 2005; **39**: 6063–6070.
56 Aalto, P., Hameri, K., Paatero, P., Kulmala, M., Bellander, T., Berglind, N., Bouso, L., Castano-Vinyals, G., Sunyer, J., Cattani, G., Marconi, A., Cyrys, J., von Klot, S., Peters, A., Zetzsche, K., Lanki, T., Pekkanen, J., Nyberg, F., Sjovall, B. and Forastiere, F. Aerosol particle number concentration measurements in five European cities using TSI-3022 condensation particle counter over a three-year period during health effects pollution on susceptible subpopulations. *Journal of the Air & Waste Management Association* 2005; **55**: 1064–1076.
57 Ketzel, M., Wahlin, P., Kristensson, A., Swietlicki, E., Berkowicz, R., Nielsen, O.J. and Palmgren, F. Particle size distribution and particle mass measurements at urban, near-city and rural level in the Copenhagen area and Southern Sweden. *Atmospheric Chemistry and Physics* 2004; **4**: 281–292.
58 Tunved, P., Hansson, H.C., Kulmala, M., Aalto, P., Viisanen, Y., Karlsson, H., Kristensson, A., Swietlicki, E., Dal Maso, M., Strom, J. and Komppula, M. One year boundary layer aerosol size distribution data from five nordic background stations. *Atmospheric Chemistry and Physics* 2003; **3**: 2183–2205.
59 Russell, M., Allen, D.T., Collins, D.R. and Fraser, M.P. Daily, seasonal, and spatial trends in PM2.5 mass and composition in Southeast Texas. *Aerosol Science and Technology* 2004; **38**: 14–26.
60 Laakso, L., Koponen, I.K., Monkkonen, P., Kulmala, M., Kerminen, V.M., Wehner, B., Wiedensohler, A., Wu, Z.J. and Hu, M. Aerosol particles in the developing world: a comparison between New Delhi in India and Beijing in China. *Water Air and Soil Pollution* 2006; **173**: 5–20.
61 Venkataraman, C., Thomas, S. and Kulkarni, P. Size distributions of polycyclic aromatic hydrocarbons – gas/particle partitioning to urban aerosols. *Journal of Aerosol Science* 1999; **30**: 759–770.
62 Venkataraman, C., Reddy, C.K., Josson, S. and Reddy, M.S. Aerosol size and chemical characteristics at Mumbai, India, during the INDOEX-IFP (1999). *Atmospheric Environment* 2002; **36**: 1979–1991.
63 He, K., Yang, F., Ma, Y., Cadle, S.H., Chan, T., Mulawa, P.A. and Yan, Y. Trends in PM2.5 concentration and its relationship with PM10 and TSP in China. In: *Proceedings of the Sixth International Aerosol Conference*. Taipei, Taiwan: 2002; pp. 257.
64 Zhang, X.Y., Cao, J.J., Li, L.M., Arimoto, R., Cheng, Y., Huebert, B. and Wang, D. Characterization of atmospheric aerosol over XiAn in the South Margin of the Loess Plateau, China. *Atmospheric Environment* 2002; **36**: 4189–4199.
65 Parekh, P.P., Khwaja, H.A., Khan, A.R., Naqvi, R.R., Malik, A., Shah, S.A., Khan, K. and Hussain, G. Ambient air quality of two metropolitan cities of Pakistan and its health implications. *Atmospheric Environment* 2001; **35**: 5971–5978.
66 Vega, E., Reyes, E., Sanchez, G., Ortiz E., Ruiz, M., Chow, J., Watson, J. and Edgerton, S. Basic statistics of $PM_{2.5}$ and PM_{10} in the atmosphere of Mexico city. *Science of the Total Environment* 2002; **287**: 167–176.
67 Moloi, K., Chimidza, S., Lindgren, E.S., Viksna, A. and Standzenieks, P. Black carbon, mass and elemental measurements of airborne particles in the village of Serowe, Botswana. *Atmospheric Environment* 2002; **36**: 2447–2457.

68 Darlington, T.L., Kahlbaum, D.F., Heuss, J.M. and Wolff, G.T. Analysis of PM10 trends in the United States from 1988 through 1995. *Journal of the Air & Waste Management Association* 1997; **47**: 1070–1078.

69 Brook, J.R., Dann, T.F. and Bonvalot, Y. Observations and interpretations from the Canadian Fine Particle Monitoring Program. *Journal of the Air & Waste Management Association* 1999; **49**: 35–44.

70 Houthuijs, D., Breugelmans, O., Hoek, G., Vaskovi, E., Mihalikova, E., Pastuszka, J.S., Jirik, V., Sachelarescu, S., Lolova, D., Meliefste, K., Uzunova, E., Marinescu, C., Volf, J., de Leeuw, F., van de Wiel, H., Fletcher, T., Lebret, E. and Brunekreef, B. PM10 and PM2.5 concentrations in Central and Eastern Europe: results from the Cesar study. *Atmospheric Environment* 2001; **35**: 2757–2771.

71 Woo, K.S., Chen, D.R., Pui, D.Y.H. and McMurry, P.H. Measurement of Atlanta aerosol size distributions: observations of ultrafine particle events. *Aerosol Science and Technology* 2001; **34**: 75–87.

72 Raes, F., Van Dingenen, R., Vignati, E., Wilson, J., Putaud, J.P., Seinfeld, J.H. and Adams, P. Formation and cycling of aerosols in the global troposphere. *Atmospheric Environment* 2000; **34**: 4215–4240.

73 Putaud, J.P., Raes, F., Van Dingenen, R., Bruggemann, E., Facchini, M.C., Decesari, S., Fuzzi, S., Gehrig, R., Huglin, C., Laj, P., Lorbeer, G., Maenhaut, W., Mihalopoulos, N., Muller, K., Querol, X., Rodriguez, S., Schneider, J., Spindler, G., ten Brink, H., Torseth, K. and Wiedensohler, A. European aerosol phenomenology-2: chemical characteristics of particulate matter at kerbside, urban, rural and background sites in Europe. *Atmospheric Environment* 2004; **38**: 2579–2595.

74 Poschl, U. Atmospheric aerosols: composition, transformation, climate and health effects. *Angewandte Chemie –International Edition* 2005; **44**: 7520–7540.

75 CAFE. Second Position Paper on Particulate Matter 2004. Available at: http://ec.europa.eu/environment/air/cafe/pdf/working_groups/2nd_position_paper_pm.pdf.

76 Herner, J.D., Ying, Q., Aw, J., Gao, O., Chang, D.P.Y. and Kleemani, M.J. Dominant mechanisms that shape the airborne particle size and composition distribution in central California. *Aerosol Science and Technology* 2006; **40**: 827–844.

77 Anlauf, K., Li, S.M., Leaitch, R., Brook, J., Hayden, K., Toom-Sauntry, D. and Wiebe, A. Ionic composition and size characteristics of particles in the Lower Fraser Valley: Pacific 2001 Field Study. *Atmospheric Environment* 2006; **40**: 2662–2675.

78 Karanasiou, A.A., Sitaras, I.E., Siskos, P.A. and Eleftheriadis, K. Size distribution and sources of trace metals and *n*-alkanes in the Athens urban aerosol during summer. *Atmospheric Environment* 2007; **41**: 2368–2381.

79 McNeill, J.R. Something new under the sun. In: *An Environmental History of the Twentieth Century World*. W.W. Norton, New York: 2001.

80 Wehner, B., Wiedensohler, A., Tuch, T.M., Wu, Z.J., Hu, M., Slanina, J. and Kiang, C.S. Variability of the aerosol number size distribution in Beijing, China: new particle formation, dust storms, and high continental background. *Geophysical Research Letters* 2004; **31**: Art. No. L22108.

Chapter 2
Nucleation

Ari Laaksonen and Kari E.J. Lehtinen

2.1 Introduction

Nucleation is the initiating step of a first-order phase transition. Tiny clusters ('nuclei') appear in a metastable mother phase that carry the basic properties of the new phase. Nucleation research has in recent years increased considerably, mainly because of the possible climatic relevance of atmospheric nucleation events [1]. In this chapter, we consider nucleation of liquid droplets from the vapour. The liquid nuclei consist typically of a few to a few hundred molecules, which means that they are very small objects from the thermodynamic viewpoint, and are not necessarily well described with macroscopic liquid properties. On the other hand, they are quite large and complex objects from the microscopic viewpoint that tries to describe them starting from the interactions between individual molecules. As a result, nucleation has been studied for more than a hundred years [2, 3] but a definitive theory that would yield accurate predictions is still not available. Since molecular-based theories or computer simulations do not yet capture the nucleation behaviour of real fluids quantitatively, we concentrate on traditional nucleation theory and its applications. We first review nucleation kinetics in one-component and binary systems. We then go on to nucleation thermodynamics, and derive the one-component and multicomponent classical nucleation theories. After that, we present the so-called nucleation theorem, which enables experimental measurements of the sizes of critical nuclei. Finally, we apply the nucleation theorem in examination of scaling properties of critical nuclei, and in deducing the critical nucleus composition in atmospheric nucleation events based on measurements of particle formation rates.

2.2 Nucleation kinetics

2.2.1 One-component systems

Consider the dynamics of g-mers, that is clusters containing g molecules. Assuming that cluster growth/shrinkage occurs by addition/removal of monomers only, the number concentration of g-mers N_g is described by the following differential equation [4]:

$$\partial N_g/\partial t = (C_{g-1}N_{g-1} + E_{g+1}N_{g+1}) - (C_g N_g + E_g N_g) \tag{2.1}$$

Here C_g denotes the rate of condensation of monomers to a g-mer, and E_g is the evaporation rate of g-mers. The net current of molecules between sizes g and $g + 1$ is

$$J_g = C_g N_g - E_{g+1} N_{g+1} \tag{2.2}$$

transforming Equation (2.1) into a convenient form for distinguishing different cases:

$$\partial N_g / \partial t = J_{g-1} - J_g \tag{2.3}$$

The case with J_g zero for all g is the equilibrium, or balanced steady-state, case. When J_g has a non-zero constant value for all g, we are considering the unbalanced steady-state case. In both steady-state cases N_g does not change with time.

A short non-steady-state period exists when a body of vapour has just been transformed, for example by adiabatic expansion. During this period the cluster distribution evolves through collisional processes according to Equation (2.1). With many substances these times are in the order of tens of microseconds, so that the transient period may be ignored, for example in expansions performed in the laboratory, which typically last for a few milliseconds [5]. However, there are some exceptions such as the binary sulphuric acid/water system, for which the transient period is longer than a typical expansion time in a cloud chamber experiment. This is because of the very low vapour pressure of sulphuric acid; the collisions of H_2SO_4 molecules to the clusters are very rare, and thus it takes a relatively long time to steady the cluster populations off at moderately large g.

At steady state, obtained by equating the left sides of Equation (2.1) to zero, we have for the (quasi-steady) nucleation rate

$$J = C_1 N_1 - E_2 N_2 = C_2 N_2 - E_3 N_3 = \cdots = C_g N_g - E_{g+1} N_{g+1} \tag{2.4}$$

This period with constant currents J_g lasts typically tens to hundreds of milliseconds, after which the vapour is depleted appreciably and the conditions defining our problem are no longer constant.

The condensation rate C_g is given by the gas kinetic theory as $4\pi R^2 P_v / (2\pi mkT)^{1/2}$, where R is the radius of the g-mer, m is the mass of the vapour molecule, k is Boltzmann's constant and T is temperature. The evaporation rate, E_{g+1}, is not dependent on the pressure of the vapour; it is solely determined by the temperature and the size of the cluster. We can therefore calculate it at thermodynamic equilibrium ($S = 1$), where $J = 0$ and the balanced steady state holds:

$$E_{g+1} n_{g+1} = C_g^e n_g = C_g n_g / S \tag{2.5}$$

The number density of g-mers (denoted by a lowercase n in order to distinguish it from the supersaturated number density) is given by the Boltzmann distribution

$$n_g = \frac{P_e}{kT} \exp(-\Delta G_e / kT) \tag{2.6}$$

ΔG_e is the Gibbs free energy difference between g molecules in the vapour phase and in a g-mer, and the prefactor of the exponential is often approximated with the monomer number density.

By inserting (2.5) into (2.4) we get

$$\begin{aligned} J &= C_g N_g - C_g^e \frac{n_g}{n_{g+1}} N_{g+1} \\ &= C_g n_g S^g \left(\frac{N_g}{n_g S^g} - \frac{N_{g+1}}{n_{g+1} S^{g+1}} \right) \end{aligned} \tag{2.7}$$

which leads into a series of equations:

$$\begin{aligned} J/C_1 n_1 S^1 &= N_1/n_1 S^1 - N_2/n_2 S^2 \\ J/C_2 n_2 S^2 &= N_2/n_2 S^2 - N_3/n_3 S^3 \\ &\vdots \\ J/C_G n_G S^G &= N_G/n_G S^G - N_{G+1}/n_{G+1} S^{G+1} \end{aligned} \tag{2.8}$$

We now assume that at some sufficiently large G ($\gg$ the critical nucleus size), $N_{G+1} = 0$. (We note here that the critical nucleus, i.e. the smallest stable cluster, can be defined kinetically as a cluster having equal growth and decay probabilities, or thermodynamically as a cluster which is in unstable equilibrium with the supersaturated vapour, distinguished as a maximum in ΔG as a function of cluster size.) Summing up the equations gives

$$\sum_{g=1}^{G-1} (J/C_g n_g S^g) = N_1/n_1 S \simeq 1 \tag{2.9}$$

By approximating the difference in Equation (2.7) with a differential

$$J \simeq -C_g n_g S^g \frac{d}{dg}(N_g/n_g S^g) \tag{2.10}$$

and integrating from $g = 1$ to G, we obtain

$$J = \left[\int_1^G dg/C_g S^g n_g \right]^{-1} \tag{2.11}$$

with $S^g n_g \simeq n_1 \exp(-\Delta G_e/kT + g \ \ln S)$. The quantity inside the exponential is just the free energy difference in the supersaturated case and can be written as

$$\Delta G = A\sigma - gkT \ \ln S \tag{2.12}$$

where A is the surface area of the g-mer, and σ is the surface free energy. To approximate the integral, we note that $S^g n_g$ has a sharp minimum at g^*, where the asterisk denotes critical nucleus. Thus, by approximating the integrand by its Taylor series around this minimum, we may expect to obtain a reasonable approximation. In addition, the condensation rate is a fairly smooth function of g, so that we may take it out of the denominator in Equation (2.11) as a constant, and evaluate it at g^*. The Taylor series expansion of ΔG at g^* is

$$\Delta G = \Delta G^* + \frac{1}{2}\left[\frac{\partial^2(\Delta G)}{\partial g^2}\right]_g^* (g - g^*) + \cdots \tag{2.13}$$

Note that the first derivative of ΔG is zero at g^* by definition. If we write the surface area in (2.12) in terms of g, the second derivative in the Taylor series can be expressed as

$$\left[\frac{\partial^2(\Delta G)}{\partial g^2}\right]_{g^*} = -(2/9)\Theta g^{*-4/3} \equiv -Q \tag{2.14}$$

where $\Theta = A\sigma/g^{2/3}$. (Note that Θ is independent of g.)

The integral in (2.11) is now

$$J^{-1} = \frac{\exp(\Delta G^*/kT)}{n_1 C_{g^*}} \int_{-g^*-1}^{G-g^*} \exp(-Qx^2/2kT)\mathrm{d}x \tag{2.15}$$

where we have changed to a new variable of integration $x = g - g^*$. To be able to perform the integration analytically, we shift the lower and upper limits of integration to $-\infty$ and $+\infty$, respectively. (This should be reasonable as long as g^* is at least a few tens of molecules, and $G > 2g^*$.) With this approximation, we finally have

$$J = (n_1 C_{g^*}/Z) \exp(-\Delta G^*/kT) \tag{2.16}$$

where the Zeldovich non-equilibrium factor $Z = (Q/2\pi kT)^{(1/2)}$.

2.2.2 Binary systems

Consider the growth and decay of binary clusters composed of g_1 molecules of substance 1 and g_2 molecules of substance 2. We denote the rates of condensation of monomers of species 1 and 2 to the cluster by $C_1(g_1, g_2)$ and $C_2(g_1, g_2)$, respectively. The evaporation rates are denoted by $E_1(g_1, g_2)$ and $E_2(g_1, g_2)$. As with the one-component system, we take into account the addition and removal of monomers only.

We denote the net current in the direction $(g_1, g_2) \rightarrow (g_1 + 1, g_2)$ by $J_1(g_1, g_2)$ and the net current in the direction $(g_1, g_2) \rightarrow (g_1, g_2 + 1)$ by $J_2(g_1, g_2)$. The total current is then a vector, $\mathbf{J} = (J_1, J_2)$. The currents can be expressed using the condensation and evaporation rates as

$$J_1(g_1, g_2) = C_1(g_1, g_2)N(g_1, g_2) - E_1(g_1 + 1, g_2)N(g_1 + 1, g_2) \tag{2.17}$$

$$J_2(g_1, g_2) = C_2(g_1, g_2)N(g_1, g_2) - E_2(g_1, g_2 + 1)N(g_1, g_2 + 1) \tag{2.18}$$

In equilibrium, the net currents are zero in all directions, and the concentrations are given by

$$n(g_1, g_2) = [n(1, 0) + n(0, 1)] \exp(-\Delta G(g_1, g_2)/kT) \tag{2.19}$$

The evaporation rates can then be expressed as

$$E_1(g_1, g_2) = \frac{C_1^e(g_1 - 1, g_2)n(g_1 - 1, g_2)}{n(g_1, g_2)} \tag{2.20}$$

$$E_2(g_1, g_2) = \frac{C_2^e(g_1, g_2 - 1)n(g_1, g_2 - 1)}{n(g_1, g_2)} \tag{2.21}$$

and the condensation rate is obtained from kinetic gas theory:

$$C_i^e(g_1, g_2) = n_i \left(\frac{3}{4\pi}\right)^{1/6} (6kT)^{1/2} \left(\frac{1}{m(g_1, g_2)} + \frac{1}{m_i}\right)^{1/2} \left(v(g_1, g_2)^{1/3} + v_i^{1/3}\right) \quad (i = 1, 2) \tag{2.22}$$

where $m(g_1, g_2)$ and $v(g_1, g_2)$ are the mass and volume of the cluster, respectively, and n_i, m_i and v_i are the concentration, mass and volume of the condensing monomer.

The evaporation rates can now be eliminated from the currents, and we have

$$J_1(g_1, g_2) = C_1(g_1, g_2)n(g_1, g_2)\left(\frac{N(g_1, g_2)}{n(g_1, g_2)} - \frac{N(g_1 + 1, g_{2+1})}{n(g_1 + 1, g_2)}\right) \tag{2.23}$$

$$J_2(g_1, g_2) = C_2(g_1, g_2)n(g_1, g_2)\left(\frac{N(g_1, g_2)}{n(g_1, g_2)} - \frac{N(g_1, g_2 + 1)}{n(g_1, g_2 + 1)}\right) \tag{2.24}$$

At the continuum limit we can replace the differences by differentials, and the vector current becomes

$$\mathbf{J} = \begin{pmatrix} J_1 \\ J_2 \end{pmatrix} = -n(g_1, g_2)\begin{pmatrix} C_1 & 0 \\ 0 & C_2 \end{pmatrix}\nabla\left(\frac{N}{n}\right) \tag{2.25}$$

where

$$\nabla = \begin{pmatrix} \frac{\partial}{\partial g_1} \\ \frac{\partial}{\partial g_2} \end{pmatrix} \tag{2.26}$$

(Note that all of the matrix elements in Equation (2.25) would be non-zero if we accounted for condensation/evaporation of n-mers.)

Assuming steady state, Equation (2.25) can be integrated, but we will not go into the details of this somewhat tedious exercise. Instead, we note that the free energy ΔG forms a saddle-shaped surface as a function of g_1 and g_2, and the free energy of the critical cluster corresponds to the saddle point. The integration involves finding the direction of the nucleation current at the saddle point and, as with the one-component case, expanding ΔG in a Taylor series in the vicinity of the saddle point. The resulting formula for the nucleation rate is [6]

$$J = N_v A B_{AV} Z \exp(-\Delta G/kT) \tag{2.27}$$

Here $N_v = N_{1v} + N_{2v}$ is the total number density of the vapour phase, A is the surface area of the nucleus, B_{AV} is the average growth rate and Z is the Zeldovich non-equilibrium factor. The average growth rate is given for non-associated vapours by

$$B_{AV} = \frac{B_1 B_2}{B_1 \sin\Theta + B_2 \cos\Theta} \tag{2.28}$$

where Θ is the angle between the g_1-axis and direction of growth at the saddle point in the three-dimensional space $(g_1, g_2, \Delta G)$. The growth rates of the single species are obtained from kinetic gas theory:

$$B_i = N_{iv}\sqrt{\frac{kT}{2\pi m_i}} \tag{2.29}$$

The Zeldovich factor is given by

$$Z = \frac{-D_{11}\cos^2\Theta + 2D_{12}\cos\Theta\sin\Theta + D_{22}\sin^2\Theta}{\sqrt{D_{12}^2 - D_{11}D_{22}}} \tag{2.30}$$

with

$$D_{ij} = \frac{1}{2}\frac{\partial^2 \Delta G^*}{\partial g_1\,\partial g_2} \tag{2.31}$$

For practical purposes, the angle Θ can be approximated by the angle of steepest descent:

$$\tan\Theta \approx \frac{g_2}{g_1} \tag{2.32}$$

The derivatives in the Zeldovich factor can be evaluated numerically only if the saddle surface is known. In practice, one usually knows only the free energy at the saddle point, ΔG^*, and an approximate form for Z is needed. Kulmala and Viisanen [7] considered a critical nucleus consisting of fictitious 'average' monomers with volume

$$v_{AV} = (g_1 v_1 + g_2 v_2)/(g_1 + g_2) \tag{2.33}$$

in which case the problem reduces to one-component nucleation:

$$Z = \sqrt{\frac{\sigma}{kT}\frac{v_{AV}}{2\pi R^{*2}}} \tag{2.34}$$

2.3 Nucleation thermodynamics

2.3.1 One-component systems

In this section we will consider the thermodynamics of cluster formation in a supersaturated vapour. The vapour phase (denoted by a subscript v) is at constant pressure and temperature, and contains in itself a 'liquid' cluster (subscript l; the quotation marks are used because we do not assume that the cluster is characterised by actual bulk liquid properties). We divide the volume of the system into two parts

$$V = V_v + V_l \tag{2.35}$$

separated by an arbitrarily chosen dividing surface (Figure 2.1). Here V_l encloses the cluster and V_v contains the vapour phase. The dividing surface is a mathematical construction introduced by Gibbs (see e.g. Ref. 8) that has a zero thickness, and the volume of the interface is consequently zero. The total number of molecules in the system g_t is then

$$g_t = g_v + g_l + g_s \tag{2.36}$$

Here g_v is the number of molecules that would occupy a volume V_v of uniform vapour phase (density ρ_v), g_l is the corresponding number for the volume of the uniform 'liquid' phase at density ρ_l (see Figure 2.1) and g_s is the number of molecules that corrects for the shape of the actual interfacial profile. The subscript s refers to surface, even though the surface volume is zero. This is because the value of g_s depends on the choice of dividing surface. The special surface at which $g_s = 0$ is referred to as the equimolar surface.

The Gibbs free energy change to create the cluster from the vapour phase is

$$\Delta G = (P_v - P_l)V_l + (\mu_l - \mu_v)g_l + (\mu_s - \mu_v)g_s + \Phi(V_l, \mu_v) \tag{2.37}$$

Here μ_l, μ_v and μ_s are the chemical potentials of the liquid, vapour and surface phases, and Φ is the total surface energy of the cluster. Note that the pressure of the surface phase does not come in, since it has no volume.

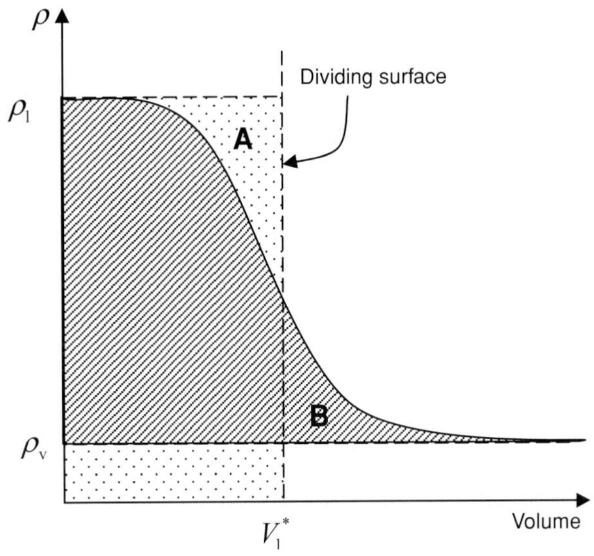

Figure 2.1 A schematic showing the density profile of a nucleus in uniform vapour. The excess number of molecules in the nucleus g^* (see Equation (2.45)) would be obtained by integrating over the smoothly varying (real) density profile of the nucleus and subtracting the uniform vapour portion (as described by the dark, shaded area). In the Gibbs model, the nucleus is taken to have a uniform density ρ_l up to the dividing surface and a volume V_l^* . The number of molecules within V_l^* is denoted by g_l. The excess number of molecules in the nucleus is thus $g_l - V_l\rho_v + g_s$, where the second term represents the uniform vapour that would be present in the nucleus volume if the nucleus were not there, and the g_s is a surface excess term that corrects for the difference between the actual nucleus and the Gibbs model, and correspods to the difference of the areas A and B.

The critical nucleus (superscript *) is in chemical and (unstable) mechanical equilibrium with the vapour. The equilibrium conditions can be written as

$$\begin{aligned} &\partial \Delta G / \partial g_l = 0 \\ &\partial \Delta G / \partial g_s = 0 \\ &\partial \Delta G / \partial V_l = 0 \end{aligned} \tag{2.38}$$

leading to

$$\mu_v = \mu_l^* = \mu_s^* \tag{2.39}$$

$$P_l^* = P_v + \frac{\partial \Phi^*}{\partial V_l^*} \tag{2.40}$$

The nucleation work $W^* = \Delta G(V_l^*, \mu_v)$ is then

$$W^* = (P_v - P_l^*)V_l^* + \Phi(V_l^*, \mu_v) \tag{2.41}$$

The above development is very general. We will next move to a somewhat more specialised direction by assuming that the surface energy is given by $\Phi(V_l^*, \mu) = 4\pi R_l^{*2}\sigma(R_l^*, g^*)$, with

$V_l^* = (4/3)\pi R_l^{*3}$. The number g^* is defined as the *excess number of molecules over the uniform vapour* (see Figure 2.1); that is it is given by

$$g^* = (\rho_l - \rho_v)V_l^* + g_s = g_l + g_s - \rho_v V_l^* \quad (2.42)$$

and is determined solely by the chemical potential of the vapour, μ_v.

With these assumptions Equation (2.40) becomes the generalised Laplace equation

$$P_l^* - P_v = 2\sigma^*/R_l^* + [\partial\sigma(R_l^*, g^*)/\partial R_l^*] \quad (2.43)$$

and the work of formation is given by

$$W^* = (P_v - P_l^*)V_l^* + 4\pi R_l^{*2}\sigma(R_l^*, g^*) \quad (2.44)$$

For a cluster with fixed g^*, the surface free energy has a minimum at the so-called Gibbs 'surface of tension' (subscript G): $[\partial\sigma(R_l^*, g^*)/\partial R_l^*]_{R_G^*} = 0$. At R_G^*, σ_G becomes effective surface tension (note that it still does not necessarily have the flat interface value because it depends on g^*), and Equation (2.43) becomes

$$P_l^* - P_v = 2\sigma_G/R_G^* \quad (2.45)$$

which is the well-known Laplace relation. The work of formation is given by

$$W^* = (P_v - P_l^*)V_G^* + 4\pi R_G^{*2}\sigma_G \quad (2.46)$$

With model fluids, for which μ and P can be calculated as a function of density, Equations (2.39), (2.45) and (2.46) are sufficient to determine the radius and work of formation of the critical cluster (provided that σ_G is known). However, with real fluids, the chemical potential of the liquid is known only at equilibrium, and thus Equation (2.39) cannot be used to find the density of the liquid coexisting with the supersaturated vapour. To overcome this restriction, we assume that the droplet consists of an incompressible liquid, that is, the molecular volume is given by $v = V_l^*/g_l^*$. The Gibbs–Duhem relation for the droplet is

$$S_l\,dT - V_l\,dP_l + g_l\,d\mu_l = 0 \quad (2.47)$$

where S_l denotes the entropy of the liquid phase. Integrating this equation at constant temperature subject to the incompressibility constraint gives

$$\mu_l(P_l) - \mu_l(P_e) = v(P_l - P_e) \quad (2.48)$$

where we make use of the fact that the pressures of the vapour and the liquid phases are equal at equilibrium.

The vapour is assumed to be ideal:

$$P_v V_v = g_v kT \quad (2.49)$$

The Gibbs–Duhem relation for the vapour is

$$S_v\,dT - V_v\,dP_v + g_v\,d\mu_v = 0 \quad (2.50)$$

The two relations can be combined to give

$$S_v\,dT - [g_v kT/P_v]dP_v + g_v\,d\mu_v = 0 \quad (2.51)$$

Integration of Equation (2.51) at constant temperature yields

$$\mu_v(P_v) - \mu_v(P_e) = kT \ln(P_v/P_e) \tag{2.52}$$

Subtracting Equation (2.48) from Equation (2.52), and making use of the fact that the liquid- and vapour-phase chemical potentials are equal both in the supersaturated state (Equation (2.39)) and at equilibrium ($\mu_v(P_e) = \mu_l(P_e)$), we have

$$kT \ln(P_v/P_e) = v(P_l - P_e) \tag{2.53}$$

Substituting this to the Laplace equation (2.45) gives

$$kT \ln S = 2\sigma_G v/R_G^* + v(P_v - P_e) \tag{2.54}$$

Usually the last term is small and can be ignored; we thus obtain the well-known Kelvin equation:

$$kT \ln S = 2\sigma_G v/R_G^* \tag{2.55}$$

which can be used to find the radius of the critical cluster at given supersaturation (provided the surface tension is known). With the same approximations, the work of formation (Equation (2.46)) is given by

$$W^* = -g_G^* kT \ln S + 4\pi R_G^{*2} \sigma_G \tag{2.56}$$

where g_G^* denotes the number of molecules enclosed by the surface of tension.

2.3.2 The classical theory

The classical nucleation theory (CNT) follows from the assumption that the density and surface tension of the droplet are equal to those of equilibrium bulk liquid. This is generally called 'the capillarity approximation'. Note that capillarity approximation for the surface tension holds in the special case that the surface of tension happens to coincide with the equimolar surface. This can be seen by writing out the Gibbs–Duhem relation for the surface phase at the Gibbs surface of tension:

$$S_s \, dT + A_G \, d\sigma_G + g_{s,G} \, d\mu_s = 0 \tag{2.57}$$

However, at the equimolar surface, $g_s = 0$, so that at constant temperature we have

$$A_G \, d\sigma_G = 0 \tag{2.58}$$

or σ = constant (i.e. not dependent on droplet size). Thus we have $\sigma_G = \sigma_\infty$ = surface tension of flat interface, $g_G^* = g^*$, and the Kelvin equation becomes

$$kT \ln S = \frac{2\sigma_\infty v}{R^*} \tag{2.59}$$

where the radius is given by $(4/3)\pi R^{*3} = g^* v$. The work of formation is correspondingly

$$W^* = -g^* kT \ln S + 4\pi R^{*2} \sigma_\infty \tag{2.60}$$

Equations (2.59) and (2.60) constitute the working formulae of the CNT for calculation of the size of the critical nucleus and the nucleation work.

2.3.3 Multicomponent systems

The thermodynamics needed to determine the work of nucleus formation can be generalised from one-component systems to multicomponent systems. We will not go into all details; only the most important results are presented below.

Consider a multicomponent supersaturated vapour at temperature T and pressure P_v with mole fractions x_{iv}, in which critical liquid nuclei are appearing. The bulk properties (pressure and composition) of the liquid phase can be determined by solving the simultaneous equations:

$$\mu_{iv}(T, P_v, x_{iv}) = \mu_{il}(T, P_l, x_{il}) \quad (i = 1, 2, \ldots) \tag{2.61}$$

where μ_{iv} and μ_{il} are the vapour- and liquid-phase chemical potentials of component i, respectively, and P_l and x_{il} denote the liquid-phase pressure and mole fractions.

Assuming that the mole fractions x_{il} in the centre of the droplet are the same as for the hypothetical macroscopic liquid at pressure P_l, and that the surface tension is known, the radius R_G^* of the nucleus (assumed spherical) can be determined from the Laplace equation:

$$P_l - P_v = 2\sigma/R_G^* \tag{2.62}$$

The reversible work of formation of a critical nucleus is then given by

$$W^* = -V_G^*(P_l - P_v) + \sigma A_G^* \tag{2.63}$$

where V^* is the volume and A^* is the surface area of the nucleus. These equations follow from a straightforward generalisation of the derivation presented above for one-component systems. Note that the droplet volume V_G^* enclosed by the surface of tension equals $\sum g_{il} v_{il}$, where v_{il} denotes the partial molecular volume of species i.

The differentials of chemical potentials and pressure in the liquid phase are related by

$$d\mu_{il} = v_{il}\, dP_l \tag{2.64}$$

Assuming incompressible liquid phase, this relation can be integrated from P_v to P_l, and making use of the equality of chemical potentials across phases, we can write

$$v_{il}\Delta P = \mu_{il}(P_l) - \mu_{il}(P_v) = \mu_{iv}(P_v) - \mu_{il}(P_v) = -\Delta\mu_i \tag{2.65}$$

and the multicomponent Kelvin equations are obtaind by substitution to the Laplace equation:

$$\Delta\mu_i = 2\sigma v_{il}/R_G^* \tag{2.66}$$

The work of nucleus formation is given by

$$W^* = \sum_i g_{il}\Delta\mu_i + \sigma A_G^* \tag{2.67}$$

Equation (2.64) can also be utilised in deriving a relation that can be used for determination of the critical nucleus composition iteratively. For example, in a binary system, we have (assuming incompressibility):

$$\Delta\mu_1/v_{1l} = \Delta\mu_2/v_{2l} \tag{2.68}$$

The equations of the classical multicomponent nucleation theory for determining the critical nucleus size and work of formation are obtained by replacing σ in Equations (2.66) and (2.67) by the flat interface surface tension σ_∞. It can be shown [9] that this is equivalent to assuming that the special equimolar dividing surface defined by $\sum g_{is}v_{il} = 0$ coincides with the surface of tension.

In the binary CNT, the total numbers of molecules in the critical nucleus can be solved within the incompressibility approximation from [10]:

$$V = g_1 v_1 + g_2 v_2 \tag{2.69}$$

$$g_1 \, \mathrm{d}\mu_{1l} + g_2 \, \mathrm{d}\mu_{2l} + A \, \mathrm{d}\sigma_\infty = 0 \tag{2.70}$$

where the last equation follows from the addition of the Gibbs–Duhem relation and the Gibbs adsorption equation.

Note that the binary nucleation theory of Doyle [11] (that used to be referred to as the classical binary nucleation theory) does not take account of adsorption, and the resulting Kelvin equations include an erroneous surface tension derivative term. An equivalent error is made when the saddle point is located numerically from a $(g_1, g_2, \Delta G)$-surface, and the thermodynamic properties of the clusters (density, vapour pressure and surface tension) are evaluated using the overall mole fraction $g_1/(g_1 + g_2)$. The reason for this is that generally the interior and surface compositions of a cluster are different, and thus the interior mole fraction (which should be used to calculate the properties) is different from the overall mole fraction. In other words, using $g_1/(g_1 + g_2)$ to determine the thermodynamic properties means that adsorption is ignored, and the erroneous surface tension derivative is implicitly included in the saddle point conditions. This is avoided if one has some means to explicitly specify the interior and surface mole fractions of a (g_1, g_2)-cluster.

2.4 The nucleation theorem

We will now derive the so-called nucleation theorem that connects the excess number of molecules in the critical cluster and the derivative of the work of formation with respect to the vapour-phase chemical potential. In order to do this, we go back to Equations (2.40) and (2.41). (Note that at this point the dividing surface is not fixed.) We take the partial derivative of Equation (2.41) with respect to the chemical potential μ_v in the vapour phase:

$$\frac{\partial W^*}{\partial \mu_v} = V_l^* \frac{\partial (P_V - P_l^*)}{\partial \mu_v} + (P_V - P_l^*) \frac{\partial V_l^*}{\partial \mu_v} + \frac{\partial \Phi^*}{\partial V_l^*} \frac{\partial V_l^*}{\partial \mu_v} + \frac{\partial \Phi^*}{\partial \mu_v} \tag{2.71}$$

where we have denoted $\Phi(V_l^*, \mu_v)$ by Φ^* This equation can be simplified by using Equation (2.40) to cancel the second and third summand and by recalling the Gibbs–Duhem relations at constant temperature:

$$V_v \, \mathrm{d}P_V = g_v \, \mathrm{d}\mu_v \tag{2.72}$$

$$V_l^* \, \mathrm{d}P_l = g_l^* \, \mathrm{d}\mu_l \tag{2.73}$$

$$\mathrm{d}\Phi^* = -g_s^* \, \mathrm{d}\mu_s \tag{2.74}$$

Together with the equality of chemical potentials across phases, these transform Equation (2.71) into

$$\frac{\partial W^*}{\partial \mu_v} = -g_l^* - g_s^* + g_v \frac{V_l^*}{V_v} \tag{2.75}$$

This equation can be rewritten by noting that the number density ρ_v in the vapour phase is

$$\rho_v = g_v / V_v \tag{2.76}$$

Equation (2.75) then becomes

$$\frac{\partial W^*}{\partial \mu_v} = -g_l - g_s + \rho_v V_l = -g^* \tag{2.77}$$

which is the nucleation theorem. It is straightforward to generalise the nucleation theorem to multicomponent systems, yielding [12] $(\partial W^* / \partial \mu_v i)_{\mu_v j} = -g_i^*$.

The nucleation theorem, which was first derived by Kashchiev [13] and later in more general ways by Oxtoby and Kashchiev [12] (whose derivation we have more or less repeated above) and others [14–16] is particularly useful, because it can be further shown that (at constant temperature)

$$\frac{d \ln J}{d \ln S} \simeq g^* + \Delta \tag{2.78}$$

where Δ is on the order of unity or smaller. In other words, the molecular content of critical clusters can be calculated from the slopes of experimental isothermal nucleation rates versus supersaturation. Measurements performed for various substances [5, 17–20] show, quite surprisingly, that the classical Kelvin equation (Equation (2.59)) predicts the size of the critical cluster very accurately as long as it contains more than 30–40 molecules. On the other hand, the nucleation rates predicted by the classical theory are generally different from the experimental rates, indicating that the classical work of formation is incorrect.

2.4.1 Scaling properties of critical clusters

What might be the reason for the puzzling success of the classical Kelvin equation that makes use of planar surface tension in describing highly curved critical clusters? Let us take a look at the work of formation of the classical nucleus following McGraw and Laaksonen [21]. By making the substitution $\Delta\mu = \mu_v(P_v) - \mu_v(P_e)$ for $kT \ln S$ (equation 2.52), and combining Equations (2.59) and (2.60), we see that

$$\frac{W^*_{CNT}}{g^*_{CNT} \Delta\mu} = \frac{1}{2} \tag{2.79}$$

A *general* expression would be, for example,

$$\frac{W^*}{g^* \Delta\mu} = \frac{1}{2} - f(g^*, \Delta\mu) \tag{2.80}$$

where $f(g^*, \Delta\mu)$ gives the departure of experimental results from classical predictions. We will now apply the nucleation theorem to Equation (2.80): (Note that the nucleation theorem can be written as $dW^*/d\Delta\mu = -g^*$, because $\mu_v(P_e)$ is a constant at constant temperature.)

$$\frac{dW^*}{d\Delta\mu} = \frac{d}{d\Delta\mu}\left(\frac{g^*\Delta\mu}{2} - f g^* \Delta\mu\right) = -g^* \tag{2.81}$$

or

$$\Delta\mu \frac{\mathrm{d}}{\mathrm{d}\Delta\mu} g^* + 3g^* = 2\frac{\mathrm{d}}{\mathrm{d}\Delta\mu}(fg^*\Delta\mu) \tag{2.82}$$

where the argument of f has been suppressed.

In the classical nucleation theory, $f = 0$ and the solution of the differential equation (2.82) is a homogeneous function of the form

$$g^*_{\mathrm{CNT}} = C(T)(\Delta\mu)^{-3} \tag{2.83}$$

where $C(T)$ is a function of temperature alone. This is in agreement with the Kelvin relation, which yields an explicit formula for $C(T)$:

$$g^*_{\mathrm{CNT}} = \frac{32\pi v^2 \sigma_\infty^3}{3}(\Delta\mu)^{-3} \tag{2.84}$$

It is well known that the nucleation rate predictions of CNT usually fail, and thus f is generally non-zero. However, Equation (2.82) cannot be solved without additional information. The simplest solutions of Equation (2.82) are homogeneous, that is those obtained when both sides of the equation are set equal to zero separately. This homogeneity ansatz produces the following generalisation of the CNT:

$$g^* = C'(T)(\Delta\mu)^{-3} \tag{2.85}$$

and

$$fg^* = D(T)(\Delta\mu)^{-1} \tag{2.86}$$

where the integration constants $C'(T)$ and $D(T)$ depend only on T.

Now, the classical Kelvin equation can be assumed to hold accurately at the planar limit; that is $g^*/g^*_{\mathrm{CNT}} \to 1$ as $\Delta\mu \to 0$. This requirement is fulfilled only if $C'(T) = C(T)$, or in other words, if g^* of the generalised theory is given by the classical Kelvin equation. Experiments show that this seems indeed to be the case, at least when g^* is greater than about 30–40, indicating that the homogeneity ansatz is a reasonable approximation in this size range.

Equations (2.80) and (2.86) have the interesting consequence that the displacement between works of nucleus formation in the classical and generalised theories is a function only of temperature,

$$W^* - W^*_{\mathrm{CNT}} = W^* - \frac{1}{2}g^*\Delta\mu = -D(T) \tag{2.87}$$

This conclusion is also supported by experiment; the slopes of measured nucleation rates versus supersaturation are generally predicted quite well by the classical theory, even though the temperature dependence is not correct (see e.g. Ref. 5).

Equation (2.87) could be useful in formulation of nucleation theories that are as easily applicable as the classical theory, but yield more accurate predictions. One could for example envision semiempirical forms that correlate $D(T)$ with molecular and critical properties of fluids. Figure 2.2 shows $D(T)$ for several different fluids as a function of the reduced temperature $T_r = T/T_c$ extracted from nucleation rate measurements. The measurements were carried out with a single type of instrument, the two-valve nucleation pulse chamber [5]. The different

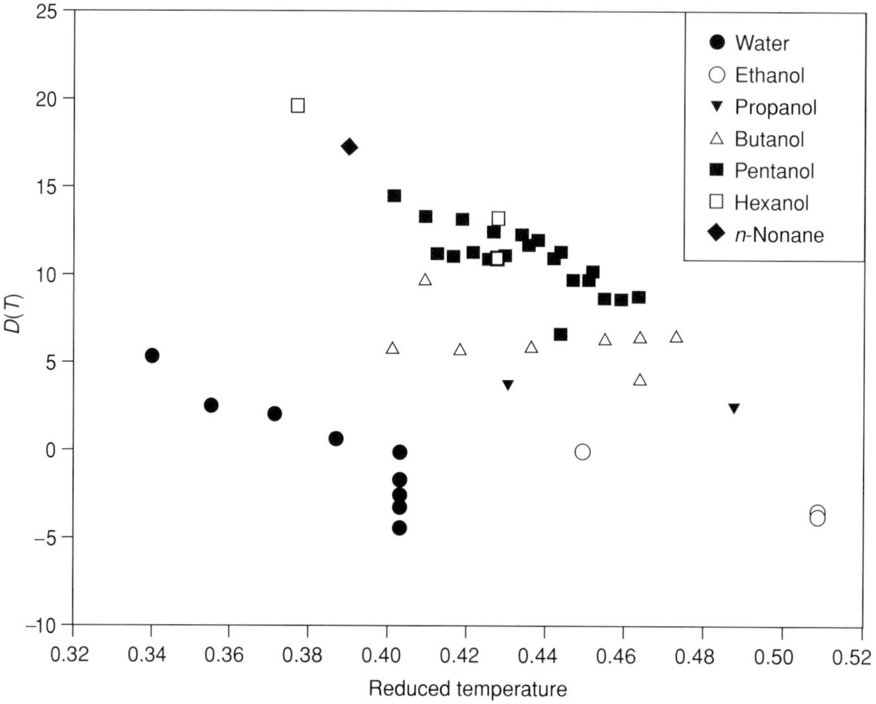

Figure 2.2 $D(T)$ as a function of the reduced temperature $T_r = T/T_c$, where T_c denotes critical temperature, for water [19, 22, 23], ethanol [22–25], *n*-propanol [23, 25, 26], butanol [23, 25], pentanol [20, 23, 27], hexanol [23–25] and *n*-nonane [25]. The experimental work of formation was extracted from nucleation rate data, assuming that the rate is adequately described with Equation (2.16). $D(T)$ in units of kT.

data sets are not in complement agreement as can be seen for example from the data points for water. (Note, however, that there is even bigger quantitative, and sometimes even qualitative, disagreement between nucleation rata data obtained using different measurement techniques [20].) However, it appears from Figure 2.2 that an experimental fit of the form $D(T) = a + bT_r$ could be made, with b negative and more or less constant for all substances, and a positive and increasing as a function of molecular weight. Whether this is really the case, though, can be decided only with a larger data set involving more data points for the individual substances, and a larger number of different substances.

So far, there are few theoretical approaches for determining $D(T)$. Talanquer [28] noted that the CNT does not predict the existence of the spinodal line of the $P - V$ phase diagram where the supersaturated vapour becomes unstable and W must tend to zero. He suggested that $D(T)$ could be obtained by determining the vapour-phase chemical potential at the spinodal using a model equation of state (e.g. van der Waals or the Peng–Robinson EoS [29]) and requiring that $W^*_{\mathrm{CNT}} = D(T)$ at the spinodal. Talanquer's theory seems to be succesful with some substances; however, it should be noted that in this theory, $D(T)$ can only be positive, wheras for many substances it can in reality be negative, depending on temperature, as shown in Figure 2.2.

2.4.2 An application to atmospheric nucleation

Atmospheric new particle formation has received considerable attention during the recent years [1]. The nature of the nucleating vapours is still to be elucidated, although sulphuric acid is strongly suspected to participate in the nucleation in most cases. One way to analyse the atmospheric particle formation, suggested by the nucleation theorem, is to consider the slope of the observed particle formation rate versus the concentration of sulphuric acid, or whatever vapour is suspected to cause the nucleation.

Applying the nucleation theorem on atmospheric particle formation measurements is not straightforward, however, since the available instrumentation does not allow us to measure nucleation directly but instead is limited to a certain size, currently at approximately 3 nm. Thus, after nucleation, the size distribution undergoes complex aerosol dynamical processes, before the particles reach the measurable range.

The measured formation rate J_m (at diameter d_m) is smaller than the actual nucleation rate J^* (at diameter d^*) and determined by the competition between condensational growth (GR = growth rate from experimental observations) and scavenging (rate proportional to condensation sink CS) [30]:

$$J_m = J^* \exp\left[\gamma(d_m^{-1} - d^{*-1})\frac{\mathrm{CS}}{\mathrm{GR}}\right] \tag{2.88}$$

where $\gamma = 0.23\,\mathrm{nm^2\,m^2\,h^{-1}}$. The condensation sink CS describes the loss rate of molecules with diffusion coefficient D onto a distribution $n(d_p)$ of existing particles:

$$\mathrm{CS} = 1/2 \int_0^\infty d_p \beta_M(d_p) n(d_p) d\, d_p \tag{2.89}$$

Here β_M is the transitional correction factor [31]. Setting the formation rate at 3 nm (J_3) and nucleation rate at 1 nm (J_1) gives

$$J_3 = J_1 \exp\left[\gamma \frac{\mathrm{CS}}{\mathrm{GR}}\right] \tag{2.90}$$

where $\gamma = -0.153\,\mathrm{nm\,m^2\,h^{-1}}$. If we assume that both the nucleation rate and the formation rate of 3-nm particles can be expressed using a power–law dependence of the vapour concentration C, J_1 being proportional to C^{n1} and J_3 to C^{n3}, we obtain, by taking logarithms on each side,

$$n_3 \ln C = n_1 \ln C - 0.153\frac{\mathrm{CS}}{\mathrm{GR}} \tag{2.91}$$

Now, in addition to the nucleation rate terms J_1 and J_3, the vapour concentration C can appear in the growth rate term GR. The formation rate dependency on C is illustrated in Figure 2.3. The curve representing J_1 is obviously a straight line, with the slope n_1 depending on the nucleation mechanism. If the effect of sulphuric acid on growth is negligible, that is the growth rate GR does not depend on C, then the term 0.153CS/GR in Equation (2.8) is constant (with respect to C). Hence the resulting apparent nucleation rate at 3 nm J_3 (Figure 2.3) has the same slope as J_1, but a lower absolute value than J_1 by magnitude 0.153CS/GR. If GR does depend on C, the exact value for the slope is not straightforward to determine. In any case, it is clear that at otherwise the same conditions, increasing C will increase GR, thus decreasing the magnitude of 0.153CS/GR. This means that J_3 will approach J_1 at high values of C, as

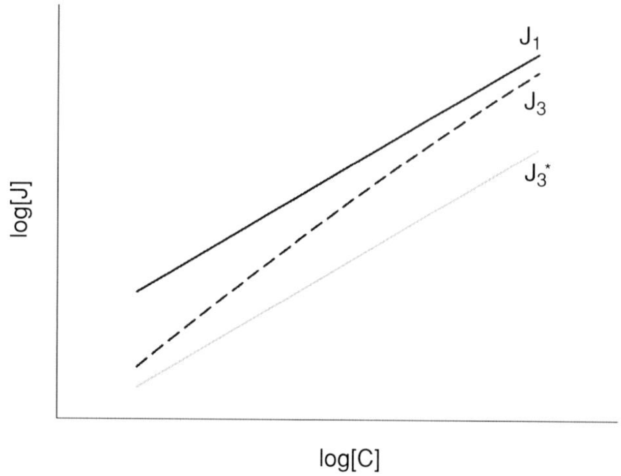

Figure 2.3 Dependences of nucleation rate at 1-nm (J_1) and 3-nm particle formation rate (J_3) on the concentration of the nucleating vapour.

indicated qualitatively (dashed line) in Figure 2.1. Thus the slope for J_3 should be steeper than the slope for J_1; that is

$$n_1 \leq n_3. \tag{2.92}$$

Several atmospheric field measurements have indicated a clear presence of sulphuric acid vapour in the particle formation process. By investigating the slopes of $\ln J$ versus $\ln C$ (with C the sulphuric acid concentration), Weber *et al.* [32] showed that the slope for J_3 was between 1 and 2 in Idaho Hill, US. Preliminary results [33] from Hyytiälä, Finland, also show that the slope is smaller than 2. Now, the above analysis shows that if n_3 is smaller than 2, n_1 should also be smaller than 2. In classical homogenous nucleation, the dependence of sulphuric acid vapour should be stronger, with n_1 at least in the range 5–10 in typical atmospheric/experimantal conditions. Thus, these studies suggest another nucleation mechanism, either kinetic barrierless nucleation, or some kind of an activation mechanism involving other vapours participating in particle growth.

References

1 Kulmala, M., Vehkamaki, H., Petäjä, T., Dal Maso, M., Lauri, A., Kerminen, V.M., Birmili, W. and McMurry, P.H. Formation and growth rates of ultrafine atmospheric particles: a review of observations. *Journal of Aerosol Science* 2004; **35**: 143–176.

2 Abraham, F.F. *Homogeneous Nucleation Theory*. Academic Press, New York: 1974.

3 Laaksonen, A., Talanquer, V. and Oxtoby, D.W. Nucleation – measurements, theory, and atmospheric applications. *Annual Review of Physical Chemistry* 1995; **46**: 489–524.

4 McDonald, J.E. Homogeneous nucleation of vapour condensation II. Kinetic aspects. *American Journal of Physics* 1963; **31**: 31–41.

5 Strey, R., Wagner, P.E. and Viisanen, Y. The problem of measuring homogeneous nucleation rates and the molecular contents of nuclei – progress in the form of nucleation pulse measurements. *Journal of Physical Chemistry* 1994; **98**: 7748–7758.

6 Stauffer, D. Kinetic theory of two-component ('heteromolecular') nucleation and condensation. *Journal of Aerosol Science* 1976; **7**: 319–333.

7 Kulmala, M. and Viisanen, Y. Homogeneous nucleation – reduction of binary nucleation to homomolecular nucleation. *Journal of Aerosol Science* 1991; **22**: S97–S100.

8 Alberty, R.A. *Physical Chemistry*, 6th edn. John Wiley & Sons, New York: 1983.

9 Laaksonen, A., McGraw, R. and Vehkamäki, H. Liquid-drop formalism and freeenergy surfaces in binary homogeneous nucleation theory. *Journal of Chemical Physics* 1999; **111**: 2019–2027.

10 Laaksonen, A., Kulmala, M. and Wagner, P.E. On the cluster compositions in the classical binary nucleation theory. *Journal of Chemical Physics* 1993; **99**: 6832–6835.

11 Doyle, G.J. Self-nucleation in the sulfuric acid-water system. *Journal of Chemical Physics* 1961; **35**: 795–799.

12 Oxtoby, D.W. and Kashchiev, D. A general gelation between the gucleation gork and the size of the nucleus in multicomponent nucleation. *Journal of Chemical Physics* 1994; **100**: 7665–7671.

13 Kashchiev, D. On the relation between nucleation work, nucleus size, and nucleation rate. *Journal of Chemical Physics* 1982; **76**: 5098–5102.

14 Viisanen, Y., Strey, R. and Reiss, H. Nucleation rates for water. *Journal of Chemical Physics* 1993; **99**: 4680–4692.

15 ten Wolde, P.R. Numerical study for homogeneous nucleation. Ph.D. Thesis. University of Amsterdam, 1998.

16 MacDowell, L.G. Formal study of nucleation as described by fluctuation theory. *Journal of Chemical Physics* 2003; **119**: 453–463.

17 Viisanen, Y. and Strey, R. Homogeneous nucleation rates for *n*-butanol. *Journal of Chemical Physics* 1994; **101**: 7835–7843.

18 Kim, Y.J., Wyslouzil, B.E., Wilemski, G., Wolk, J. and Strey, R. Isothermal nucleation rates in supersonic nozzles and the properties of small water clusters. *Journal of Physical Chemistry A* 2004; **108**: 4365–4377.

19 Wölk, J. and Strey, R. Homogeneous nucleation of H_2O and D_2O in comparison: the isotope effect. *Journal of Physical Chemistry B* 2001; **105**: 11683–11701.

20 Iland, K., Wedekind, J., Wölk, J., Wagner, P.E. and Strey, R. Homogeneous nucleation rates of 1-pentanol. *Journal of Chemical Physics* 2004; **121**: 12259–12264.

21 McGraw, R. and Laaksonen, A. Scaling properties of the critical nucleus in classical and molecular-based theories of vapour-liquid nucleation. *Physical Review Letters* 1996; **76**: 2754–2757.

22 Viisanen, Y., Strey., R., Laaksonen A. and Kulmala, M. Measurement of the molecular content of binary nuclei. II. Use of the nucleation rate surface for water-ethanol. *Journal of Chemical Physics* 1994; **100**: 6062–6072.

23 Strey, R., Viisanen, Y. and Wagner, P.E. Measurement of the molecular content of binary nuclei. III. Use of the nucleation rate surfaces for the water-*n*-alcohol series. *Journal of Chemical Physics* 1995; **101**: 4333–4345.

24 Strey, R. and Viisanen, Y. Measurement of the molecular content of binary nuclei. Use of the nucleation rate surface for the ethanol-hexanol. *Journal of Chemical Physics* 1993; **99**: 4693–4704.

25 Viisanen, Y., Wagner, P.E. and Strey, R. Measurement of the molecular content of binary nuclei. IV. Use of the nucleation rate surfaces for the *n*-nonane-*n*-alcohol series. *Journal of Chemical Physics* 1998; **108**: 4257–4265.

26 Viisanen Y. Experimental study of binary nucleation in the water *n*-propanol vapour mixture. Ph.D. Thesis. University of Helsinki, 1991.

27 Hruby, J., Viisanen,Y. and Strey, R. Homogeneous nucleation rates for *n*-pentanol in argon: determination of the critical cluster size. *Journal of Chemical Physics* 1996; **104**: 5181–5187.

28 Talanquer, V. New phenomenological approach to gas-liquid nucleation based on the scaling properties of the critical nucleus. *Journal of Chemical Physics* 1997; **106**: 9957–9960.

29 Peng, D.-Y. and Robinson, D.B. New two-constant equation of state. *Industrial & Engineering Chemical Fundamentals* 1976; **15**: 59–64.

30 Kerminen, V.-M. and Kulmala, M. Analytical formulae connecting the real and apparent nucleation rate and the nuclei number concentration for atmospheric nucleation events. *Journal of Aerosol Science* 2002; **33**: 609–622.

31 Fuchs, N.A. and Sutugin, A.G. Highly dispersed aerosol. In: G.M. Hidy and J.R. Brock, eds. *Topics in Current Aerosol Research*. Pergamon, New York: 1971, pp. 1–60.

32 Weber, R.J., McMurry, P.H., Eisele, F.L. and Tanner, D.J. Measurement of expected nucleation precursor species and 3-500-nm diameter partciles at Mauna Loa Observatory, Hawaii. *Journal of Atmospheric Sciences* 1995; **52**: 2242–2257.

33 Kulmala, M., Lehtinen, K.E.J. and Laaksonen, A. Cluster activation theory as an explanation of the linear dependence between formation rate of 3 nm particles and sulphuric acid concentration. *Atmospheric Chemistry and Physics Journal* 2006; **6**: 787–793.

Chapter 3
Mass Transfer to Aerosols

Charles F. Clement

3.1 Introduction

Environmental aerosols consist of many types, ranging from water clouds and mists extending over large areas to localised pollution aerosols. They are formed by direct emissions into the atmosphere, by water condensation, from wind entrainment (dust, sand and sea spray), volcanic activity, smoke from burning, spores from plants and by gas-to-particle conversion in the atmosphere. All of these aerosols are transitory in nature and eventually evaporate or fall back to earth with lifetimes varying from hours, days or weeks in the troposphere to years in the stratosphere. In this chapter we are concerned not with the direct emission processes, but with the interchange of material between aerosol and the surrounding gases in the atmosphere. In terms of mass, the largest interchange by far is that of water in condensation and evaporation processes, but important mass transfer processes also take place from trace gases in the atmosphere and can lead to formation or growth of new aerosol or to chemical reactions in existing aerosol.

Several physical and chemical processes cause mass transfer to take place to atmospheric aerosol:

(a) *Phase change.* When the temperature is lowered, vapour can become supersaturated and condense into a liquid or solid form. This principally, but not exclusively, affects water in the atmosphere and leads to mists and clouds.

(b) *Production of involatile material from gases in the atmosphere (gas-to-particle conversion).* Principal examples are the production of sulphuric acid by photochemical reactions with SO_2 and possible subsequent reaction with ammonia to produce ammonium sulphate. If there is not enough existing aerosol present on which to condense the material, new aerosol will be nucleated. Involatile or semi-volatile organics are also produced by oxidising reactions in the atmosphere, and condense on aerosol.

(c) *Uptake of trace atmospheric gases into cloud droplets or aerosols, in which they are absorbed or undergo chemical reactions.* SO_2 is partly removed from the atmosphere by this process.

Important processes are summarised in Table 3.1. The formation of stratospheric clouds, which leads to ozone depletion and enhancement of harmful UV light reaching the earth's surface, is discussed in Chapter 7. The production processes lead to increases in mass and number

Table 3.1 Atmospheric mass transfers to aerosol

Process	Cause	Effects
Cloud and mist formation	Cooling and phase change	Condensation of water on cloud condensation nuclei (CCN)
Nucleation events	Involatile material production	Sulphuric acid, ammonium sulphate, iodine compounds
Aerosol acidification	Acid production and condensation	Condensation increases aerosol hygroscopicity and number of CCN
Organics condensation	Production of semi-volatile material	Increases aerosol mass and number of CCN
Stratospheric polar cloud production	Phase change and chemistry at low *T*	Removes nitric acid onto aerosol and leads to ozone hole
Trace gas processing in clouds	Absorption into cloud droplets plus chemistry	Produces acid from SO_2 and other gases

concentrations of atmospheric aerosol with adverse health effects and changes to the radiative properties of the atmosphere. They also increase concentrations of cloud condensation nucleus (CCN) in the atmosphere. These changes to CCN concentrations, together with the process of cloud formation, affect the radiative properties of clouds as well as their longevity. Thus mass transfer processes to aerosol in the atmosphere play an important role in the characterisation of the indirect effect of aerosols on the climate as well as on the direct radiative effect.

There have been few reviews or summaries of work on aerosol condensation and growth, a notable one being that of Wagner [1]. In this chapter, we explain and summarise the important features of the mass transfer process as they are currently understood. By now there is a large literature on the subject, and it would not be possible to describe it all in detail here. In particular, the topic of initial mass transfer involving aerosol nucleation is omitted and left to Chapter 2 on the subject. Instead, we concentrate on the growth process, first as it applies to a single particle and then the aerosol as a whole. More emphasis is placed on the key points of the theory involved in describing it, rather than on experimental observations and techniques, although we refer to what is probably the best technique used in the laboratory for measuring nucleation and growth of aerosol, which is due to Wagner and his collaborators [2, 3].

In Section 3.2, we give a description of the transfer from the surrounding gas to an individual aerosol particle. As the process basically involves the motion of molecules, the fundamental properties of their motion in a gas are briefly given, and how the process of diffusion arises and differs between diffusion in gases and that in condensed media. This leads to simple models of molecular transfer to a particle by gas kinetic motion onto the surface and by diffusion through a boundary layer around a particle, and how these two models may be combined. The method used is that of combining resistances analogous to those in an electrical circuit, thus illustrating the basic method used by Pöschl *et al.* [4] to construct a much more elaborate model framework for transmission and chemical reactions of molecules within aerosol particles and droplets.

In Section 3.3, we then consider mass transfer between the gas and an aerosol as a whole, which is a collection of droplets or particles distributed in space. These individual objects are never identical and will always have a distribution of sizes, and possibly also shapes and compositions. The key feature that allows us to make a relatively simple description of the transfer process is that molecules in the gas move around fast enough to keep their concentrations uniform in spatial regions containing many particles. Thus we can average over the aerosol in determining these concentrations. The same averaging applies to heat transfer and the temperature of the gas. This feature has been implicitly assumed for a long time in the literature, but only recently was it proved [5] and, in analogy to a similar averaging used in other areas of physics, termed mean field theory. The proof is briefly reproduced here. For a given aerosol, and knowledge of the transfer mechanism, the averaging enables us to calculate the timescale for mass transfer between the surrounding gas and the aerosol. This usually depends on just two moments of the aerosol size distribution, which are its surface area and mean size, both per unit volume of the gas. The timescales are calculated for a representative empirical collection of regional aerosols given by Jaenicke [6] by an improved method over that for previous results [7]. Also, in Section 3.3, we describe redistribution processes that redistribute mass in an aerosol between droplets of different sizes, or different phases. These occur particularly with water aerosols; water clouds can be turned into ice clouds at low temperatures.

Before giving these descriptions of mass transfer, we have to discuss just exactly what it is that drives the process. Is there an equilibrium state in which no transfer takes place?

3.1.1 Equilibration between aerosol and gas

A dispersion of particles or droplets in a gas is not a thermodynamic state. For a fixed volume, the stable state for all the atoms and molecules present would be a single condensed solid and/or liquid mass in contact with the gas. We do not consider the important aerosol process of coagulation, which could lead to this state, in this chapter, except in Section 3.3.6 where we point out that it can contribute significantly to the apparent growth rate of an aerosol. By equilibration we therefore mean that there is no net transfer of molecules between any of the particles or droplets making up the aerosol and the surrounding gas. The conditions for this equilibrium at the surfaces of the two media are well known (e.g. [8]):

(a) Temperature, T, is uniform throughout the system.
(b) Chemical potentials of all molecular species are equal across all interfaces.

Condition (a) might be thought to be easily satisfied by sensible heat transfer leading to an overall equilibrium. However, in the atmosphere, environmental aerosols are always subject to long-range radiative heat transfer, particularly from the sun. This strongly affects carbonaceous aerosols, which are strong absorbers of solar radiation, and also water clouds, which can be evaporated by absorption of sunlight and also radiate at long wavelengths into space when high in the atmosphere. Water clouds can also cool and grow from radiation into space. External radiation can be important in driving aerosols away from equilibrium with surrounding air.

The condition on chemical potentials can concern the coexistence of the same molecule in two phases, or the interaction, or chemical reaction, between molecules in the gas and different molecules on the surface of, or within, the aerosol particles. Other chapters deal extensively with equilibrium reached by such processes, and the corresponding mass transfer chemical problems can be very complicated, as shown in the work of Pöschl *et al.* [4]. We only touch on them here in Section 3.2.9.

For a single substance, the chemical potential equilibrium requirement reduces to that the vapour pressure of the substance in the gas be equal to an equilibrium value, $p_{ve}(T)$, which, for a plane surface, depends only on temperature. However, aerosol droplets have spherical curved surfaces, and the Kelvin effect states that the equilibrium vapour pressure over a curved liquid surface, radius R, is increased over that for a plane surface, so that

$$p_{ve}(R, T)/p_{ve}(T) = \exp(R_\gamma/R) \approx 1 + R_\gamma/R \tag{3.1}$$

where R_γ is defined in terms of the liquid surface tension, γ_L, and liquid density, ρ_L, by

$$R_\gamma = 2\mu_v\gamma_L/(\rho_L R_G T) \tag{3.2}$$

where μ_v is the molecular weight.

For water droplets, R_γ is 1.071 nm at 20°C decreasing to 0.696 nm at 100°C so that the expanded form is generally adequate in Equation (3.1). Because the vapour pressure varies with R, a cloud of water droplets is never in thermodynamic equilibrium and water will evaporate from smaller size droplets to condense on larger sizes, the *Ostwald ripening process* discussed in Section 3.3.5.

Convenient parameterisations of the equilibrium vapour pressure for water are given in Appendix A, and also a vapour pressure for ice at low temperatures. Because that over water is greater than that over ice, water droplets in a mixed cloud will evaporate and condensation take place on ice particles. Again, no equilibration is possible in a mixed water–ice cloud. It should also be noted that there are many different forms of ice particles whose condensation is favoured at different temperatures.

3.2 Transfer to a particle

In this section we consider molecular motion from the surrounding gas to an isolated particle. The concentration of these molecules in the gas a long way from the particle is regarded as given, and any direct effect of other aerosol particles is ignored. This is an excellent approximation as long as the aerosol is dilute, as is the case for practically all environmental aerosols.

3.2.1 Molecular motion

Mass transfer takes place by the motion of molecules, and this forms the basis of the description adopted here. We have to consider motion in the gas surrounding the aerosol droplets or particles, and, in several cases when there are different molecules in the aerosol, motion within the droplets or particles. As a general rule, molecular motion in gases is much better known and described than in solids or liquids, especially motion in surfaces of tiny objects, so that there is still much to learn about molecular motion within the aerosol.

The motion of a molecule depends on its energy and thus the local temperature. The temperature is not necessarily uniform in space when mass transfer is taking place, because energy can be released when molecules move into a different state from that in the gas. In the case of the condensation of the same type of molecule this energy is the latent heat corresponding to the change of state from a gas to a liquid or solid. In the case of a chemical reaction between the incoming molecule and a different one in the aerosol (e.g. chemisorption on the surface), a considerable amount of energy may also be released. Because they are small and have little thermal capacity, aerosol droplets or particles can then heat up when enough incoming molecules are arriving. They may not have time to cool by their collisions with inert gas molecules for the temperature to remain the same. The cooling of the aerosol by heat transfer can then provide the determining rate for mass transfer. The main, and possibly only, occurrence in environmental aerosols of when mass transfer is controlled by heat transfer is in the condensation of water except at low temperatures, and we shall describe this in detail in Section 3.2.7.

3.2.2 Motion in gases: kinetic theory

An extensive practical description of molecular motion in gases is given by Hirschfelder *et al.* [9]. More mathematical treatments of the underlying physics are given by Chapman and Cowling [10] and Lifshitz and Pitaevskii [11]. In low-density uncharged gases, molecules interact strongly with each other by collisions only when they are in close proximity. The distribution $f(\mathbf{r}, \mathbf{v}, t)$ of molecular positions $\mathbf{r}$ and velocity $\mathbf{v}$ then satisfies the Boltzmann equation, whose equilibrium solution is the well-known Maxwell–Boltzmann distribution:

$$f(v_x, v_y, v_z) = c(\beta_\mathrm{T}/\pi)^{3/2} \exp\left[-\beta_\mathrm{T}\left(v_x^2 + v_y^2 + v_z^2\right)\right] \tag{3.3}$$

where c is the molecular number concentration and, with m, the molecular mass, and k_B, Boltzmann constant,

$$\beta_\mathrm{T} = m/(2k_\mathrm{B}T) \tag{3.4}$$

The mean molecular velocity is

$$\begin{aligned} v_\mathrm{T} &= \int v f(v) d^3v/c = 2[2k_\mathrm{B}T/(\pi \mathrm{m})]^{1/2} \\ &= 145.5[T(K)/\mu]^{1/2}\ \mathrm{m\,s^{-1}} \end{aligned} \tag{3.5}$$

where μ is the molecular weight of the molecule concerned.

At $T = 20°$C and $\mu = 100$, $v_\mathrm{T} = 249.1\ \mathrm{m\,s^{-1}}$, which suggests that mass transfer associated with free molecular velocities would be very fast. The mass flux onto a surface (mass per unit area per unit time) associated with the free molecular distribution, (3.3), is obtained by integrating cmv_x with positive v_x over the distribution:

$$\begin{aligned} \nu_\mathrm{v}(T) &= m\int_0^\infty v_x\,\mathrm{d}v_x \int \mathrm{d}v_y \int \mathrm{d}v_z f(v) = mcv_\mathrm{T}/4 \\ &= [m/(2\pi k_\mathrm{B})]^{1/2} p/T^{1/2} \end{aligned} \tag{3.6}$$

where the pressure, p, associated with the molecules concerned is related to the concentration, c, by the gas law,

$$p = ck_\mathrm{B}T \tag{3.7}$$

When aerosol particles are small enough not to change the free molecular distribution of molecules around them, the flux (3.6) gives the mass flux of the molecules striking their surfaces, and is thus a vital ingredient in determining mass transfer to aerosol. This is usually termed the *kinetic regime*, and a measure for its description is given by the *Knudsen number*,

$$\mathrm{Kn} = \frac{\lambda_m}{R} \tag{3.8}$$

where R is the radius of the aerosol particle and λ_m is the molecular mean free path in the gas which may be thought of as the average distance travelled by a molecule between collisions with other molecules (see Hirschfelder *et al.* [9] for a fuller discussion). For consistency with the kinetic limit of the Fuchs–Sutugin relation for mass transfer to an aerosol particle (see Section 3.2.6), we relate λ_m to the diffusivity, D, of the molecule in the gas by

$$\lambda_m = 3D/(m\pi/8R_G T)^{1/2} \tag{3.9}$$

where R_G is the gas constant. Similar relations, but with different numerical constants, may be found in the literature, but these give small numerical discrepancies in the kinetic limit.

The total current of molecules striking the surface is then given by multiplying the kinetic flux (3.6) by its assumed spherical area, $4\pi R^2$. Not all molecules striking the surface will be absorbed, and some will be reflected immediately, so that we must multiply the total striking current by a *sticking probability*, S_p, to obtain the actual absorbing mass current. This constant has been given various different names in the literature with slightly different meanings, and we discuss this matter further together with values for S_p in Section 3.2.10. Molecules will also be generally emerging or evaporating from the droplet at a rate determined by its temperature and size and any other variables which would put it into equilibrium with a concentration, c_e, in the surrounding gas. The principle of detailed balance states that, in equilibrium, the emerging current must be equal to the incident absorbing current. Thus we can write the net mass current onto the droplet in a Hertz–Knudsen form:

$$J_K = 4\pi R^2 S_p\{[mk_B T/(2\pi)]^{1/2} c(R) - [mk_B T_d/(2\pi)]^{1/2} c_e(R, T_d)\} \tag{3.10}$$

where T_d is the droplet temperature, and c_e may be replaced by $p_{ve}/(k_B T_d)$, as in Equation (3.7). This form shows the effect of allowing the droplet to be heated up by the release of latent heat of condensation or exothermic chemical reactions.

This form is based on the molecule in question having a Maxwell–Boltzmann distribution adjacent to the surface, a situation which must apply at small enough R, but there is no requirement that it should apply at larger R. Alternative theories have been used at this point, as we show in Section 3.2.6. When there is little or no temperature rise, $T_d = T$, and the current becomes

$$J_K = 4\pi R^2 S_p[mk_B T/(2\pi)]^{1/2}[c(R) - c_e(R)] \equiv [c(R) - c_e(R)]/R_K \tag{3.11}$$

where we define R_K as the kinetic resistance.

At normal temperatures and pressure, the mean free path is about 0.1 μm, so that the *kinetic regime* where $\mathrm{Kn} \gg 1$ corresponds to $R \ll 0.1$ μm. For $R \gg 0.1$ μm where $\mathrm{Kn} \ll 1$, any incoming molecule will undergo many collisions in the vicinity of the particle, and its motion must then be described as *diffusion*, which describes the motion of a molecule subject to many random collisions. The macroscopic diffusion equation for a continuous molecular concentration is then used to describe molecular motion, and the motion when $\mathrm{Kn} \ll 1$ is in the *continuum regime*. We describe diffusion below. In the kinetic and continuum regimes the interactions and exchanges between an aerosol particle and the surrounding gas are easier to understand and calculate than in the *transition regime*, where $\mathrm{Kn} = O(1)$. In the simpler regimes we can use continuum theories and associated empirical constants, for example vapour diffusivity in the gas, or molecular gas theory based on the Boltzmann equation to describe the problem. In the transition regime, more complicated theory is required; whilst solutions of the basic Maxwell–Boltzmann equation in principle describe the transition regime, they are often very complicated and the problem arises of reconciling basic molecular collision integrals with the empirical continuum constants.

3.2.3 Diffusion

Diffusion is the process by which molecules are displaced in space through their random thermal motion induced by interactions with neighbouring molecules. Diffusion is usually very different in nature in gases from that in liquids and solids because the neighbouring molecules are far apart, and molecules move freely between successive collisions between them. The theory of diffusion in gases, which must be modified if the gas is not at a uniform temperature, is described in Hirschfelder *et al.* [9], and has a simple form if the molecules in question are a small fraction of a large background gas, air in the case of the atmosphere. This arises because collisions with air molecules predominate and collisions between trace molecules in the atmosphere can be neglected. The concentration, c, of a trace molecule then obeys the transport equation:

$$\partial c/\partial t + \nabla \cdot (c\,\mathbf{v} - D\nabla c) = S(\mathbf{r}, t) \qquad (3.12)$$

where $\mathbf{v}$ is the local gas velocity and S is a source term, describing the formation of the molecules i.

The molecular *diffusivity*, D, is specified in a two-molecular collision model with hard spheres [9] with a pressure and temperature dependence,

$$D \sim T^{3/2}/p \qquad (3.13)$$

An empirical form for D [12] can be used for water vapour in air:

$$D = (2.23/p)(T/273)^{1.75} 10^{-5}\ \mathrm{m^2\ s^{-1}} \qquad (3.14)$$

where p (bar) is the total pressure.

A typical value for heavier molecules in air, which is often used for sulphuric acid at normal temperature and pressure, is

$$D = 10^{-5}\ \mathrm{m^2\ s^{-1}} \qquad (3.15)$$

Diffusive processes may occur once a molecule reaches a liquid aerosol droplet or a solid particle. Because molecules are closely surrounded by other molecules, diffusion within a liquid medium is very much slower than in a gas to pass close neighbouring molecules. In a solid, motion generally involves penetration of an energy barrier, E_b, and the diffusivity will then usually take on an *Arrhenius form*:

$$D = D_0 \exp(-E_b/k_B T) \qquad (3.16)$$

In solids, diffusivities of all except small atoms or molecules are very small at normal temperatures. A process that may be important in some types of mass transfer is *surface diffusion*, in which atoms or molecules hop around the surface of a particle. Surface diffusivities are very much larger than diffusivities for motion within a medium.

3.2.4 Diffusion to an aerosol particle

For diffusion to a small aerosol particle or droplet, there is no gas velocity and the diffusion equation (3.12) with no sources in a spherical geometry reduces to

$$\partial c/\partial t = (D/r^2)(\partial/\partial r)(r^2\,\partial c/\partial r) \tag{3.17}$$

Transient time-dependent solutions for diffusion to a droplet radius R (e.g. [1]) exist to this equation with a timescale

$$\tau_{\mathrm{D}} = R^2/D = 10^{-7}\,R(\mu\mathrm{m})^2\ \mathrm{s} \tag{3.18}$$

for $D = 10^{-5}\ \mathrm{m}^2\,\mathrm{s}^{-1}$.

These timescales are much shorter than time changes relevant to aerosols in environmental atmospheric situations, so that it is always justified to use 'quasi-stationary' solutions to Equation (3.17), which nevertheless can refer to concentrations at boundaries that are time dependent on a much longer timescale. Then we can take $\partial c/\partial t = 0$ and obtain the original solution of Maxwell [13] to the equation, with c depending on $1/r$ and a constant total molecular mass current to the droplet of

$$J_{\mathrm{D}} = 4\pi mDr^2\,\mathrm{d}c/\mathrm{d}r = 4\pi mDR[c(\infty) - c(R)] \tag{3.19}$$

If there is no resistance to molecules reaching equilibrium at the droplet surface, $c(R) = c_{\mathrm{e}}(R, T)$ and J_{D} gives the total net mass current to the droplet. However, we do not have to make this assumption and can easily combine the diffusive and kinetic resistances to mass transfer to a droplet.

3.2.5 Combined diffusive and kinetic model

The procedure adopted here is a simple illustration of the procedure used widely in the chemical engineering literature to describe either mass transfer or heat transfer through a series of resistances. It was in effect used by Barrett and Clement [14] to describe mass transfer to aerosol droplets when heat transfer is also involved, and some of their results are given in Section 3.2.7. Recently, it has been used extensively by Pöschl *et al.* [4] to develop a kinetic model framework for aerosol and cloud surface chemistry and gas–particle interactions. When transfer rates and removal rates are linear in the concentration of the molecule described, the theory is exactly analogous to the resistance model that is applied to electrical currents in a network. Resistances can occur in series or in parallel, and the rules for combining them are well known and are based on the conservation of currents.

In our case, we have a diffusive current, J_{D}, given by Equation (3.19), which applies up to the surface of an aerosol droplet, and we wish to combine it with the same current, J_{K} (by conservation), which crosses the surface and is given by (3.10).

Then

$$J = J_{\mathrm{D}} = J_{\mathrm{K}} = (c(\infty) - c(R)/R_{\mathrm{D}} = [c(R) - c_{\mathrm{e}}(R)]/R_{\mathrm{K}} \tag{3.20}$$

where the diffusive resistance is

$$R_{\mathrm{D}} = 1/(4\pi mDR) \tag{3.21}$$

The resistances add in series so that the total mass current is given by

$$J = [c(\infty) - c_e(R)]/R_T \tag{3.22}$$

where

$$R_T = 1/\{4\pi R^2 S_p[mk_B T/(2\pi)]^{1/2}\} + 1/(4\pi mDR) \tag{3.23}$$

This result is exact in the limit of large R, where the diffusive resistance dominates, and at small R where there is really a Maxwell–Boltzmann distribution for the impinging molecules. This is true in our three-dimensional (3D) case of a small particle surrounded by a gas even when the sticking probability is less than 1 because reflected molecules are converted back into the distribution before they can strike the particle again. However, we note that the situation is very different for absorption on a plane surface where a 1D current is incident, and the question of the distribution adjacent to the surface is much more complicated [15].

It is convenient to express the mass current in this simple *interpolation model* as a ratio to its value in the diffusive limit:

$$J = 4\pi mDRF(\text{Int})[c(\infty) - c_e(R)] \tag{3.24}$$

where

$$F(\text{Int}) = 1/[1 + (D/RS_p)(2\pi m/k_B T)^{1/2}] \tag{3.25}$$

The alternative form where J_D is expressed as a ratio to its value in the kinetic limit, and which is used in Section 3.3.2, is

$$J = 4\pi m R^2 S_p v_m[c(\infty) - c_e(R)]/(1 + \alpha R) \tag{3.26}$$

where $v_m = v_T/4(v_T$ given by (3.5)) and

$$\alpha = \frac{S_p v_m}{D} \tag{3.27}$$

3.2.6 Fuchs mass transfer model

More elaborate theories have been constructed to describe mass transfer in the *transition regime* between the *kinetic regime* for very small particles and the *diffusive regime* for large particles. For mass transfer alone, theories have been constructed which modify the continuum growth rate according to the value of Kn. Fuchs and Sutugin [16] introduced a correction factor F to the diffusive mass transfer (MD) limit which, in a modified form due to Hegg *et al.* [17] and Kreidenweis *et al.* [18], is

$$F(F - S) = F(\text{Kn})/\{1 + (4/3)\text{Kn}F(\text{Kn})[1/S_p - 1]\} \tag{3.28}$$

$$F(\text{Kn}) = (1 + \text{Kn})/(1 + 1.71\text{Kn} + 1.33\text{Kn}^2) \tag{3.29}$$

Again, a sticking probability, S_p, occurs in this formula. The relation of this definition to other similar parameters that are used in the literature is covered in Section 3.2.10, but, in the past, a wide range of values for S_p have been suggested to apply to water condensation on environmental aerosols, ranging from its maximum, $S_p = 1$, down to 0.02. In Figure 3.1, we show a comparison between $F(I)$ and $F(F - S)$ for water molecules in air at 1 atm pressure (1.013 b) and 20°C

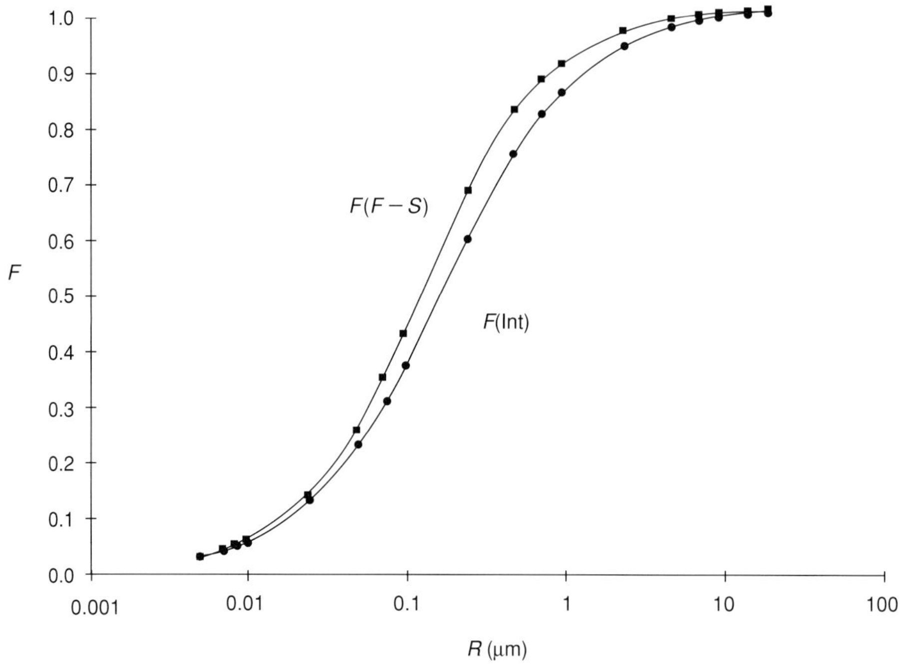

Figure 3.1 Factors $F(\mathrm{Int})$ for the interpolation model and $F(F - S)$ for the Fuchs–Sutugin model as functions of particle radius which reduce mass transfer rate for particle growth from 1 in the diffusive limit for a sticking probability $S_p = 1$.

for $S_p = 1$ and a wide range in droplet size. The difference is small and it would be difficult for experiments to distinguish between the two functions. The variability with S_p is much larger, and in Figure 3.2 we show values $F(F - S)$ over the same size range for three values of S_p. As S_p reduces from 1, differences between $F(I)$ and $F(F - S)$ become even smaller than those shown for $S_p = 1$. The straight-line regions in the logarithmic plots for $F(F - S)$ in Figure 3.2 show that the kinetic region has been reached where the mass current is proportional to R^2 and thus the surface area of the aerosol particle. At high enough values of R, the diffusive limit of $F = 1$ is always reached, but if S_p is as small as 0.01, this is not reached for cloud droplets of $R = 3$ μm and the mass current would be significantly reduced even at larger sizes.

3.2.7 Heat and mass transfer

As noted in Section 3.2.1, a mass current, J_D, onto a droplet carries enthalpy (heat content) which can be released as sensible heat if the molecules change their state or undergo chemical reactions in the droplet. In the case of condensation, this heat per unit mass is the latent heat of condensation, which, in the case of water, is very large. The heat capacity of droplets or particles is very small so that, when a steady-state mass current is incident onto them, any heat released must be removed by a steady-state heat current. The processes of heat and mass transfer occurring with a growing droplet are shown in Figure 3.3, including the interaction of radiation with the droplet that can transfer heat to or from a droplet.

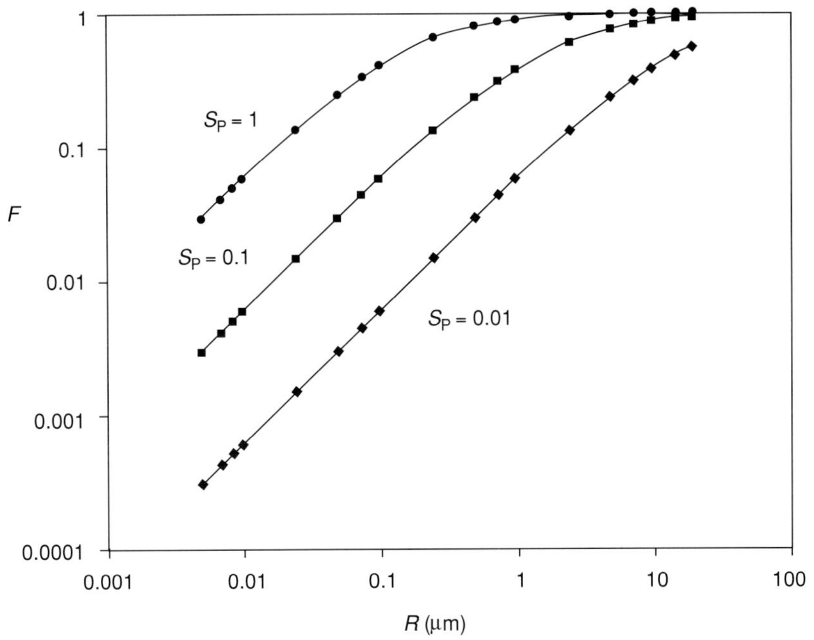

Figure 3.2 Factor F for the Fuchs–Sutugin model as a function of particle radius which reduces growth rates from their diffusive limit for sticking probabilities, $S_p = 1$, 0.1 and 0.01.

The continuum diffusive mass flux and conductive heat flux are based on molecular gas theory [9] and there are small additional fluxes from Stefan flow, thermal diffusion and the Dufour effect whose contributions to droplet growth in the continuum regime have been explored in detail by Kulmala and Vesala [19]. They are generally small and we neglect them here.

As illustrated in Figure 3.3, the droplet growth rate can then be limited by the heat transfer mechanisms of molecular heat transfer at the surface and conduction through a boundary layer as well as by the two mass transfer mechanisms we have considered already. This limitation was emphasised by Wright [20, 21] and examined by Wagner [1], but we make here a simplified and improved derivation of the results of Barrett and Clement [14], who also included radiation as a possible heat transfer mechanism. We limit the discussion to the case of vapour condensation, when the heat release is given by

$$L(T_d) = h_v(T_d) - h_c(T_d) \tag{3.30}$$

which is the difference between the enthalpies per unit mass in the vapour and in the condensed state. The appropriate generalisation to chemical cases is to replace L by this enthalpy difference.

Mathematically, the above currents, (3.11), (3.19), (3.20) and (3.24), must be modified because there is difference between the droplet temperature, T_d, and $T(R)$ in the gas adjacent to the droplet, and between $T(R)$ and $T(\infty)$. When these temperature differences are small, as will be the case invariably in the atmosphere, the equilibrium concentrations, c_e, or equivalently the equilibrium vapour pressures, p_{ve}, which appear in the currents, (3.10) and (3.20), can be

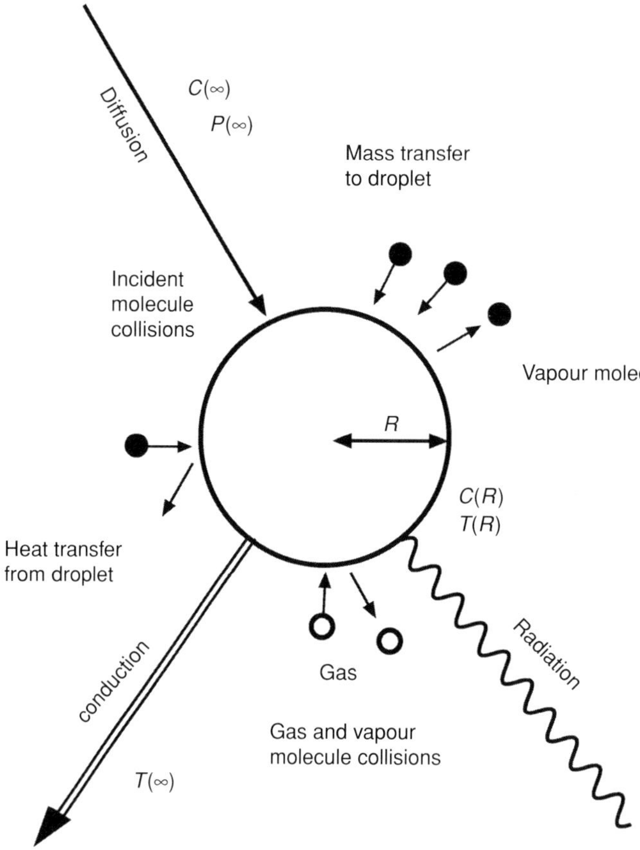

Figure 3.3 The mass and heat transfer processes which occur during droplet growth with associated release of latent heat. Radiation may add or remove heat from a droplet.

expanded in terms of the equilibrium vapour pressure at $T(\infty)$, using the Clausius–Clapeyron relation:

$$(\mathrm{d}p_{\mathrm{ve}}/\mathrm{d}T)/p_{\mathrm{ve}} = \beta/T(\infty)^2 \tag{3.31}$$

where

$$\beta = \mu_{\mathrm{v}}L/R_{\mathrm{G}} \tag{3.32}$$

When supersaturations and/or temperature differences are large, the expansions are no longer valid and some results are given by Barrett and Clement [14] and by Williams [22].

The heat and mass transfers are illustrated in Figure 3.3, where the total inward mass flux is taken to be J as before, and a radiative heat flux, Q_{R}, is included in the corresponding total outward enthalpy or heat flux, Q, from the droplet. Then Q in a conductive boundary layer is

$$Q = 4\pi kR(T(R) - T(\infty)) - h_{\mathrm{v}}(T)J + Q_{\mathrm{R}} \tag{3.33}$$

where k is the thermal conductivity, and it was shown previously [14] that we can ignore temperature differences in $h_v(T)$. The heat flux crossing the kinetic region adjacent to the droplet was shown to be

$$Q = 4\pi R^2 H(T_d - T(R)) - (h_v(T) - \tfrac{1}{2} R_G T_d/\mu_v)J + Q_R \tag{3.34}$$

where the heat transfer coefficient is

$$H = \alpha_g c_{bg} \nu_g(T_d) + \alpha_v c_{bv} \nu_g(T_d) \tag{3.35}$$

where $c_b T = h_v(T) - \tfrac{1}{2} R_G T/\mu_v$ and α is an effective thermal accommodation coefficient (in practice close to unity). Again, there is a negligible correction in changing the argument of the fluxes, ν, given by (3.6) from T_d to $T(\infty)$. The outgoing enthalpy flux is also given by

$$Q = -h_c(T_d)J \tag{3.36}$$

which is negative because the droplet is growing.

The equations (3.33), (3.34), (3.36) and (3.30) determine the temperature differences in terms of J, Q_R, L and

$$L_m = L - \tfrac{1}{2} R_G T_d/\mu_v \tag{3.37}$$

The results are

$$LJ = 4\pi k R(T_d - T(\infty)) + Q_R \tag{3.38}$$

$$L_m J = 4\pi H R^2(T_d - T(R)) \tag{3.39}$$

An incorrect extra term in (3.38) was included by Wagner [1], but was later removed [23].

Because there is now a temperature difference between R and ∞, Equation (3.19) is no longer an accurate expression for the diffusive flux, and should be replaced by

$$J_D = 4\pi m R D(T(\infty))(p_v(\infty) - p_v(R))/(k_B T(\infty)) \tag{3.40}$$

As pointed out earlier [14], this is a good approximation to account for the increase in the diffusivity, D, with temperature. In terms of the pressures, the kinetic flux becomes exactly

$$J_K = 4\pi R^2 S_p[mk_B/(2\pi)T(R)]^{1/2}[p_v(R) - p_{ve}(R, T_d)((T(R)/(T_d)^{1/2}] \tag{3.41}$$

We now expand the final term from its value at $T(\infty)$ in temperature differences, $T_d - T(\infty)$ and $T_d - T(R)$, use (3.38) for the derivative of p_{ve}, eliminate p_v between Equations (3.39) and (3.41) and substitute for the temperature differences from (3.31) and (3.40). The result is expressed in terms of the asymptotic vapour saturation,

$$S = p_v(\infty)/p_{ve}(T(\infty)) \tag{3.42}$$

and the droplet growth rate,

$$\rho_d \, dR/dt = J/(4\pi R^2) \tag{3.43}$$

where ρ_d is the droplet density.

$$dR/dt = [S - p_{ve}(R, T(\infty))/p_{ve}(T(\infty)) + [\text{RAD}]]/(\rho_d R_T) \tag{3.44}$$

where there are four contributions to the total resistance, R_T, from diffusive mass transfer (MD), kinetic mass transfer (MK), conductive heat transfer (HC) and kinetic heat transfer (HK):

$$R_{MD} = R/D\rho_{ve}(T(\infty)) \tag{3.45}$$

$$R_{MK} = 1/(S_p \nu_{ve}(T(\infty)) \tag{3.46}$$

$$R_{HC} = \beta L R/kT(\infty)^2 \tag{3.47}$$

$$R_{HK} = L_m/(HT(\infty)(\beta/T(\infty) - 1/2) \tag{3.48}$$

and the radiative driving term is

$$[\mathrm{RAD}] = Q_R/(4\pi R^2 T(\infty))[(\beta/T(\infty)(R/k) + (1/H)(\beta/T(\infty) - 1/2)] \tag{3.49}$$

These resistances are slightly more accurate than the expressions given by Barrett and Clement [14], as they give the correct limits in the four cases of large or small R, and large or small vapour density, ρ_{ve}, or, equivalently pressure, p_{ve}. However, we have not included here correction terms arising when vapour pressures or densities are not very small compared to those of the gas, air in our case, as they are not generally relevant to environmental problems. The dimensionless ratios of the conductive heat current to the latent heat carried by the mass current have been termed *surface condensation numbers* [24] in the continuum region, and are given by the inverse of the corresponding resistance ratios:

$$\mathrm{Cn_s} = R_{MD}/R_{HC} = kT(\infty)^2/(LD\beta\rho_{ve}(T(\infty)) \tag{3.50}$$

In the continuum region for large R, (3.44) becomes [14] a more accurate version of the Mason equation [25]:

$$\frac{dR}{dt} = \frac{D\rho_{ve}(T(\infty))}{\rho_d R}(S - 1 + [\mathrm{RAD}])\frac{\mathrm{Cn_s}(T(\infty))}{\mathrm{Cn_s}(T(\infty)) + 1} \tag{3.51}$$

which clearly shows the role of $\mathrm{Cn_s}(T_\infty)$ in determining whether mass or heat transfer controls the growth rate. A similar expression can be defined in terms of $\mathrm{Cn_k}(T_\infty)$ for the growth rate in the kinetic regime.

Calculations have been performed for water condensation to see where in the temperature–droplet size plane the four processes, MD and HC in the continuum regime and MK and HK in the kinetic regime, limit water droplet growth [14]. The results depend strongly on the value chosen for S_p. As the pressure reduces, the $\mathrm{Cn_s} = 1$ line moves down to lower T from a value of about 4°C at sea-level atmospheric pressure, as is shown in Figure 3.4 for continuum size droplets that make up mists and clouds in the atmosphere. Where heat from condensation is transferred to a surface, as in the initial formation of ground mist in the atmosphere for subsequent removal by radiation, it was shown by Clement [24] and Clement and Ford [26] that maximum aerosol formation takes place when $\mathrm{Cn_s} \approx 1$. This is responsible for the misty conditions seen in parts of the world, such as the British Isles, where winter temperatures are frequently near 4°C.

To estimate the importance of radiation on droplet growth, we examine the simple case of a single large droplet at temperature, T_d, with surface emissivity ε radiating into a black cavity with wall temperature, T_w, when

$$Q_R = 4\pi R^2 \varepsilon\sigma\left(T_d^4 - T_w^4\right) \tag{3.52}$$

where σ is the Stefan–Boltzmann constant.

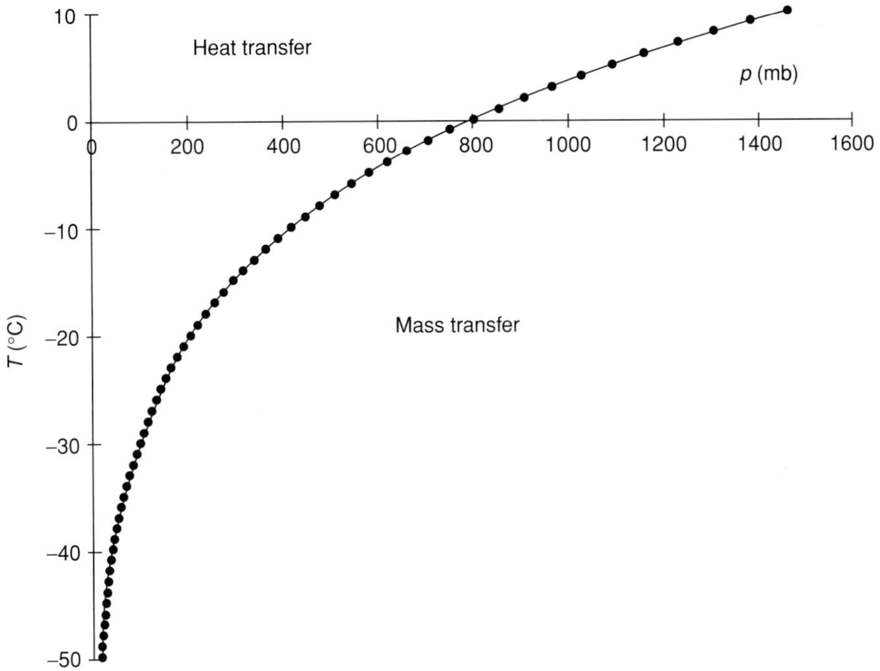

Figure 3.4 The line in the pressure, p, temperature, T, plane where the condensation number, $Cn_s = 1$, showing where resistances to particle growth by water condensation in the atmosphere from heat transfer and mass transfer are equal in the continuum region. Above and below the line, the resistances dominate from heat transfer and mass transfer, respectively.

If $T_w \approx T_d$, as would be the case for a droplet well within a cloud, Q_R is small and it has been pointed out by several authors, including Wagner [1] and Barrett and Clement [14], that [RAD] will be negligible. However, if the droplet is at the top edge of a cloud, particularly high in the atmosphere, and half of it 'sees' outer space where $T_w \approx 0$, [RAD] can be equivalent to a fraction of 1% supersaturation. Its importance for droplet growth in these circumstances was pointed out by Barkstrom [27], and is realised for growing the top of ground mists and for cloud formation at the top of hurricane eyewalls.

Formally, a much better expression than (3.52) is needed for the interaction with radiation, and Q_R can be generally expressed in terms of the droplet's efficiency factor, $Q_a(R, \lambda)$, for absorption of radiation of wavelength $\lambda$and the radiation intensity, $I_\lambda(s, \Omega)$, which depends on position s along a path with Ω as the propagation direction [28,29]:

$$\begin{aligned} Q_R &= \pi R^2 \int_0^\infty d\lambda\, Q_a(R, \lambda) \left[4e_{b\lambda}(T_d) - \int_{4\pi} I_\lambda(s, \Omega) d\Omega \right] \\ &= \pi R^2 < Q_a(R) > \left[4\sigma T_d^4 - \int_{-1}^{1} < I(x, \mu) > d\mu \right] \end{aligned} \tag{3.53}$$

Here $e_{b\lambda}(T_d)$ is the Planck black body function at temperature T, and the final form, with brackets denoting wavelength averages, is obtained in the grey medium approximation for a 1D radiation field along distance x with μ as the cosine of the angle to the x-direction. The two final terms in (3.53) have the physical interpretation as corresponding to droplet emission and absorption, respectively.

3.2.8 Multicomponent droplets and binary growth

The most important examples of multicomponent droplets are water droplets containing soluble material that may (sulphuric acid or ammonium nitrate), or may not (inorganic salts), have a significant vapour pressure. In the latter case just water vapour transport to the droplet occurs as before, and the theory only needs modification of the equilibrium water vapour pressure over the droplet. According to a modified Raoult law, the vapour pressure is proportional to the mole fraction, $X_L = M_L\mu_s/(M_L\mu_s + M_s\mu_L)$ of the solvent, where M_L and M_s are the liquid and solute masses [30], so that Equation (3.1) is replaced by

$$p_{ve}(R, T, X_L)/p_{ve}(T) = \Gamma_L X_L \exp(R_\gamma/R) \tag{3.54}$$

where Γ_L is the *activity coefficient*. Such a description applies in general to droplets which are mixtures of liquids, and activities have generally to be found from experiment, and, for a much more comprehensive description of the thermodynamics of such droplets, see Chapter 6. We confine the discussion here to a brief description of how solvent affects water condensation. For dilute solutions ($M_s \ll M_L$), the ratio may be rewritten as [31]

$$S_e(R_s, R) = 1 + R_\gamma/R - b(R_s/R)^3 \tag{3.55}$$

where the exponential has been expanded as in Equation (3.1), R_s is the radius of a sphere corresponding to the mass M_s in the droplet at solid density, ρ_s, and

$$b = i\mu_L\rho_s/(\mu_s\rho_L) \tag{3.56}$$

where i is the van't Hoff factor to allow for dissociation in the liquid, a quantity which is formally related to the activity coefficient.

In the above expressions, (3.44) and (3.51), for droplet growth, the factor $S - 1$ should be replaced by $S - S_e(R_s, R)$ for droplets containing soluble material. This implies that droplets grow if $S > S_e(R_s, R)$, but evaporate if $S < S_e(R_s, R)$. Plots of $S_e(R_s, R)$ show the equilibrium saturation corresponding to R_s and R and are known as *Köhler curves* and have maxima at a critical radius, R_c, and saturation, S_c,

$$R_c = \left(3bR_s^3/R_\gamma\right)^{1/2} \tag{3.57}$$

$$S_c = 1 + [4/(27b)^{1/2}](R_\gamma/R_s)^{3/2} \tag{3.58}$$

For $S < S_c$ and $R < R_c$, the droplets have a stable equilibrium value on the curves, a fact responsible for haze droplets in the atmosphere. For $R > R_c$, a droplet grows indefinitely if $S > S(R)$, but evaporates to reach its stable value if $S < S(R)$. A nucleus is said to be *activated* once $R > R_c$, and forms a CCN for the given saturation, S_c.

For droplets containing two or more vaporisable materials, for example ammonia–water or sulphuric acid–water, a theoretical description becomes much more complicated, even if, as is usually the case in the atmosphere, concentrations of both the vapours are small so that they have uncoupled diffusive mass currents to a droplet. Considerable progress has been made in obtaining useful expressions for binary growth and evaporation [32–35]. Models and numerical procedures for binary particles are given by Vesala *et al.* [36]. Such binary and ternary aerosols (e.g. including nitric acid as well as water and sulphuric acid) are particularly important in the stratosphere [37], where they play a large role in the chemical reactions leading to holes in the ozone layer (see Chapter 7).

3.2.9 Chemical model framework

Much of the chemical conversion of SO_2 to sulphuric acid by oxidation takes place by chemical reactions inside cloud droplets (see, e.g., [38]), and only aqueous phase oxidation growth in cloud droplets can explain observed atmospheric aerosol growth in short timescales (10–30 min) [39]. Many other reactions take place between trace atmospheric species and atmospheric aerosol particles and droplets. An extensive kinetic framework has been constructed by Pöschl *et al.* [4] to describe molecular mass transport to the particles and chemical reactions taking place on aerosol surfaces and within the aerosol. Example calculations using this framework are reported by Ammann and Pöschl [40]. It is not possible to describe the details here, and we discuss only some of the main features of the framework.

The main aim of the work is to understand the loss rate of molecules from the atmosphere by being able to calculate the *uptake coefficient*, Y, which is defined to be

$$Y = \frac{\text{Net number of molecules taken up per unit time by a particle}}{\text{Number of molecular collisions with a particle per unit time}} \tag{3.59}$$

The same quantity has also been called the *attachment coefficient* [41,42], and could assume negative values to allow for formation of the molecule. If chemical reactions and desorption of reaction products follow, an effective adsorption or reaction coefficient for each product, r, may be defined as advocated by Baltensperger *et al.* [41]; for example

$$Y_r = \frac{\text{Number of molecular surface collisions resulting in emission of } r}{\text{Total number of molecular surface collisions}} \tag{3.60}$$

In order to relate Y to the molecular concentration, $c(\infty)$, in the gas, Pöschl *et al.* [4] consider the kinetic (MK) and diffusive (MD) resistances discussed here, including the Fuchs modifications, but do not insist, as we have done here, that the kinetic flux always reduces to the kinetic limit (3.11) for small R. They do not include limitations from heat transfer, so that the model framework [4] does not apply to water condensation in general. However, it would apply to the attachment of water molecules associated with the attachment of hygroscopic material. In this important case, which includes the condensation of sulphuric acid, ammonium sulphate and other salts on aerosol, water molecules will almost simultaneously condense to reach equilibrium according to the local humidity of the atmosphere. Because this condensation occurs at a slow rate determined by that for the hygroscopic material, no limitation from heat transfer arises.

To describe what happens to molecules once they reach the particle surface, Pöschl *et al.* [4] then introduce models for the layers shown in Figure 3.5 whose characteristics are briefly summarised in Table 3.2. The currents operating between the layers shown in Figure 3.5 differ for

Table 3.2 Layers in the kinetic model framework

Layer	Description
Gas phase	Bulk gas – asymptotic concentration $c(\infty)$
Near-surface gas phase	Gas adjacent to aerosol – within mean free path concentration $c(R)$
Sorption layer	1 molecule thick – adsorbed volatile or semi-volatile molecules
Quasi-static surface layer	1 molecule thick – non-volatile molecules
Near-surface bulk	Several molecules thick ≈1 nm – not directly exposed to sorption layer but affects the quasi-static surface layer
Bulk	Mainly homogeneous liquid or solid, but could have slow diffusion of trace species into it

the three types of molecules considered: non-volatile material (no molecules in the gas phase), which is confined to the particle; semi-volatile material (this can exist as vapour or condensate); volatile material (e.g. gases which dissolve in the particle). In each layer, molecules are allowed to have production (P) and loss (L) rates from chemical reactions with other molecules. Adsorption sites on the surface and their coverage can be treated. We discuss adsorption rates in relation to the sticking probability, S_p, we have used in the next section.

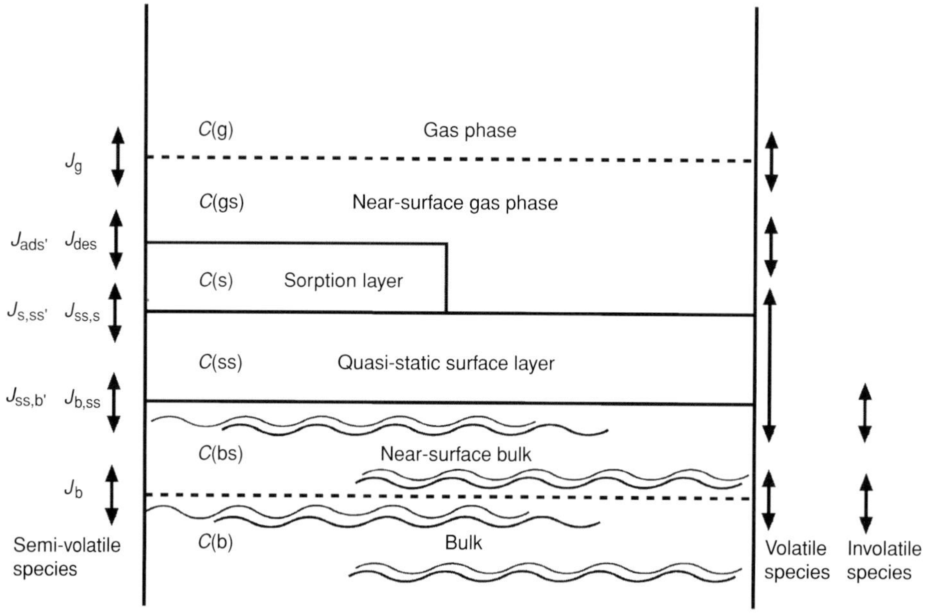

Figure 3.5 Layers surrounding a particle or droplet in the kinetic model framework for aerosol and cloud surface chemistry of Pöschl *et al.* [4]. Mass currents that are taken into account between the layers are indicated for involatile, semi-volatile and volatile molecules.

One achievement of the models included is to explain how the temporal evolution of gas uptake coefficients can have different behaviour at short and long times (see figures in [42]). Whether it can cope with even more complex behaviour associated with the condensation of organics, for example adding chemical reactions to the equilibrium-partitioning concept [43–46], remains an open question. Also, the model does not treat explicitly the averaging over the aerosol in mass transfer that we discuss here in Section 3.3, which includes the possibility of slow redistribution processes similar to those described in Section 3.3.6.

3.2.10 Kinetic coefficients

A bewildering differing nomenclature has been used in the literature for coefficients related to molecules interacting with surfaces, and wide ranges of values have been quoted for some of the coefficients, for example values between 0.02 and 1 for sticking probabilities for sulphuric acid and water molecules. The values are important, as they can affect the numbers of CCN in the atmosphere [47]. The following brief discussion attempts to remove some of the confusion about the subject.

There are two types of accommodation coefficients, although Pöschl *et al.* [4] appear to lump them together (p. 2130).

3.2.10.1 Thermal accommodation coefficients

These are usually denoted by α_T and give the extent to which a molecule *immediately* rebounding from a collision with a surface molecule acquires the mean thermal energy (temperature) of the surface molecules [48]. If $\alpha_T = 1$, the molecule is equilibrated at the same temperature. When the molecule has internal degrees of freedom, it is possible to define separate coefficients for the different translational, rotational and vibrational degrees of freedom.

3.2.10.2 Mass accommodation coefficients

In this and their following different forms, the coefficient is the probability that the molecule colliding with the surface is neither *immediately* reflected nor has an immediate chemical reaction. This statement agrees with that of Pöschl *et al.* [4, p. 2130], and implies adsorption at the surface, but not necessarily in a single-molecule-surface layer as in Pöschl *et al.* [4]. This distinction is meaningful only if surface molecules are different from interior molecules, and if transfers between the surface layer and interior are treated separately. The coefficient is often denoted by α_M. Alternative similar definitions are as follows:

Surface accommodation coefficient. The name given by Pöschl *et al.* [4] for adsorption in the surface layer, and denoted by S_{Xi} for a molecule Xi.

Condensation coefficient. The probability of no immediate reflection and usually refers to condensation of a single substance. Widely used in chemical engineering literature and by Pruppacher and Klett [49]. Denoted by σ_C. There is a corresponding *evaporation coefficient*, σ_E, for a condensable molecule which appears in the outgoing flux from a surface, and which, by detailed balance, is equal to σ_C at the same temperature.

Sticking probability. Denoted by S_p and used here and widely in the literature for no immediate reflection. It does not prejudice the later fate of the molecule concerned.

Uptake coefficient. Denoted by Y, and as defined by Equation (3.59) it is different from mass accommodation, as it refers to the net flux onto the surface, and involves the probabilities of removal or change of the molecule by chemical reaction and lack of re-emission.

A theoretical discussion of the likely values of the sticking probability or mass accommodation coefficients for atmospheric aerosols has been given by Clement *et al.* [50]. Consideration was given to experimental evidence from aerosol growth experiments, molecular beam experiments and molecular dynamics simulations. The classical physics of momentum and energy transfer relevant to surface collisions was reviewed. The conclusion was that S_p cannot be much less than unity in atmospheric conditions. This does not mean, however, that uptake coefficients are near unity or close to their maximum value for a given supersaturation because molecules may have to penetrate a layer of different surface molecules to remain on a particle or droplet. For water, once condensation is initiated on a CCN, it is unlikely that such a layer will exist to prevent a maximum condensation rate. For much of the atmospheric aerosol that is initially repellent to water molecules, subsequent ageing involving the deposition of sulphuric acid converts it into potential CCN.

Many older experiments, which appeared to give values of much less than unity for S_p for water or sulphuric acid, were misinterpreted, as heat removal was limiting the mass transfer rate. The most reliable experiments for the measurement of accommodation coefficients have been performed in an expansion cloud chamber in which an aerosol is nucleated, nucleation is cut off, and then the aerosol grows at a well-defined supersaturation. The size of the growing droplets is observed by the constant-angle Mie scattering technique as a function of time [2]. Some results of recent experiments on the growth of water droplets [51] are shown in Figure 3.6. The authors conclude that their experiments exclude values below 0.85 for α_T and values below 0.5 for α_M over the temperature range 250–290 K, and that both coefficients are likely to be 1 for all studied conditions. This conclusion is in accord with the study [50] that S_p for sulphuric acid condensation in the atmosphere should also be close to unity.

3.3 Transfer to an aerosol

3.3.1 Aerosol description

So far we have considered mass transfer to an isolated droplet or particle, but an actual aerosol consists of a large number of droplets and/or particles in a region of space.

Whilst a gas or vapour has identical molecular constituents, this is never true of an aerosol, whose constituents may differ in size, shape and chemical composition. Where a constituent contains different chemicals it may be *internally mixed*, that is a uniform molecular mixture except possibly at the surface, such as a cloud droplet with dissolved material, or *externally mixed*, that is different molecules in different spatial regions, such as a black carbon particle coated with acid or coagulated with a dust particle. We shall only consider composition here in connection with internally mixed droplets, and assume only spherical particles whose number concentration (number per unit volume) at a point $\mathbf{r}$ and whose radius lies between R and $R + \mathrm{d}R$ is $n(\mathbf{r}, R)\mathrm{d}R$. The size distribution is obtained experimentally by observing the aerosol

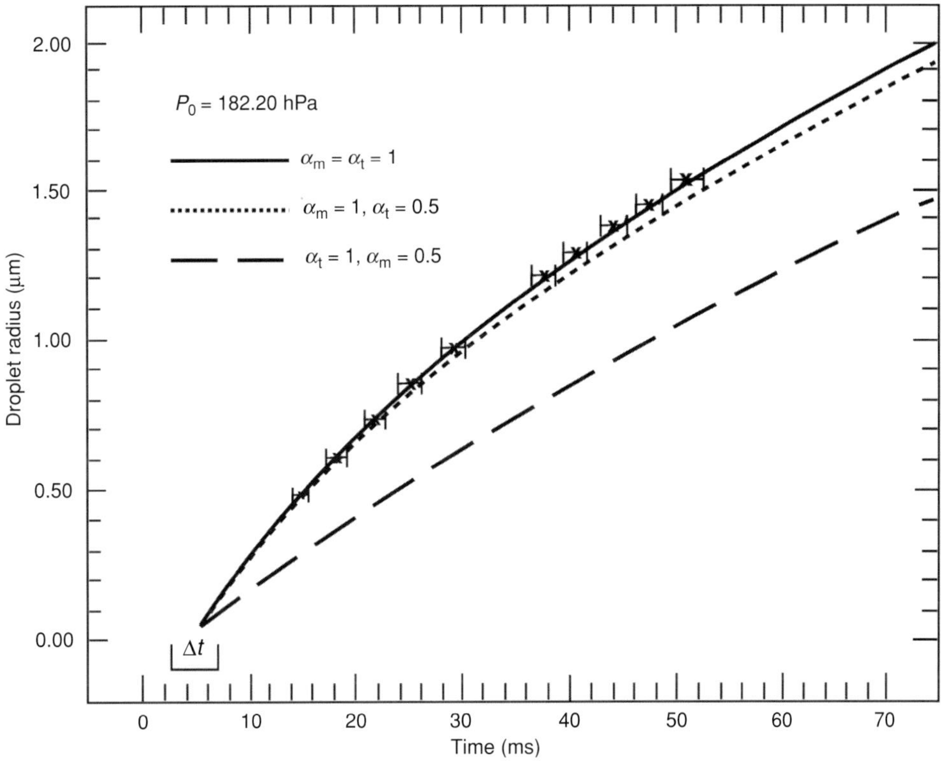

Figure 3.6 Droplet radius as a function of time during the condensation of water vapour as measured by Winkler *et al.* [51] compared to calculations assuming the mass accommodation coefficients, α_m, and thermal accommodation coefficients, α_t, shown. At the end of expansion in the experiments, the saturation $S_0 = 1.49$, and temperature $T = 282.1$ K. The start of the growth process was in the experimental interval, Δt. (Reprinted with permission from *Physical Review Letters.* Copyright 2004 by the American Physical Society.)

in a limited volume centred at **r**, which is nevertheless large enough to contain many particles. Important physical properties of the aerosol then deduced by summing or averaging over this size distribution in terms of its moments are as follows:

$$M_n(\mathbf{r}) = \int R^n n(\mathbf{r}, R) \mathrm{d}R \tag{3.61}$$

The most important moments are

Number concentration: $N(\mathbf{r}) = M_0$ (3.62)

Mean radius: $R_m(\mathbf{r}) = M_1/M_0$ (3.63)

Aerosol surface area concentration: $A(\mathbf{r}) = 4\pi M_2$ (3.64)

Aerosol mass concentration: $\rho_a(\mathbf{r}) = (4\pi/3)\rho_d M_3$ (3.65)

For aerosols consisting of other simple basic shapes, such as rods or ellipsoids, these moments are easy to modify, and are still integral moments, but for fractal aerosols, the moments would generally be non-integral in terms of a 'radius' for the aerosol.

For a vapour–gas–aerosol mixture, basic thermodynamic quantities will be expressed in terms of the moments. For a gas, the pressure is related to the molecular number concentration, c, by Equation (3.7), but aerosol particles do not exert the same pressure as gas molecules on walls, as they will often stick to them rather than bounce off them, altering the momentum transfer. However, since N will always be many orders of magnitude less than c, any contribution of an aerosol to the pressure can always be neglected. We are concerned with dilute aerosols, which means that the aerosol volume fraction, $M_3 \ll 1$, but we cannot always neglect the contribution of the aerosol to the total mass concentration or density of the mixture. Furthermore, if the aerosol consists of a vaporisable substance like water, it can contribute greatly to the enthalpy or heat content of the mixture. To evaporate a water aerosol concentration of ρ_a, the enthalpy concentration of the air must be raised by an amount $L\rho_a$, which in practice will often be equivalent to a rise in temperature of several degrees.

Each individual droplet or particle interacts with its surrounding gas, but does it want to equilibrate only with its immediate surrounding volume, as has been suggested in the literature [52]? This would mean that the values of $c(\infty)$ and $T(\infty)$ would initially change to equilibrate locally and then, as they would be different for different particles, diffusion and heat conduction between the volumes around different particles would take place to equilibrate the gas–aerosol mixture as a whole. We now reproduce an argument given by Clement [5] to show that this is not the case, and that, in the exchange of mass and heat between aerosol and gas, $c(\infty)$ and $T(\infty)$ maintain average values that slowly change in response to the average aerosol properties such as are given by the moments of the aerosol size distribution.

3.3.2 Mean field approximation

As illustrated in Figure 3.7, we consider an aerosol to be made up of droplets or particles each surrounded by a sphere of radius, R_p, which is the volume per particle and is thus related to the number concentration, N, by

$$1/N = (4/3)\pi R_p^3 \tag{3.66}$$

We now determine the time taken for the mean molecular vapour concentration, c, within this sphere to equilibrate with a particle of radius R at its centre. If the vapour in the sphere around our particle does not exchange with neighbouring vapour, its concentration can change only with the mass loss rate to the particle, J_D, and a possible molecular production rate, P, giving for c

$$(4/3)\pi m R_p^3 \, dc/dt = mP - J_D \tag{3.67}$$

We take the loss rate from Equation (3.26) and, since $R_p \gg R$, an excellent approximation is $c = c(\infty)$, so that this equation becomes

$$dc/dt + 3 S_p v_m R^2 (c - c_e)/\left[R_p^3 (1 + \alpha R)\right] = P/\left[(4/3)\pi R_p^3\right] \tag{3.68}$$

The final dividing factor for P is inadvertently omitted in Equation (8) of Clement [5].

The inclusion of P would give an approximate quasi-steady-state solution with

$$c = c_E + P(1 + \alpha R)/(4\pi S_p v_m R^2) \tag{3.69}$$

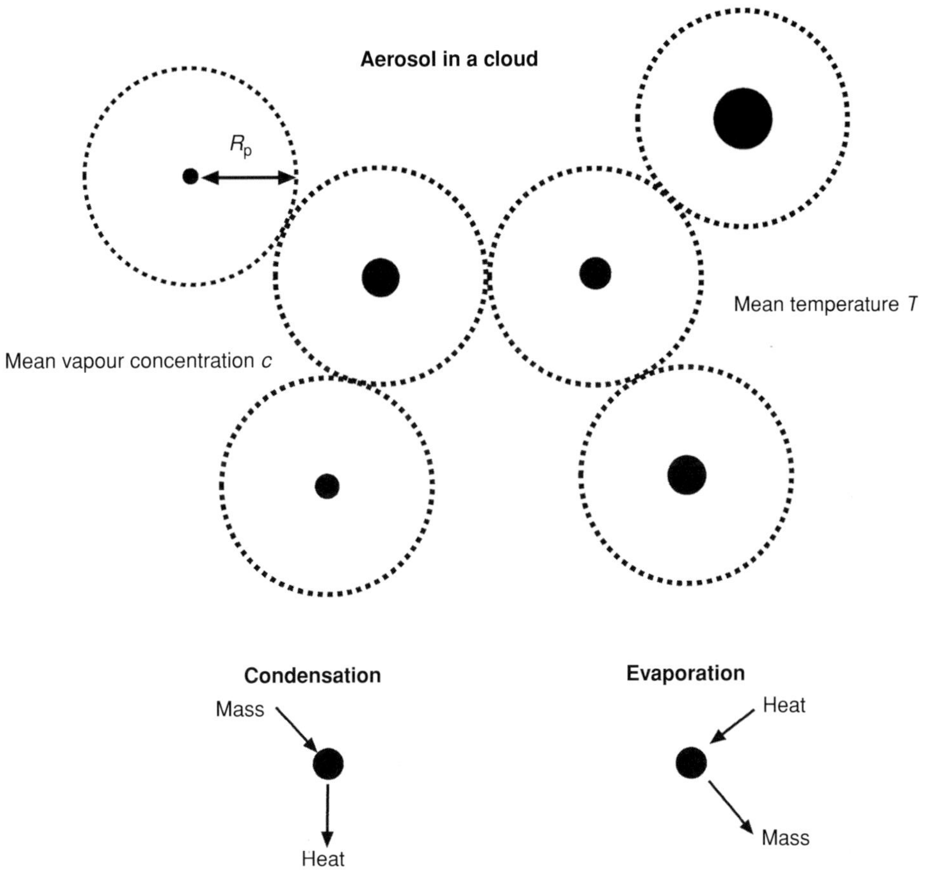

Figure 3.7 Schematic diagram showing aerosol particles in a cloud at the centres of volumes of average radius, R_p, with mean vapour concentration and temperature uniform away from the particles.

in which R is changing with time. The formal solution without P is

$$c(t) = c_e + (c_0 - c_e) \exp \left\{ - \int_0^t dt'/\tau_E(R(t')) \right\} \tag{3.70}$$

where τ_E is a timescale given by

$$\tau_E(t) = R_p^3(1 + \alpha R)/(3S_p v_m R^2) \tag{3.71}$$

Wu and Biswas [53] obtain analytic solutions to these equations in the limits when $\alpha = 0$ (free molecular regime) and the αR term dominates (continuum regime). In general, it was shown in Appendix A of Clement [5] that, in any growth by condensation from R_1 to R_2, the exponent in (3.70) can be replaced by $-t/\tau_E(R)$, where $R_1 < R < R_2$, so that appropriate choices for R in Equation (3.71) give all possible timescales for approach to equilibrium.

Table 3.3 Ratio of diffusion timescale to condensation equilibrium timescale for aerosol number concentrations, *N*, and radii, *R*

N (cm^{-3}) / R (μm)	100	1000	10 000	100 000	1 000 000
0.001	1.39×10^{-8}	3.00×10^{-8}	6.45×10^{-8}	1.39×10^{-7}	3.00×10^{-7}
0.01	1.32×10^{-6}	2.84×10^{-6}	6.11×10^{-6}	1.32×10^{-5}	2.84×10^{-5}
0.1	8.62×10^{-5}	1.86×10^{-4}	4.00×10^{-4}	8.62×10^{-4}	1.86×10^{-3}
1	$1.94 \times^{-3}$	4.17×10^{-3}	8.99×10^{-3}	1.94×10^{-2}	4.17×10^{-2}
10	0.0221	0.0476	0.103	0.221	0.476

To estimate these timescales we take $S_p = 1$ and a velocity of typical molecules of interest in the atmosphere with a molecular weight of 100 and $T = 293$ K, when we have $v_m = 62.3$ m s^{-1}. For the diffusivity we take $D = 10^{-5}$ m^2 s^{-1} and obtain the numerical estimate:

$$\tau_E = 5.3510^{-9} R_p^3(\mu m)(1 + 6.23 R(\mu m))/R(\mu m)^2 \text{ s} \tag{3.72}$$

These timescales are inversely proportional to the number concentration and depend strongly on the size of the particle. In the actual atmosphere, the differences that would arise between molecular concentrations around different particles could be removed by molecular diffusion between spheres around different particles (see Figure 3.7). To see whether this is possible within the equilibration time, we must evaluate the time taken for molecules to diffuse the typical distance needed, that is R_p. This diffusive equilibration timescale is given by

$$\tau_D = R_p^2/D = 10^{-7} R_p^2(\mu m) \text{ s} \tag{3.73}$$

using our numerical value for D. The ratio of τ_D to the equilibration timescale (3.71) is

$$\tau_D/\tau_E = 18.7 R(\mu m)^2/[R_p(1 + 6.23 R(\mu m))] \tag{3.74}$$

Values for this ratio covering the likely range of atmospheric aerosols with $R = 1$ nm – 10 μm and $N = 10^2$ cm^{-3} – 10^6 cm^{-3} are shown in Table 3.3 taken from Clement [5]. In the continuum limit for large R, the ratio given in (3.74) is independent of the diffusivity, D, since both timescales are then proportional to $1/D$. Except for dense aerosols of very large size, the ratio is very much smaller than 1, showing that vapour diffusion in the interstitial gas between particles is much faster than condensation on the particles. This means that the vapour concentration will be kept uniform throughout the aerosol by diffusion.

This result extends to the case when vapour production, P, occurs. The solution (3.69), which depends on radius R, is invalid, and the steady-state vapour concentration is determined by the balance between the production rate and the *average* removal rate onto the aerosol (see Section 3.3.4 below).

The same result applies to the temperature of the vapour–gas mixture between particles in the aerosol. This is determined by the average heat transfer to or from the aerosol associated, respectively, with evaporation or condensation and the latent heat absorbed or released, respectively. The rates of heat transfer by conduction or from molecules at liquid surfaces are similar to those for mass transfer (see, e.g., Clement [24] and Barrett and Clement [14]). Thus there is no need to repeat the above argument with the vapour diffusivity replaced by the thermometric conductivity of the mixture. The ratio of the latter to the diffusivity is the dimensionless Lewis number which is of the order of unity. This discussion has therefore rigorously established the existence of mean fields, vapour concentration, c, and temperature, T, with which an aerosol interacts during condensation, evaporation or indeed any mass transfer to and from the atmosphere.

3.3.3 Growth equations

As a result of the discussion above, we can assume that a single molecular concentration, c, interacts with the aerosol. However, molecules join or leave particles one at a time, and we should justify a continuous description of the aerosol growth process. If c_i is the concentration of particles containing i molecules, the basic discrete growth equations are

$$dc_i/dt = J(i-1) - J(i), \quad n > 1 \tag{3.75}$$

where the currents which add to or subtract from the concentration are

$$J(i) = \beta(i)\, c_i - \gamma(i+1)c_{i+1} \tag{3.76}$$

and the growth and decay probabilities for adding or subtracting one molecule are β and γ, respectively. The $\beta(i)$ are proportional to $c_1 = c$ and the $\gamma(i)$ are determined from detailed balance when $c = c_e$. These equations, with extra terms allowing coagulation between clusters, were used by McMurry and Friedlander [54], McMurry [55, 56] and Rao and McMurry [57] to describe collision-controlled formation and growth of aerosols when there is no significant nucleation barrier. The macroscopic limit of the equations, when i (or the corresponding radius, R) becomes a continuous variable, was explored by Clement and Wood [58, 59], and the extra real-size diffusion in the aerosol size distribution which occurs with the discrete equations has been examined by Clement *et al.* [60]. The spreading resulting from the diffusion is only likely to be significant when supersaturations are small, but might apply to CCN in atmospheric water clouds. For supersaturations of greater than 10%, it was found that the largest spreading could be only 1–2 nm, so that, outside the nucleation region, it is always justifiable to use the continuum equation for the growth of a size distribution, $n(R, t)$:

$$\partial n(R)/\partial t + \partial(n\, dR/dt)/\partial R = S_c(R,\ t) \tag{3.77}$$

where S_c is a source term from nucleation, and the growth rate, dR/dt, given by (3.44), has a general R-dependence of the form

$$dR/dt = \mu_G(S(t) - S_e(R,\ R_s))/(1 + \alpha_G R) \tag{3.78}$$

where we have omitted [RAD] but have allowed a droplet to contain salts with $S_e(R, R_s)$ given by Equation (3.55). The functions μ_G and α_G depend on thermodynamic and kinetic quantities, but are independent of R:

$$\mu_G = [\rho_d(R_{MK} + R_{HK})]^{-1} \tag{3.79}$$

$$\alpha_G = (R_{MD} + R_{HC})/[R(R_{MK} + R_{HK})] \tag{3.80}$$

General solutions of Equation (3.78) with source terms are available [61], and also, with S independent of R, explicit solutions for the evolving distribution (see, e.g., Barrett and Clement

[62]). Here, we evaluate the increase in the aerosol mass concentration, $\rho_a(t)$, of droplets over a nucleated size, R_N, resulting from growth:

$$\begin{aligned} d\rho_a/dt &= (4\pi/3)\rho_d \int (\partial n/\partial t) R^3 \, dR \\ &= (4\pi/3)\rho_d R_N^3 n(R_N) R(R_N) + 4\pi \rho_d \int R^2 (dR/dt) n \, dR \end{aligned} \quad (3.81)$$

where the integrals run from R_N to ∞, and the contribution of the time derivative to the integral has been evaluated using (3.77) and partial integration. The first term in (3.81), which is the additional mass from freshly nucleated droplets, is usually negligible, but the corresponding contribution, $n(R_N)R(R_N)$, to the total number concentration derivative, dN_a/dt, obtained by performing the same procedure with (3.62), gives the entire result for the increase in number concentration. The final term in (3.81) is, as expected, the mass flux from particle growth summed over the size distribution. From the form of dR/dt given by (3.78) with $S_e(R, R_s)$ approximated by unity, which is in the kinetic regime when $\alpha_G R \ll 1$ and the continuum regime when $\alpha_G R \gg 1$, the limiting mass transfer rates from vapour to the aerosol cloud are

$$d\rho_a/dt = 4\pi \rho_d \mu_G (S(t) - 1) M_2, \quad \text{kinetic regime} \quad (3.82)$$

$$= 4\pi \rho_d (\mu_G/\alpha_G)(S(t) - 1)\, M_1, \quad \text{continuum regime} \quad (3.83)$$

For tiny particles, the surface area of the aerosol controls the mass transfer rate, but the rate becomes relatively smaller for large particles where diffusion and conduction through boundary layers result in M_1 as the controlling moment. Since $S = c_v/\underline{c}_{ve}(T)$, and mass conservation means $m_v c_v + \rho_a$ is constant, these equations are essentially differential equations for $\rho_v = m_v c_v$. In the continuum case, taking μ_G/α_G from Equation (3.51) and neglecting [RAD], Equation (3.83) becomes

$$d\rho_v/dt + 4\pi D_v M_1 [\mathrm{Cn}_s/(\mathrm{Cn}_s + 1)](\rho_v - \rho_{ve}) = 0 \quad (3.84)$$

If mass transfer controls aerosol growth, this immediately gives the equilibration timescale between a continuum-sized aerosol and its surrounding mixture,

$$t_e = (4\pi D_v M_1)^{-1}, \qquad \mathrm{Cn}_s \gg 1 \quad (3.85)$$

In the opposite limit where heat transfer is controlling growth, the temperature is changing so that Cn_s is a function of time, and it is necessary to consider the equation for T in addition. A derivation in this case [63] gives, neglecting small terms,

$$\begin{aligned} t_e &= (4\pi D_v M_1)^{-1} (\mathrm{Cn}_s + 1)(\mathrm{Cn}_s + Le)^{-1} \\ &\approx [4\pi M_1 k/(\rho c_p)]^{-1}, \qquad \mathrm{Cn}_s \ll 1 \end{aligned} \quad (3.86)$$

These timescales are relevant to water clouds and condensation, where, for most of the growth, the droplets are in the continuum regime at sizes of 1 μm or more. To obtain a numerical estimate, we take the kinetic factor to be 2.10^{-5} m^2 s^{-1} and M_1 to be the product of the number of CCN, N_0 in units of cm^{-3} and a radius R in μm. The timescale is then

$$t_e \approx 4000/(N_0(\mathrm{cm}^{-3}) R_0(\mu\mathrm{m}))\ \mathrm{s} \quad (3.87)$$

N_0 is less than 100 only in clean parts of the atmosphere, so that most cloud formation timescales are of the order of seconds rather than minutes. This is responsible for the rapid appearance of clouds following adiabatic expansion in the atmosphere. The most dramatic evidence for the short timescale are the apparently stationary clouds over mountains, such as Table Mountain above Cape Town in South Africa, which appear and disappear suddenly in airflows which reach over 50 km h^{-1}.

3.3.4 Empirical atmospheric timescales

Apart from the condensation of water, most mass transfer to aerosols in the atmosphere results from gas-to-particle conversion, for example the oxidation of SO2 to sulphuric acid. Any non-volatile material produced will then condense on existing aerosol, and, if the removal rate from this condensation is not large enough, the supersaturation will increase and a nucleation event will occur. Usually, the existing aerosol will not be changed much by the condensation, so that the value of the removal rate, or its inverse, the equilibration timescale, is a critical factor in determining whether nucleation will occur or not. One example is that the occurrence of Finnish forest nucleation events depends on the size of the radiation intensity, which produces condensable species, divided by the removal rate exceeding a certain value [64]. It is therefore highly desirable to be able to calculate removal rates and timescales for empirical atmospheric aerosols. This was done approximately [7] using mass transfer timescales in the continuum region (Equation (3.85)) and the kinetic region for the different types of global aerosols proposed empirically by Jaenicke [6], which are combinations of lognormal distributions. We repeat the calculation here, but with an analytic method which gives accurate results across both regions. A preliminary account was given of this method [65].

Omitting the nucleation term, and keeping only the mass transfer resistances in dR/dt and α_G given by Equation (3.80), Equation (3.81) reduces to

$$d\rho_a/dt = 4\pi\rho_{ve}S_p(S-1)(R_G T)/(2\pi\mu_v)^{1/2}\int dR\,n(R)R^2/(1+\alpha_G R) \qquad (3.88)$$

For an aerosol that is not much changed by the growth process so that $n(R)$ can be regarded as fixed, the timescale for molecular mass equilibration between the aerosol and the vapour is obtained in a similar way as the continuum timescale (3.85) is derived from (3.84):

$$t_e = \rho_{ve}(S-1)(d\rho_a/dt)^{-1} \qquad (3.89)$$

From Equation (3.88) this equilibration timescale is inversely proportional to the integral

$$I = \int dR\,n(R)R^2/(1+\alpha_G R) \qquad (3.90)$$

Lognormal distributions are specified by the radius R_a and width parameter σ:

$$n(R) = (2\pi\nu)^{-1/2}(N/R)\exp\{-\tfrac{1}{2}[\ln(R/R_a)]^2/\nu\} \qquad (3.91)$$

where N is the total number concentration and, for convenience, we define

$$\nu = (\ln\sigma)^2 \qquad (3.92)$$

$$\mu = \alpha_G R_a \qquad (3.93)$$

The integral to be calculated becomes

$$I = NR_a^2 I(\mu,\ \nu) \qquad (3.94)$$

where, with $x = \ln(R/R_a)$,

$$I(\mu, \nu) = (2\pi\nu)^{-1/2} \int_{-\infty}^{\infty} dx \exp(-x^2/2\nu + 2x)/(1+\mu \exp x) \quad (3.95)$$

An analytic and numerically accurate result is then obtained by the method of steepest descent [65]:

$$I(\mu, \nu) = \{[1 + \mu \exp(\beta\nu)][1 + \mu \exp(\beta\nu) + \nu(2 - \beta)]\}^{-1/2} \exp[\nu\beta(2 - \beta/2)] \quad (3.96)$$

where β is a solution of the non-linear equation:

$$\beta\nu - \ln\{(2 - \beta)/[\mu(\beta - 1)]\} = 0 \quad (3.97)$$

and lies between 1 in the continuum limit as $\mu \to \infty$ and 2 in the kinetic limit when $\mu = 0$:

$$I(\mu, \nu) = \exp(\tfrac{1}{2}\nu)/\mu, \quad \mu \gg 1. \quad (3.98)$$

$$I(0, \nu) = \exp(2\nu), \quad \mu = 0 \quad (3.99)$$

Approximate analytic forms for β are obtained by iterating solutions near these limits:

$$\beta = 2 - (1 + \nu + \exp(-2\nu)/\mu)^{-1}, \quad \mu < \exp(-2\nu)/(1 + \nu) \quad (3.100)$$

$$\beta = 1 + (1 + \nu + \mu \exp \nu)^{-1}, \quad \mu > (1 + \nu) \exp(-\nu) \quad (3.101)$$

In practice, we find that taking one of these forms, or their mean if neither condition is satisfied, gives sufficiently accurate values for the integral that it is generally not necessary to use the Newton–Raphson procedure to solve Equation (3.97) for β.

Calculations were performed for the global aerosols of Jaenicke [6] corresponding to condensation of sulphuric acid at $T = 10°C$ with $\mu_v = 98.08$ g and $D = 10^{-5}$ m^2 s^{-1}and two values of the sticking probability, S_p, which have been proposed for sulphuric acid. The results are shown compared to the previous values [7] in Table 3.4. Usually, only one of the three lognormal distributions given for each region dominates the result. Jaenicke (private communication) is no longer confident that the distributions he gave for the polar aerosols are representative of the region. We believe that $S_p = 1$ is much more likely to give realistic times in Table 3.4, so that

Table 3.4 Equilibration times, t_e, for atmospheric aerosols

	$S_p = 1$		$S_p = 0.02$	
Aerosol	Previous[a]	Present	Previous[a]	Present
Polar	73–103 min	181 min	19–65 h	70 h
Background	3.7 min	5.2 min	4.9–35 min	44 min
Maritime	10.6 min	20.3 min	83 min–5.2 h	8.2 h
Remote continental	41–78 s	123 s	55–61 min	63 min
Desert dust storm	74 s	112 s	4.4 min	9.0 min
Rural	1.2–2 min	3.6 min	99 min	112 min
Urban	0.2–15 s	22 s	12 min	12 min

[a]Clement and Ford [7].

removal timescales are very short except in remote parts of the atmosphere. In reality, cleaner conditions occur everywhere following precipitation and removal of aerosol by rainout, but it is important to know threshold values of the removal integral or t_e which signal the onset of nucleation events. The above formulae can be used to calculate t_e for any lognormal fit to any observed atmospheric aerosol without performing numerical integration.

3.3.5 Redistributive processes

We have just seen that aerosols are likely to equilibrate rapidly with surrounding condensable vapours in the sense that no further overall mass transfer takes place and $d\rho_v/dt = 0$. This does not mean, however, that the diverse individual droplets and particles are all in equilibrium, and in many cases when the condensable material is volatile, mass will continue to be redistributed over an aerosol size distribution. This is particularly true for water aerosols. Several mechanisms that can cause this redistribution are illustrated in Figure 3.8, where the transfers between different sizes are pictorially represented. We summarise some existing results on the subject here.

We first include the radiation term [RAD] in the growth rate (3.44) as $G(R)$ to emphasise its dependence on R, in addition to the solute term, $S_e(R, R_s)$, and describe results of their effects obtained in studies by [63, 66]. In an expanded form, the simplified growth rate (3.78) then becomes

$$dR/dt = \mu_G(S - 1 - R_\gamma/R + b(R_s/R)^3 + G(R))/(1 + \alpha_G R) \tag{3.102}$$

Starting from an arbitrary value of S, there will be a rapid equilibration, as described in the previous section, followed by slower change to the size distribution. This may or may not involve further total mass change with a finite $d\rho_a/dt$, depending on whether heat is being transferred in or out of the vapour–gas–aerosol mixture, by radiation for example. In either case, the growth rate can be re-expressed in terms of $d\rho_a/dt$ and additional terms using the integral, $4\pi\rho_p \int R^2(dR/dt)n\,dR$, in (3.81). The result appears in terms of moments of the distribution and in the continuum regime takes the relatively simple form,

$$\begin{aligned} dR/dt = {} & (d\rho_a/dt)/(4\pi\rho_p M_1 R) \quad \text{overall growth} \\ & + (\mu_G/\alpha_G)(R_\gamma/R)(N_a/M_1 - 1/R) \quad \text{Ostwald ripening} \\ & - (\mu_G/\alpha_G)(bR_s^3/R)(M_{-2}/M_1 - 1/R^3) \quad \text{solvent redistribution} \\ & + (\mu_G/\alpha_G)(1/R)[G(R) - < RG(R) > /M_1], \quad \text{radiative redistribution} \end{aligned} \tag{3.103}$$

where $<RG(R)>$ represents the moment of the size distribution and is proportional to the moment of Q_R.

The three final terms involve mass redistribution because they change sign across the size distribution. For Ostwald ripening the sign is positive for large R, so that large droplets grow at the expense of small ones, making a cloud of pure drops basically unstable. The phenomenon also occurs in colloids and has an extensive literature, including the existence of an asymptotic theory for $R \gg R_\gamma$ [67, 68] with an asymptotic distribution and timescale [63].

$$t_{OR} = 2.25\alpha_G R_m^3/(\mu_G R_\gamma) = 17.4 R_m^3(\mu m)\ \text{s} \tag{3.104}$$

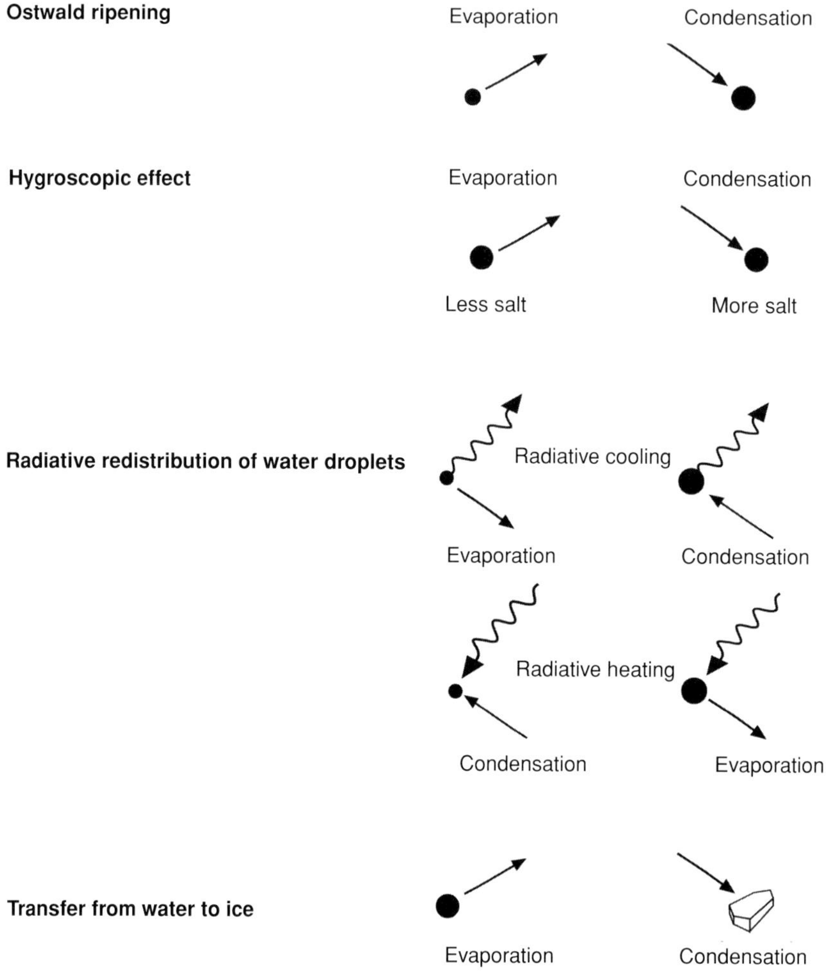

Figure 3.8 Processes which redistribute mass in a water cloud: for pure water droplets (Ostwald ripening), in response to hygroscopic material content, from the interaction with radiation.

for water droplets at 20°C, where R_m is the mean aerosol radius given by (3.63). In practice small water droplets in the atmosphere do not disappear, but become stabilised according to their solute content (see Section 3.2.8). The relatively short timescale indicates that a pure water aerosol makes a highly unstable aerosol.

Solvent redistribution can occur following rapid condensation at high S. Each droplet with a given R_s tends to a stable size as large ones evaporate and small ones grow. Some calculations also indicating relatively short timescales were performed by Barrett *et al.* [63]. As acids or salts are formed in atmospheric cloud droplets by chemical reactions, the water content of the cloud will adjust rapidly to the changes by redistributing itself between droplets. This redistribution would provide the basis for the following reported phenomena.

Table 3.5 Percentage reductions in equilibrium vapour pressure from that over water to that over ice

T (°C)	−5	−10	−15	−20	−25	−30	−35	−40	−45	−50
Reduction (%)	5	10	14	18	22	25	29	32	36	39

The practice of inducing rainout from clouds by seeding them with silver iodide crystals was once practiced in the US but activity stopped because of legal actions. It has continued elsewhere in the world. There have also been claims that mists can be cleared and thunderstorms induced to precipitate before producing damaging hail by seeding them with fairly large very hygroscopic particles. To the author's knowledge, there are no definitive calculations that validate these claims, although the effect of hygroscopicity on cloud droplet formation has been examined [69]. Existing estimates of water redistribution timescales certainly indicate the possibility of being able to make large-scale changes in cloud water content.

For radiative redistribution, the case of pure radiative cooling, when the temperature falls, and the situation with $d\rho_a/dt = 0$, corresponding to the radiative interaction of a cloud whose temperature is maintained as constant, were considered by Barrett and Clement [66] and Barrett *et al.* [63]. In the cooling case, the conditions for redistribution to occur are that Q_R is not proportional to R, which is satisfied for water droplets for which $G(R) = \text{constant}\, R\{1 - \exp[-0.3(\mu m^{-1})R(\mu m)]\}$, and the Lewis number $Le < 1$ (0.85 for water vapour in air). For a water droplet cloud interacting with radiation alone, we can have the interesting situations illustrated in Figure 3.8. Whilst the bulk of the size distribution is growing or evaporating, very small droplets may be doing the opposite. The effect arises because rates of heat and mass transfer through the boundary layers differ ($Le < 1$) and the relative amounts of heat radiated at different sizes do not correspond to the same relative amounts conducted through boundary layers.

At temperatures below 0°C, the equilibrium water vapour pressure over ice drops below that over water, as shown by the equilibrium pressures given in the Appendix and in Table 3.5.

Until temperatures are very low, water clouds mostly form in preference to ice or snow clouds. As temperatures fall, ice nucleation occurs in the water droplets. In mixed clouds of water and ice, the large difference in vapour pressures shown in Table 3.5 means a large driving force for water molecules to evaporate from water droplets and condense on ice. However, until temperatures are very low, particularly at low pressures high in the atmosphere, a considerable heat current must pass in the opposite direction. The problem therefore involves both heat and mass transfer, two phase droplets with ice and nucleation events. Calculations showing its considerable complexity are reported by Hienola *et al.* [70]. Our formulation here and the Mason equation fail, especially at lower temperatures, because values of $S - 1$ are not small. However, the authors report that an analytic formulation of the mass flux by Lehtinen *et al.* [35], which goes to second order in the expansion of the equilibrium vapour pressure in temperature differences, works well and is much faster to calculate than use of a purely numerical calculation scheme [36].

3.3.6 *Nucleation mode mass transfer*

So far, we have concentrated on mass transfer to environmental aerosols solely by molecules. However, it is becoming apparent that an important mechanism exists in which molecules

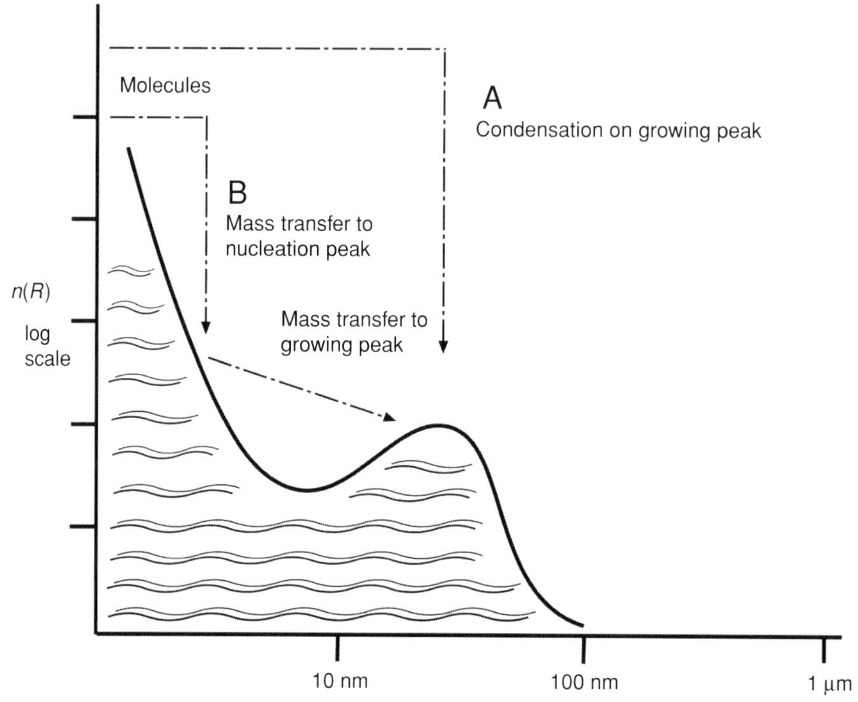

Figure 3.9 Schematic diagram of an aerosol size distribution, $n(R)$, as a function of radius R, in which mass transfers are taking place both from molecular condensation and also by coagulation with small particles in the nucleation peak.

nucleate to tiny particles in the nucleation mode, which then rapidly coagulate to a growing aerosol. This mechanism competes with direct condensation on the aerosol, as shown in Figure 3.9. In several cases, it has been found that sulphuric acid aerosols have an observed growth rate larger by a factor of 2–3 than the maximum predicted by condensation from the observed concentration of sulphuric acid molecules [71]. In addition to possible contributions from other molecules, nucleation mode coagulation can contribute to closing the discrepancy.

Analyses [72, 73] have been made of the formation of a sulphuric acid aerosol that was observed in the upper troposphere [74]. A basic assumption of the analytic model introduced by Clement *et al.* [72] is that all the acid being produced is condensing on a monodisperse aerosol initiated by a burst of nucleation, so that

$$d\rho_a/dt = mP \tag{3.105}$$

where P is the production rate of sulphuric acid molecules.

Numerical calculations [73] based on the full sectional model of Pirjola [75] show that nucleation continues well after the burst in the analytic model has finished, but that the tiny particles produced after a short initial time period are all coagulating with a growing peak in the size distribution as shown in Figure 3.9. The number of growing particles in both models is practically the same. Thus the basic assumption of the analytic model is justified that the acid production rate determines the mass transfer rate to the aerosol and, together with coagulation, determines its growth rate. However, the existence of the nucleation peak in the numerical model shows that the acid monomer concentration is reduced over that, which would produce the same growth rate from monomer condensation alone. This effect may partly explain why some observed growth rates are larger than those predicted from the monomer concentration.

The conclusion is that mass transfer to an aerosol from production of a relatively non-volatile species can be determined directly from the production rate of the species. The relative magnitude of the two mass transfer mechanisms of direct monomer condensation or nucleation mode coagulation will depend on how close the nucleation rate is to its barrierless limit, and this is presently being investigated [76]. We already know that nucleation mode coagulation is larger at the barrierless limit.

3.3.7 Source terms

We conclude by briefly discussing the source terms for mass transfer to environmental aerosols. We are not concerned here with direct not localised emissions into the atmosphere, such as from vehicle exhausts, burning, fires and volcanoes, where condensable material immediately turns into aerosol, but transfers which can take place in the bulk of the atmosphere. They may be divided into two classes.

3.3.7.1 Non-volatile materials

These must be produced by chemical reactions to appear in the atmosphere; examples are sulphuric acid, ammonium sulphate and some organics that are practically involatile at atmospheric temperatures.

3.3.7.2 Volatile and semi-volatile materials

These may also be produced by chemical reactions, but also change from vapour to aerosol when atmospheric temperature falls and the vapour becomes supersaturated. The principal example is water, but the class also includes ammonium nitrate and many organics.

From the point of view of calculations, the practical difference between the two classes is that, for non-volatile materials, it is possible to neglect the equilibrium vapour pressure, or molecular concentration when dealing with growth rates such as (3.44). In this case, the spatial extent of the mass transfer is determined solely by the rates of chemical or photochemical reactions that change gaseous substances into non-volatile ones. For the oxidation of SO_2 to sulphuric acid these are quite well known, so that the whereabouts of acid production can be calculated by following the movement of SO_2 in the atmosphere. At present, this is not the case for organics whose characterisation and chemical reactions are much less known. How far from their sources do organics contribute to the growth of aerosol? As production processes are not expected to have a large local, spatial variability in the atmosphere, the mass transfers to aerosols are generally expected to be uniform in space.

Volatile materials are conserved in the atmosphere, so that their movement can be followed, but their transformation into aerosols is very dependent upon temperature that has large fluctuations in the atmosphere. One has to only look at clouds to see the highly localised aerosols that can result from this variability. The variability is not adequately addressed in theories and codes that attempt to calculate aerosol production of these materials, including water and the process of cloud formation. The origin of this variability is atmospheric mixing during which vapour diffusion and sensible heat conduction take place. Atmospheric turbulent velocities alone cannot be responsible, as they transport vapour concentration and heat simultaneously, as was pointed out by Clement [24]. To see this explicitly, we give the following equation for the difference between the actual vapour concentration and its equilibrium value at the local temperature,

$$u = c - c_e(T) \tag{3.106}$$

Corrections of the order of the vapour concentration divided by the gas concentration are omitted from the full equation [77, 78] that was derived from the separate equations for c and T given by Clement [24] augmented by additional terms for pressure change and interaction with radiation:

$$\begin{aligned} \partial u/\partial t + \mathbf{v}\cdot\nabla u + \nabla\cdot(-Du) = {} & Dc_e''(\nabla T)^2 - (\partial p/\partial t - \nabla\cdot\mathbf{q})/(L\mathrm{Cn}) - c_e'(1 - Le^{-1}) \\ & \times(\partial T/\partial t + \mathbf{v}\cdot\nabla T) - (1+\mathrm{Cn})^{-1}(\mathrm{d}\rho_a/\mathrm{d}t)/\rho_d \end{aligned} \tag{3.107}$$

where c_e' is $\mathrm{d}c_e/\mathrm{d}T$, $\mathbf{q}$ is the radiation current interacting with the aerosol and Cn is the bulk condensation number:

$$\mathrm{Cn} = k/(L D\rho c_e'(T)) \tag{3.108}$$

The LHS terms in (3.107) are the familiar transport and diffusive terms so that it is the terms on the RHS which lead to possible supersaturation and mass transfer to aerosols, the final term being the actual transfer rate, if it occurs, which is itself proportional to $u = c - c_e$. The first term on the RHS is proportional to the second derivative of c_e. It is essentially positive and its non-linearity shows why supersaturations arise in temperature gradients. The proportionality to $-\partial p/\partial t$ shows the effect of reduced pressure in rising air columns to produce supersaturation, but the radiative term can have either sign in adding or removing heat from the aerosol. Of great interest is the term depending on the Lewis number, Le, which is the ratio of the thermometric conductivity to the vapour diffusivity:

$$\mathrm{Le} = k/(D\rho c_p) \tag{3.109}$$

For water vapour in air, Le is about 0.85, showing that water vapour diffuses faster than heat travels in air. The term with Le is therefore non-zero, and its sign depends on whether the temperature of the transported air at the point in question is rising or falling. In any mixing situation in the atmosphere, air packets will exist, which have opposite signs. This will lead to highly local fluctuations in the water content of clouds as well as the evident fluctuations in the existence of the cloud that are seen at cloud edges. An example of the locality of the aerosol formation produced is seen in the calculations for cooled laminar flow in a tube performed by Barrett and Fissan [79], but calculations for actual turbulent flows in the atmosphere are lacking.

The same physics will apply to any volatile material that forms aerosol when the temperature is lowered in the atmosphere. The fluctuations that arise will depend on the Lewis number for vapour transport of the material in air. Most organics are expected to have relatively slow diffusion in air with Le > 1. Much work remains to be done to characterise the formation of aerosols by mass transfer of volatile material in the atmosphere, and the fluctuations in aerosol density that result from this process.

Acknowledgements

The author thanks Dr J.C. Barrett for carefully checking the manuscript, Brian Ruddock for his help with the diagrams and Dr U. Pöschl for providing copies of his papers.

Appendix: water data

The equilibrium vapour pressure of water can be calculated from either of two convenient empirical forms given by (a) Richards [80] or (b) Bolton [81]:

(a) $p_{ve}(T) = 1007.84 \exp(13.31 T_r - 1.976 T_r^2 - 0.6445 T_r^3 - 0.1299 T_r^4)$ mb
where
$T_r = (T(K) - 373)/T(K)$

(b) $p_{ve}(T) = 6.112 \exp[17.67\ T(°C)/(T(°C) - 243.5)]$ mb

These formulae give values differing by not more than 0.25% from each other over the range, $-45°C < T < 45°C$, which covers most atmospheric situations.

An analytic model fitted to the data for the equilibrium vapour pressure over ice over the range $-50°C < T < 0°C$ [70] is

$$p_{ve}(T) = 10^{-2} \exp(60.83 - 6947.18/T(K) - 5.64 \ln T(K) + 0.0097T).$$

Nomenclature

A	area
b	dimensionless coefficient (3.56)
c	molecular concentration (number per unit volume)
c_b	modified enthalpy for heat transfer ($= h_v - R_G T/\mu_v$)
c_p	specific heat at constant pressure
Cn	bulk condensation number (3.108)
Cn_s	surface condensation number (3.50)
D	molecular diffusivity
E	energy
F	factor reducing flux from diffusive limit
$G(R)$	radiative driving term [RAD] as a function of R
h	enthalpy or heat content per unit mass
H	heat transfer coefficient (3.35)
i	van't Hoff factor (3.56)
I	integral (3.90)
I_λ	radiation intensity
J	mass flux or current

$J(ni)$	cluster concentration growth current (3.76)
k	thermal conductivity
k_B	Boltzmann constant
Kn	Knudsen number (3.8)
L	latent heat of vapourisation
L_m	modified latent heat (3.37)
Le	Lewis number (3.109)
m	molecular mass
M_L	liquid mass in droplet
M_n	moment of aerosol size distribution (3.61)
M_s	solute mass in droplet
$n(R)$	aerosol size distribution
N	aerosol number concentration
p	pressure
$\mathbf{q}$	radiation current
Q_R	radiative heat flux from droplet
$\mathbf{r}$	space coordinate
r	radial coordinate
R	aerosol radius
R_a	lognormal radius parameter
R_c	critical radius for nucleation
R_m	mean aerosol radius
R_G	gas constant
R_s	radius of solute spherical mass in droplet
R_γ	Kelvin radius (3.2)
R	subscripted with D, K, T, MK, MD, HC, HK: resistances to mass current
S	saturation
$S(\mathbf{r}, t)$	source term
S_c	critical saturation (3.58)
S_e	equilibrium saturation for droplet containing solute (3.55)
S_p	molecular sticking probability
t	time
t_{OR}	Ostwald ripening timescale (3.104)
T	temperature
u	vapour concentration difference (3.106)
v	velocity
v_m	$= v_T/4$ in Equation (3.26)
v_T	molecular thermal velocity
V	volume
X_L	solvent mole fraction
Y	uptake coefficient (3.59)
Y_r	reaction coefficient (3.60)
α	thermal accommodation coefficients (subscripted)
α_G	coefficient in aerosol growth rate (3.27) (3.78)
β	coefficient in Clausius–Clapeyron Equation (3.32)
β_T	constant in Boltzmann distribution (3.4)

$\beta(n)$	n-cluster growth probability
ε	surface radiation emissivity
γ_L	surface tension
$\gamma(n)$	n-cluster decay probability
Γ_L	activity coefficient
λ	wavelength
λ_m	molecular mean free path
μ	$= \alpha_G R_a$ (3.93); molecular weight (subscripted v or L)
μ_G	growth rate coefficient (3.79)
ν	$= (\ln \sigma)^2$ (3.92)
$\nu_v(T)$	molecular kinetic mass flux at temperature T
ρ	density
σ	width parameter for lognormal distribution; Stefan–Boltzmann constant
σ_c	condensation coefficient
σ_E	evaporation coefficient
τ_D	diffusion timescale (3.18), (3.73)
τ_D	equilibrium growth timescale (3.71)
Ω	solid angle

Subscripts

a	pertaining to aerosol
d	pertaining to droplet
D	pertaining to diffusion
e	pertaining to equilibrium
H	pertaining to heat
K	pertaining to kinetic
L	pertaining to liquid
M	pertaining to mass
N	pertaining to nucleation
T	pertaining to thermal
v	pertaining to vapour
x, y, z	pertaining to Cartesian coordinate directions

References

1 Wagner, P.E. Aerosol growth by condensation. In: W.H. Marlow, ed. *Aerosol Microphysics II*. Springer-Verlag, Berlin: 1982; pp. 129–178.

2 Wagner, P.E. A constant-angle Mie scattering method (CAMS) for investigation of particle formation processes. *Journal of Colloid and Interface Science* 1985; **105**: 456–467.

3 Strey, R., Wagner, P.E. and Viisanen, Y. The problem of measuring homogeneous nucleation rates and the molecular contents of nuclei: progress in the form of nucleation pulse measurements. *Journal of Physical Chemistry* 1994; **98**:7748–7758.

4 Pöschl, U., Rudich, Y. and Amman, M. Kinetic model framework for aerosol and cloud surface chemistry and gas-particle interactions: Part I – general equations, parameters and terminology. *Atmospheric Chemistry and Physics Discussions* 2005; **5**: 2111–2191.

5 Clement, C.F. Mean field theory for condensation on aerosol and applications to multi-component organic vapours. *Journal of Aerosol Science* 2003; **34**: 27–40.
6 Jaenicke, R. Tropospheric aerosols. In: *Aerosol-Cloud-Climate Interactions*. Academic Press, New York: 1993; pp. 1–31.
7 Clement, C.F. and Ford, I.J. Gas-to-particle conversion in the atmosphere: I. Evidence from empirical atmospheric aerosols. *Atmospheric Environment* 1999; **33**: 475–487.
8 Landau, L.D. and Lifshitz, E.M. *Statistical Physics*, 3rd edn. Pergamon Press, Oxford: 1980.
9 Hirschfelder, J.O., Curtiss, C.F. and Bird, R.B. *Molecular Theory of Gases and Liquids*. John Wiley, New York: 1954.
10 Chapman, S. and Cowling, T.G. *The Mathematical Theory of Non-Uniform Gases*. Cambridge University Press, Cambridge: 1970.
11 Lifshitz, E.M. and Pitaevskii, L.P. *Physical Kinetics*. Pergamon, Oxford: 1981.
12 Dorsey, N.E. *Properties of Ordinary Water Substance*. Reinhold, New York: 1940.
13 Maxwell, J.C. Diffusion. *Encyclopaedia Brittanica* 1877; **2**: 82. Reprinted in *The Scientific Papers of James Clerk Maxwell* (W.D. Niven, ed.), Vol. 2. Cambridge University Press, Cambridge: 1890; p. 625.
14 Barrett, J.C. and Clement, C.F. Growth rates for liquid drops. *Journal of Aerosol Science* 1988; **19**: 223–242.
15 Barrett, J.C. and Clement, C.F. Kinetic evaporation and condensation rates and their coefficients. *Journal of Colloid and Interface Science* 1992; **150**: 352–364.
16 Fuchs, N.A. and Sutugin, A.G. *Highly Dispersed Aerosols*. Ann Arbor Science, Ann Arbor, Michigan: 1970.
17 Hegg, D.A., Radke, L.F. and Hobbs, P.V. Measurements of Aitken nuclei and cloud condensation nuclei in the marine atmosphere and their relation to the DMS-cloud-climate hypothesis. *Journal of Geophysical Research* 1991; **96**: 18727–18733.
18 Kreidenweis, S.M., Yin, F., Wang, S.C., Grosjean, D., Flagan, R.C. and Seinfeld, J.H. Aerosol formation during photooxidation of organosulfur species. *Atmospheric Environment* 1991; **25**A: 2491–2500.
19 Kulmala, M. and Vesala, T. Condensation in the continuum regime. *Journal of Aerosol Science* 1991; **22**: 337–346.
20 Wright, P.G. On the discontinuity involved in diffusion across an interface (the Δ of Fuchs). *Discussions Faraday Society* 1960; **30**:100–112.
21 Wright, P.G. The effect of the transport of heat on the rate of evaporation of small droplets. Part 1. Evaporation into a large excess of gas. *Proceedings of the Royal Society of Edinburgh* 1962; **66A**: 65–80.
22 Williams, M.M.R. Growth rates of liquid drops for large saturation ratios. *Journal of Aerosol Science* 1995; **26**: 477–487.
23 Wagner, P.E. Energy balance during condensational drop growth. *Journal of Aerosol Science* 1991; **22**: 789–791.
24 Clement, C.F. Aerosol formation from heat and mass transfer in vapour-gas mixtures.*Proceedings of the Royal Society, London A* 1985; **398**: 307–339.
25 Mason, B.J. *The Physics of Clouds*. Oxford University Press, Oxford: 1971.
26 Clement, C.F. and Ford, I.J. Maximum aerosol densities from evaporation and condensation processes. *Journal of Aerosol Science* 1989; **20**: 293–302.
27 Barkstrom, B.R. Some effects of 8–12 μm radiant energy transfer on the mass and heat budgets of cloud droplets. *Journal of Atmosphere Science* 1978; **35**: 665–673.
28 Özisik, M.N. *Radiative Transfer and Interaction with Conduction and Convection*. John Wiley, New York: 1973.
29 Barrett, J.C. The interaction of radiation with aerosol-vapour-gas mixtures. Thesis. Queen Mary College, London and AERE-TP 1171, 1985.
30 Seinfeld, J.H. *Atmospheric Chemistry and Physics of Air Pollution*. Wiley Interscience, New York: 1986.

31 Manton, M.J. The physics of clouds in the atmosphere. *Reports of Progress in Physics* 1983; **46**: 1393–1444.
32 Kalkkinen, J., Vesala, T. and Kulmala, M. Binary droplet evaporation in the presence of an inert gas: an exact solution of the Maxwell-Stefan equations. *International Communications in Heat and Mass Transfer* 1991; **18**: 117–126.
33 Kulmala, M. Vesala, T. and Wagner, P.E. An analytical expression for the rate of binary condensational particle growth. *Proceedings of the Royal Society, London A* 1993; **441**: 589–605.
34 Kulmala, M., Laaksonen, A., Korhonen, P., Vesala, T., Ahonen, T. and Barrett, J.C. The effect of atmospheric nitric acid vapour on cloud condensation nucleus activation. *Journal of Geophysical Research* 1993; **98** (D12): 22949–22958.
35 Lehtinen, K.E.J., Kulmala, M., Vesala, T. and Jokiniemi, J.K. Analytic methods to calculate condensation rates of a multicomponent droplet. *Journal of Aerosol Science* 1998; **29**: 1035–1044.
36 Vesala, T., Kulmala, M., Rudolph, R., Vrtala, A. and Wagner, P. Models for condensational growth and evaporation of binary aerosol particles. *Journal of Aerosol Science* 1997; **28**: 565–598.
37 Meilinger, S.K., Koop, T., Luo, B.P., Huthwelker, T., Carslaw, K.S., Krieger, U., Crutzen, P.J. and Peter, T. Size-dependent stratospheric droplet composition in lee wave temperature fluctuations and their potential role in PSC freezing. *Geophysical Research Letter* 1995; **22**: 3031–3034.
38 Langner, J. and Rodhe, H. A global three-dimensional model of the tropospheric sulfur cycle. *Journal of Atmospheric Chemistry* 1991; **13**: 225–263.
39 Smith, M.H., O'Dowd, C.D. and Lowe, J.A. Observations of cloud-induced aerosol growth. In: M. Kulmala and P.E. Wagner, eds. *Nucleation and Atmospheric Aerosols 1996.* Pergamon, Elsevier Science, Oxford: 1996; pp. 937–940.
40 Ammann, M. and Pöschl, U. Kinetic model framework for aerosol and cloud surface chemistry and gas-particle interactions: Part 2 – exemplary practical applications and numerical simulations. *Atmospheric Chemistry and Physics Discussions* 2005; **5**: 2193–2246.
41 Baltensperger, U., Ammann, M., Kalberer, M. and Gäggeler, H.W. Chemical reactions on aerosol particles: concept and methods. *Journal of Aerosol Science* 1996; **27**: S651–S652.
42 Porstendörfer, J., Röbig, G. and Ahmed, A. Experimental determination of the attachment coefficients of atoms and ions on monodisperse aerosols. *Journal of Aerosol Science* 1979; **10**: 21–28.
43 Pankow, J.F. An absorption model of gas/particle partitioning of organic compounds in the atmosphere. *Atmospheric Environment* 1994; **28**: 185–188.
44 Pankow, J.F. An absorption model of gas/particle partitioning involved in the formation of secondary organic aerosol. *Atmospheric Environment* 1994; **28**: 189–193.
45 Odum, J.R., Hoffmann, T., Bowman, F., Collins, D., Flagan, R.C. and Seinfeld, J.H. Gas/particle partitioning and secondary organic aerosol yields. *Environmental Science and Technique* 1996; **30**: 2580–2585.
46 Bowman, F.M., Odum, J.R., Seinfeld, J.H. and Pandis, S.N. Mathematical model for gas-particle partitioning of secondary organic aerosols. *Atmospheric Environment* 1997; **31**: 3921–3931.
47 Pandis, S.N., Russel, I.M. and Seinfeld, J.H. The relationship between DMS flux and CCN concentrations in remote marine regions. *Journal of Geophysical Research* 1994; **99**: 16945–16957.
48 Knudsen, M. *Kinetic Theory of Gases.* Methuen, London: 1950; p. 46.
49 Pruppacher, H.R. and Klett, J.D. *Microphysics of Clouds and Precipitation.* D. Reidel, Dordrecht, Holland: 1978.
50 Clement, C.F., Kulmala, M. and Vesala, T. Theoretical consideration on sticking probabilities. *Journal of Aerosol Science* 1996; **27**: 869–882.
51 Winkler, P.M., Vrtala, A., Wagner, P.E., Kulmala, M., Lehtinen, K.E.J. and Vesala, T. Mass and thermal accommodation during gas-liquid condensation of water. *Physics Review Letter* 2004; **93**: 075701.
52 Anttila, T. and Kerminen, V.-M. Condensation growth of atmospheric nuclei by organic vapours. *Journal of Aerosol Science* 2002; **34**: 41–61.

53 Wu, C.-Y. and Biswas, P. Particle growth in a system with limited vapour. *Aerosol Science and Technology* 1998; **28**: 1–20.
54 McMurry, P.H. and Friedlander, S.K. New particle formation in the presence of an aerosol. *Atmospheric Environment* 1979; **13**: 1635–1651.
55 McMurry, P.H. Photochemical aerosol formation from SO_2: a theoretical analysis of smog chamber data. *Journal of Colloid and Interface Science* 1980; **78**: 513–527.
56 McMurry, P.H. New particle formation in the presence of an aerosol: rates, time scales, and sub-0.01 μm size distributions. *Journal of Colloid and Interface Science* 1983; **95**: 72–80.
57 Rao, N.P. and McMurry, P.H. Nucleation and growth of aerosol in chemically reacting systems: a theoretical study of the near collision-controlled regime. *Aerosol Science and Technique* 1989; **11**: 120–132.
58 Clement, C.F. and Wood, M.H. Equations for the growth of a distribution of small physical objects. *Proceedings of the Royal Society, London A* 1979; **368**: 521–546.
59 Clement, C.F. and Wood, M.H. Moment and Fokker-Planck equations for the growth and decay of small objects. *Proceedings of the Royal Society, London A* 1980; **371**: 553–567.
60 Clement, C.F., Lehtinen, K.E.J. and Kulmala, M. Size diffusion for the growth of newly nucleated aerosol. *Journal of Aerosol Science* 2004; **35**: 1439–1451.
61 Clement, C.F. Solutions of the continuity equation. *Proceedings of the Royal Society, London A* 1978; **364**: 107–119.
62 Barrett, J.C. and Clement, C.F. Aerosol concentrations from a burst of nucleation. *Journal of Aerosol Science* 1991; **22**: 327–335.
63 Barrett, J.C., Clement, C.F. and Ford, I.J. The effect of redistribution on aerosol removal rates. *Journal of Aerosol Science* 1992; **23**: 639–656.
64 Clement, C.F., Pirjola, L., dal Maso, M., Mäkelä, J.M. and Kulmala, M. Analysis of particle formation bursts in Finland. *Journal of Aerosol Science* 2001; **32**: 217–236.
65 Clement, C.F. and Clement, R.A. Molecular growth integrals for lognormal size distributions for aerosols. *Journal of Aerosol Science* 1999; **30**: S415–S416.
66 Barrett, J.C. and Clement, C.F. Growth and redistribution in a droplet cloud interacting with radiation. *Journal of Aerosol Science* 1990; **21**: 761–776.
67 Lifshitz, E.M. and Slezov, V.V. Kinetics of diffusive decomposition of supersaturated solid solutions. *Soviet Physics JETP* 1959; **8** (35): 331–339.
68 Lifshitz, E.M. and Slezov, V.V. The kinetics of precipitation from supersaturated solid solutions. *Journal of Physics and Chemistry of Solids* 1961; **19**: 35–50.
69 Kulmala, M., Korhonen. P., Vesala, T., Hansson, H.C., Noone, K. and Svenningsson. B. The effect of hygroscopicity on cloud droplet formation. *Tellus* 1996; **48**: 347–360.
70 Hienola, J., Kulmala, M. and Laaksonen, A. Condensation and evaporation of water vapor in mixed aerosols of liquid droplets and ice; numerical comparisons of growth rate expressions. *Journal of Aerosol Science* 2001; **32**: 351–374.
71 Stolzenburg, M.R., McMurry, P.H., Sakurai, H., Smith, J.N., Lee Maudlin, R., III, Eisele, F.L. and Clement, C.F. Growth rates of freshly nucleated particles in Atlanta. *Journal of Geophysical Research – Atmospheres* 2005: **110**; 2005JD005935R.
72 Clement, C.F., Ford, I.J. Twohy, C.H., Weinheimer, A. and Campos, T. Particle production in the outflow of a midlatitude storm. *Journal of Geophysical Research* 2002; **107**: 1029/2001IJD001352.
73 Clement, C.F., Pirjola, L., Twohy, C.H., Ford, I.J. and Kulmala, M. Analytic and numerical calculations of the formation of sulphuric acid aerosol in the upper troposphere. *Journal of Aerosol Science* 2006; **37**: 1717–1729.
74 Twohy, C.H., Clement, C.F., Gandrud, B.W., Weinheimer, A.J., Campos, T.L., Baumgardner, D., Brune, W.H., Faloona, I., Sachse, G.W., Vay, A.S. and Tan, D. Deep convection as a source of new particles in the midlatitude upper troposphere. *Journal of Geophysical Research* 2002; **107**: D21 4560, doi: 1029/2001IJD001352.

75 Pirjola, L. Effects of the increased UV radiation and biogenic VOC emissions on ultrafine sulphate aerosol formation. *Journal of Aerosol Science* 1999; **30**: 355–367.

76 Clement, C.F., McMurry, P.H. and Pirjola, L. Aerosol formation and sulphuric acid mass transfers in the atmosphere. Paper presented at the Aerosol Society Annual meeting, Manchester, 2004.

77 Clement, C.F. The supersaturation in vapour-gas mixtures condensing into aerosols. UKAEA Harwell Report AERE-TP 1223, 1987.

78 Clement, C.F. Condensation and evaporation in clouds. In: M. Kulmala and K. Hämeri, eds. *Workshop on Condensation, Helsinki 1991*, Report Series in Aerosol Science No. 17. Finnish Association for Aerosol Research, Helsinki: 1991, 3–13.

79 Barrett, J.C. and Fissan, H. Wall and aerosol condensation during cooled laminar flow. *Journal of Colloid and Interface Science* 1989; **130**: 498–507.

80 Richards, J.M. Simple expression for the saturation vapour pressure of water in the range −50–140°C. *Journal of Physics D – Applied Physics* 1971; **4**: L15–L18.

81 Bolton, D. The computation of equivalent potential temperatures. *Monthly Weather Review* 1980; **108**: 1046–1053.

Chapter 4
Organic Aerosols

Mihalis Lazaridis

4.1 Introduction

Airborne particulate matter (PM) is a complex mixture of different chemical components and is a challenging issue to study their dynamics in the atmosphere. An important aspect in the study of PM is its complexity of the physicochemical characteristics, their multiple sources, morphology and their dynamics that are correlated with the size of the particles. PM has very diverse effects ranging from human health to climate forcing. In recent years, extensive research effort has been invested in examining the relationship between exposure to air pollution (especially PM and photo-oxidants) and resulting health effects [1, 2].

A significant fraction of the ambient PM is secondary PM in the form of sulphate, nitrate, ammonium and organic aerosol particles formed by the oxidation of sulphur dioxide, nitrogen oxides and organic gaseous species. PM is not a single pollutant and its mass includes a mixture of many pollutants distributed differently at different particle sizes.

Organic compounds contribute significantly to the total aerosol mass through the mass transfer of low-vapour-pressure organic substances and of primary organic compounds (e.g. polycyclic aromatic hydrocarbons, PAHs) to the atmospheric particle phase. The organic fraction of aerosols in urban air generally contributes 30–60% to the total fine PM in the atmosphere [3–6], and thereby has significant impact on atmospheric aerosol properties.

The term 'carbonaceous aerosol' refers to the carbonaceous component of the atmospheric aerosols since carbon components are combined with other inorganic chemical components in the atmosphere. Carbon is usually the dominant elemental fraction of atmospheric aerosols. The total amount of carbon in the PM can be determined by elemental analysis. Sampling artefacts can lead to considerable underestimation or overestimation of the carbon mass. There is a distinction between elemental (EC) and organic carbon (OC) for the carbonaceous aerosols. The term elemental carbon refers to the component of the carbonaceous aerosols that consist entirely of carbon. EC is present in the form of chain aggregates of small soot globules and is responsible for the light absorption of the material collected on filters. The transition between EC and OC is gradual since the carbon/hydrogen ratio reaches infinity within the homologous series of polynuclear aromatic hydrocarbons [3]. Thermal and wet measurement techniques can be used for the EC/OC determination. There is also an alternative methodology to characterise each compound individually. However, there are thousands of organic compounds and this methodology has not led to the determination of the majority of the organic mass in aerosols.

In addition to the EC/OC terminology there is also the term black carbon (BC) that is mainly applied to the component which is related to the absorption of visible light and optical methods are used for its determination. However, this light absorption depends on the size distribution of the soot particles and on the association of the soot particles with other substances in the aerosol particles and on sample filters. Optical methods to determine EC are therefore only semi-quantitative, and calibration factors may vary from one situation to another (e.g. Liousse *et al.* [7]). The black colour in the BC is due to the graphitic structure. Furthermore, the broad-band light absorption is caused by this structure and is due to the broad bands and small band gap, which permit the electrons to take up any energy of visible and near-IR and UV light.

There is no clear distinction between the terms EC and BC in the scientific literature and the term EC can be used as a synonym of BC [8]. Comprehensive reviews on organic atmospheric aerosols can be found in the articles by Kanakidou *et al.* [9], Jacobson *et al.* [10] and Saxena and Hildemann [11].

Absorption of solar radiation by BC heats the atmosphere, whereas the contrary results from the majority of the organic aerosol components [12]. Climate studies have also revealed the significant role of organic aerosols in the direct and indirect aerosol forcing [9, 13]. In addition, the potential of the atmospheric organics to serve as cloud condensation nuclei (CCN) has been examined in a recent article [14, 15]. Another issue that is important but not studied in great detail is the heterogeneous reactions on carbonaceous aerosols (e.g. Chughtai *et al.* [16]).

While considerable progress has been made in the understanding of physical and chemical properties of inorganic aerosols [4], the knowledge of organic aerosols remains limited due to their chemical complexity [9]. As a result experimentally determined fractional aerosol yields [17] are mainly used to describe the incorporation of organic matter in the particle phase. Aerosol yield is defined as the fraction of a reactive organic gas that is transformed to aerosol. There have been several methods for modelling the partitioning of organic matter into the aerosol phase. Pandis *et al.* [17] partitioned the condensable organic material to the aerosol phase when its gas-phase concentration exceeds its equilibrium vapour pressure. For a negligibly small saturation vapour pressure resulted to condense the majority of the condensable organic matter is in the particulate phase. Pankow [18] suggested that even undersaturated organic gaseous matter can participate in the aerosol phase using the mechanisms of adsorption and absorption processes for determining the gas–particle partitioning of organic matter in the atmosphere.

The condensable organic species are formed in the gas phase by the reaction of volatile organic compounds (VOCs) with the principal atmospheric oxidising agents, that is ozone, OH radicals and, to a small extent, NO_3 radicals. It is known that VOCs (alkanes, alkenes, aromatics and carbonyls) are emitted into ambient air from anthropogenic activities, such as transport and industry. However, recent data point to a large contribution of biogenic compounds (such as terpenes) to the formation of tropospheric ozone and secondary organic aerosols (SOAs) [9, 19, 20].

Anthropogenic VOCs (e.g. alkanes, alkenes and aromatics) are emitted from activities such as road traffic and combustion processes. Some of these compounds will readily form SOAs, especially when the background aerosol level is high. Also, biogenic VOCs are present in urban air and contribute significantly to the total PM. Studies by Anderson-Sköld and Simpson [21] and Tsigaridis and Kanakidou [22] conclude that SOA formation from biogenic precursors is greater than the contribution from anthropogenic SOAs in Nordic countries. However, there is still a lack of knowledge regarding SOA formation and aerosol size distribution in urban areas.

Finally, there is the introduction of the term 'humic-like substances' (HULIS) or 'organic macromolecules' that have tried to resolve the unexplained mass of particulate OC [3, 23]. These substances arise from heterogeneous or multiphase reactions, such as various acid-catalysed reactions.

This chapter aims to offer a condensed overview of the physicochemical properties and dynamics of ambient carbonaceous aerosol and the recent literature in the area. The chapter covers issues such as the carbonaceous aerosol sources, ambient levels and chemical processes in the atmosphere as well as sampling and analysis methodologies. Furthermore, the contribution from biogenic sources to the ambient levels of organic aerosols has been studied and a new parameterisation for SOA formation from biogenic precursors has been proposed.

4.2 Carbon-containing aerosols and their sources

The carbonaceous aerosols in the atmosphere can be identified as primary organic aerosols (POAs) and as secondary organic aerosols (SOAs) if they are primary emitted or secondary formed in the atmosphere, respectively. The SOA forms as an oxidation product of gaseous organic precursors with low vapour pressure.

As discussed previously, the distinction between EC and OC is operationally defined and arbitrary. There are several methodologies for the determination of EC and OC that result in an uncertainty in the distinction of these forms of carbonaceous aerosols. The determination of the organic aerosols based at a molecular level is also a challenging task since thousands of organic compounds are present in the PM.

Soot is the only observed carbon particle form that includes all types of organic matter and is a synonymous of combustion-generated primary carbonaceous aerosols. It consists of an array of organic compounds that are soluble in organic solvents and an insoluble part which is named EC or BC [3, 24]. The morphology of soot particles is in the form of aggregates that consist of several clusters. Combustion sources such as diesel engines emit soot particles at irregular agglomerate structures with an average diameter of primary spherules close to 22.6 ± 6.0 nm [25]. Biomass combustion produces, during the smouldering phase, spherical particles that are compact and more stable than the irregular agglomerate structures, as shown in Figure 4.1 [26]. The presence of soot particles in the atmosphere is described in detail by Gelencsér [3].

The chemical composition of soot depends on its sources that can range from almost pure EC to a considerable contribution from organic matter. Measurements using analytical techniques proposed a structure for hexane soot [27]. It appears that combustion aerosols have a graphitic structure based on the C—O bonding and with hydrocarbon segments at the particle surface. The transition between OC and BC fractions can be deduced from Figure 4.2. There is a graphitic backbone and a gradual shift to OC. The main problem is to identify this shift from BC to OC.

4.2.1 *Elemental carbon – primary organic carbon*

EC has a chemical structure similar to impure graphite and is emitted as primary particles mainly during combustion processes (wood burning and diesel engines) [1]. In Western Europe the contribution of diesel emissions to EC concentrations is estimated to be between 70 and 90% [28]. EC both absorbs and scatters light and contributes significantly to the total light extinction. Much higher concentrations of EC are found in urban areas compared to rural and remote locations. In rural and remote locations the EC concentration can vary between 0.2 and

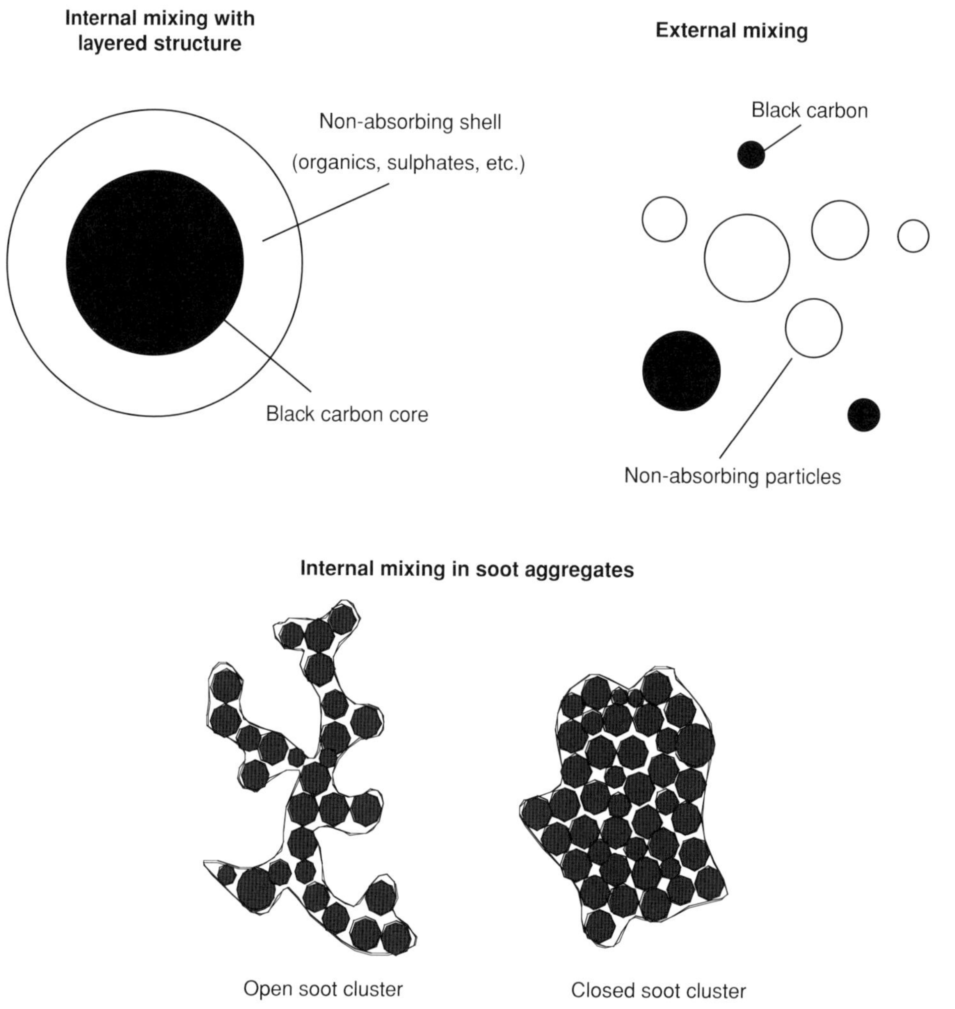

Figure 4.1 Diagram on possible combinations between BC and non-absorbing materials in soot particles. (Based on [26].)

2.0 μg m^{-3} [1] and between 1.5 and 20 μg m^{-3} in urban areas, whereas the concentration of EC in remote ocean areas is in the range 5–20 ng m^{-3} [1]. Studies concerning the size distribution and characteristics of EC aerosols are summarised in a report by EPA [1]. In polluted areas EC presents a bimodal size distribution with the first mode between 0.05 and 0.12 μm and the second mode between 0.5 and 1.0 μm. Average EC concentrations are around 1.3 and 3.8 μg m^{-3}for rural and urban sites, respectively, in the US. The ratio of EC to total carbon is in the range 0.15–0.20 in rural areas and 0.2–0.6 in urban areas.

The OC is a complex mixture of thousands of different organic compounds [17, 29, 30] and a very small portion of its molecular structure has been characterised (around 10%). Combustion sources emit particles in the fine mode and the mass distribution has a mode at 0.1–0.2 μm. Gas

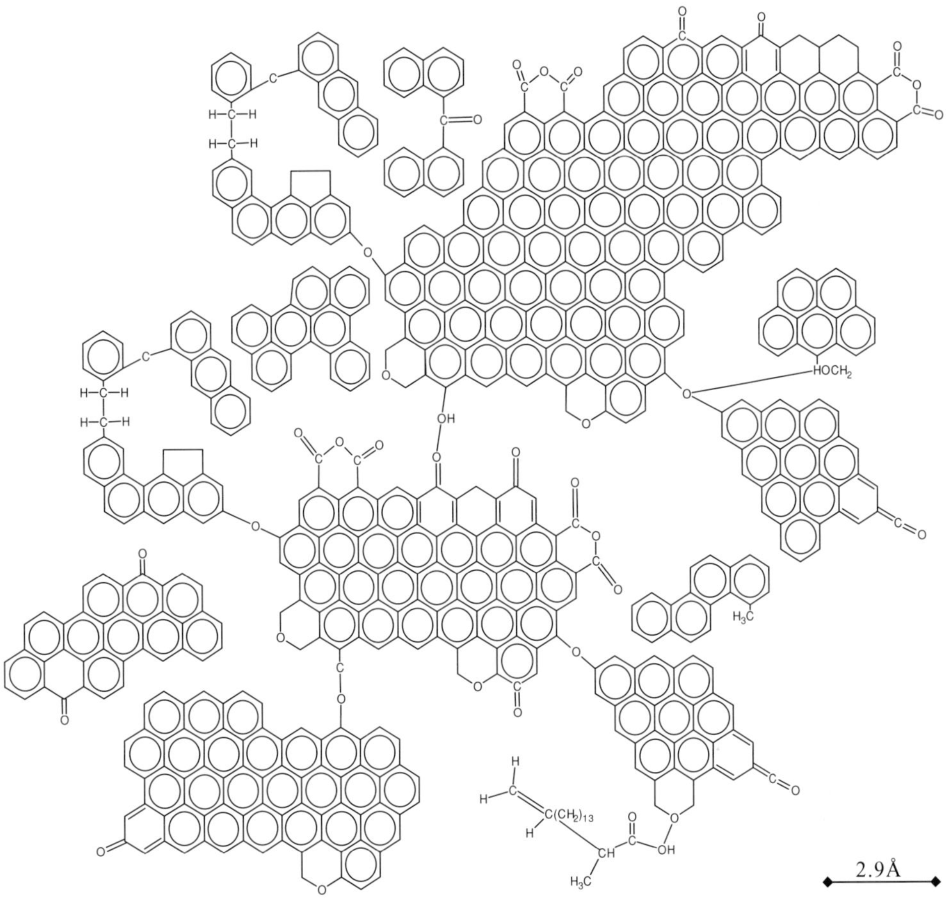

Figure 4.2 Structure of hexane soot. (Based on [27].)

chromatography combined with mass spectroscopy (GC/MS) as well as liquid chromatography combined with mass spectroscopy (LC/MS) has not, to date, provided a satisfactory solution for the molecular characterisation of the major part of the ambient organic matter. Organic compounds that have been characterised include, amongst others, *n*-alkanes, *n*-alkanoic acids and polycyclic aromatic compounds. Due to the difficulty in measuring organic compounds our current knowledge about organic matter is limited and incomplete. Primary emission sources for OC include combustion processes, geological (fossil fuels) and biogenic sources. POA components include surface-active organic matter on sea-salt aerosol.

4.2.2 Secondary organic matter formation (secondary organic carbon)

An important part of secondary aerosol particles in the atmosphere is composed by secondary formed organic matter produced from oxidation of organic compounds. Partitioning of gas–particle organic compounds in the atmosphere is an important task for determining their association with the fine PM. Understanding the mechanisms that control the conversion of organic matter from the vapour phase to PM will provide valuable information for determining future control strategies to reduce the partitioning of organic matter in the particulate phase. However, there is a great complexity of the number of different chemical forms of organic matter and the absence of direct chemical analysis has resulted in the use of experimentally determined fractional aerosol yields, fractional aerosol coefficients (FACs) and adsorption/absorption methodologies [17, 29–34] to describe the incorporation of organic matter in the aerosol phase.

An important pathway for secondary organic particle formation arises from biogenic hydrocarbons. There are very large quantities of biogenic hydrocarbons that are globally emitted, which are also highly reactive [19]. The chemical structures of some biogenic hydrocarbons relevant to SOA formation are shown in Figure 4.3. SOA formed from the oxidation of VOCs produces highly oxidised compounds that are difficult to measure with current methods. Annual global emissions of biogenic hydrocarbons are estimated to be between 825 and 1150 Tg C (per year), whereas the anthropogenic emissions are estimated to be less than 100 Tg C (per year). A detailed overview of the formation of organic aerosols from biogenic hydrocarbons is reviewed by Hoffmann *et al.* [19].

Even though considerable progress has been made concerning the characterisation of SOA in the atmosphere, there are several limitations and uncertainties connected with SOA modelling and understanding of the organic matter formation in the atmosphere. Seigneur *et al.* [36] summarised the existing uncertainties into five categories:

- Uncertainties that are inherent to the definition and use of the FAC. This is mainly due to the fact that FAC is a single, constant parameter assigned to each VOC. (*Note*: FAC (dimensionless) = [SOA] ($mg\,m^{-3}$)/$[VOC]_o$ ($mg\,m^{-3}$), where $[VOC]_o$ is the initial VOC concentration.)
- Uncertainties in the gas-phase chemistry mechanisms used in air-quality models. Some of these uncertainties have a direct impact on the description of the SOA formation.
- Weaknesses in the treatment of SOA in air-quality models. For example an important uncertainty arises since the SOA yield approach is limited to the OH–VOC reaction and ignores the ozone–VOC reaction.
- Gaps in emission data for those VOCs that are SOA precursors.
- Lack of experimental data for evaluation of model performance.

Finally, several recent modelling studies have been performed to study the transport and formation of organic aerosols in the atmosphere using state-of-the-science information on the organic aerosol chemical/physical dynamics (e.g. Johnson *et al.* [37], Pun *et al.* [38], O'Neill *et al.* [39]).

4.2.3 Biological aerosols

Biological aerosols may also have a considerable influence in the mass size distribution. Recent work in Mainz indicated that biological aerosols can constitute close to 30% of the total aerosol mass [40]. These results are valid for an urban/rural-influenced region. Results from

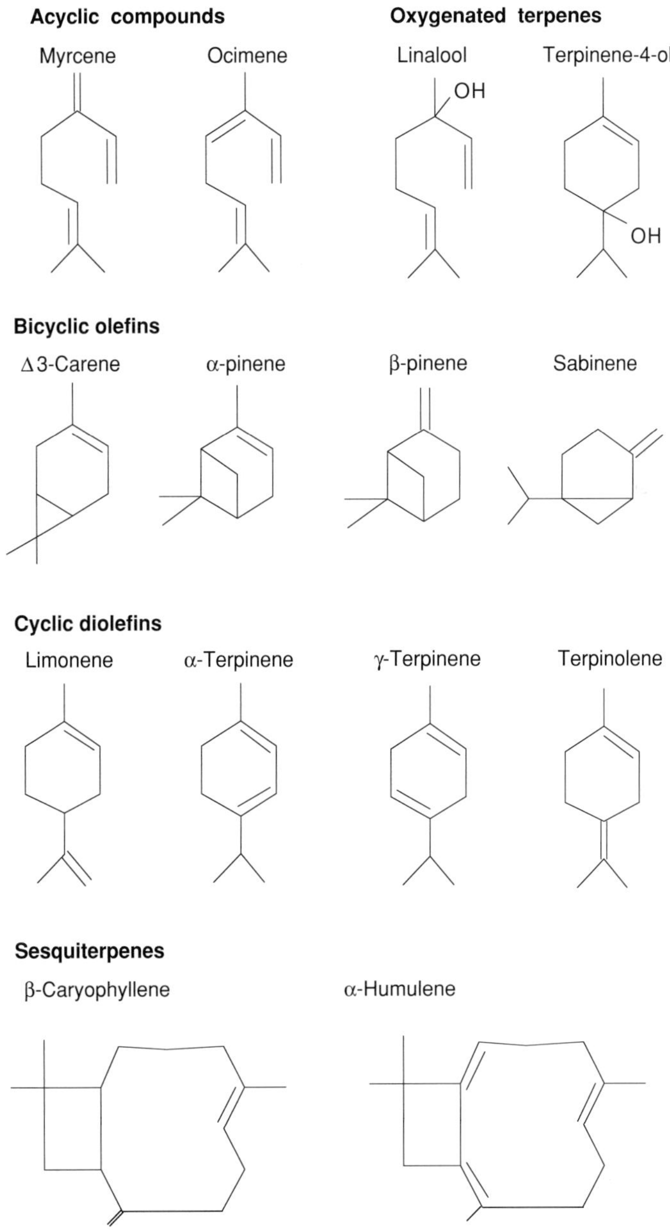

Figure 4.3 Biogenic hydrocarbons influencing SOA formation in the atmosphere. Carbon atom bonds are shown with vertices, whereas the hydrogen atoms are not shown. (Reproduced from [35] by permission of American Geophysical Union. Copyright 1999 American Geophysical Union.)

measurements on the Atlantic Ocean show a percentage of biological aerosols to total aerosol of only 14% in number and 8% in volume [41].

Primary biological aerosol particles (PBAPs) describe airborne solid particles (dead or alive) that are or were derived from living organisms, including micro-organisms and fragments of all varieties of living things. PBAP includes viruses ($0.005\ \mu m < r < 0.25\ \mu m$) ($r$ refers to particle radius), bacteria ($r \geq 0.2\ \mu m$), algae, spores of lichen mosses, ferns and fungi ($r \geq 0.5\ \mu m$), pollen ($r \geq 5\ \mu m$), plant debris such as leaf litter, part of insects and human and animal epithelial cells (usually $r > 1\ \mu m$) [41]. Most of biological aerosols arise from plants but secondary sources include industry (textile mills), agriculture (fertilising) and sewage plants.

The shape of biological aerosols can be very diverse ranging from spherical particles to needlelike and flakes. Furthermore, many viruses can change their shape and can have many different morphologies.

The composition of the PBAP is changing through the year as following:

- *spring*: micro-organism, pollen, some spores, a few fragments;
- *summer*: micro-organism, pollen, spores, a few fragments;
- *autumn*: micro-organism, fragments, spores, a few pollen;
- *winter*: micro-organism, fragments, spores, some pollen.

4.2.4 Biogenic aerosols

VOCs are emitted in large quantities from anthropogenic sources such as transport and industrial activities and their significant role as aerosol precursors has been demonstrated in the scientific literature [3, 9, 42, 43]. Natural sources of VOC arise from plant foliage, soil, oceans and fresh water as well as from wood. For example in most Mediterranean countries, the burden of biogenic emissions is comparable with that from anthropogenic sources. In particular, in summer periods, when conditions of intense sunlight and relatively high temperatures prevail, biogenic emissions may even exceed the anthropogenic ones [44]. Therefore the biogenic emissions govern the air chemistry and SOA formation in remote areas. In addition, biogenic sources have been documented to significantly affect air pollution in urban and near-urban environments with an estimated contribution of 10–60% of the total PM amount.

Photo-oxidation of aromatics has been identified to be the main source of secondary organic PM from anthropogenic VOCs such as, for example, gasoline [45]. The initial reactions are driven by OH chemistry and the formation of first-generation products has been identified [46]. Photo-oxidation of biogenic VOCs after reaction with ozone or OH radicals has been the focus of recent studies. A number of products have been identified in the gas and particle phase. It has been shown that the cyclic terpenes are not only the most abundant biogenic VOCs emitted in Europe [44] but also the ones with the highest aerosol formation potential [19]. The sum of the following four chemical classes of oxidation products from these terpenes constitutes more than 50% of the total aerosol mass formed from monoterpene oxidation, that is C_8–C_{10} dicarboxylic acids, oxo-carboxylic acids, hydroxy-keto-carboxylic acids and higher carbonyls. An important fact that has to be considered is that in order to extrapolate aerosol mass yields and product distributions obtained in the laboratory in atmospheric conditions, literature data need to be complemented with additional experimental data on the influence of fundamental parameters for aerosol formation, such as VOC/NO_X ratios, temperature and humidity.

4.2.5 Modelling biogenic aerosol formation

SOAs are formed when VOCs are oxidised in the atmosphere, creating semi-volatile products that can partition from the gas to the aerosol phase. The processes can be viewed as occurring sequentially: (1) emission of gases, (2) gas-phase chemistry, (3) gas–particle partitioning and (4) aerosol-phase chemistry. All these steps have to be modelled to represent SOA formation quantitatively. Several chemical mechanisms have been proposed in the scientific literature for the simulation of SOA [9]. The gas-phase chemical mechanisms included in air-quality models do not include an explicit reaction scheme for all chemical reactions but instead use 'lumped' categories, such as the SAPRC (Statewide Air Pollution Research Center) and RADM (random-walk advection and dispersion model) gas-phase chemical mechanisms (including lumped mechanisms for alkanes, alkenes, etc.) and the CBM-IV (carbon bond mechanism – IV) mechanism (lumped bonds such as C—C and C=C) [47].

Another widely used gas-phase chemical mechanism is the Statewide Air Pollution Research Center Mechanism (SAPRC-97) (available at http://pah.cert.ucr.edu/~carter/) [35, 48, 49]. The model includes 21 aerosol species (including 8 groups of SOAs), mass-transfer-limited condensation/evaporation, true aqueous-phase SO_2 oxidation, ad hoc nucleation scheme for H_2SO_4 and a hybrid fixed/moving aerosol bin structure. The SAPRC-97 mechanism is designed for assessing the impact of individual VOCs in the atmosphere and estimates the reactivity for more than 500 VOC categories. In the model the gases are lumped into 64 groups of compounds with similar reactivity. Eight of the lumped species are categorised as condensable organic compounds (COCs). These are semi-volatile VOC products formed from oxidation processes, which readily form SOAs. VOC oxidation occurs primarily through reaction with the hydroxyl radical (OH). In addition, if the VOC contains certain structural characteristics, oxidation also occurs via reaction with nitrate radical (NO_3) or ozone (O_3). The key reactions controlling the SOA formation in the SAPRC-97 mechanism are summarised in Table 4.1.

Table 4.1 COCs and SOA yields for the SAPRC-97 mechanism

Species	Reactants	Description of VOCs	SOA yield[a]
COC_1	CRES+OH	Cresols and other alkyl phenols	1.4
COC_2	ALK_1+OH	Alkanes with kOH $< 10^4$ ppm^{-1} min^{-1}	144
COC_3	ALK_2+OH	Alkanes with kOH $> 10^4$ ppm^{-1} min^{-1}	402
COC_4	ARO_1+OH	Aromatics with kOH $< 2 \times 10^4$ ppm^{-1} min^{-1}, e.g. toluene	416
COC_5	ARO_2+OH	Aromatics with kOH $> 2 \times 10^4$ ppm^{-1} min^{-1}, e.g. xylene	221
COC_6	OLE_1+OH, O_3, NO_3, O	Alkenes with kOH $< 7.5 \times 10^4$ ppm^{-1} min^{-1}	9.5
COC_7	OLE_2+OH, O_3, NO_3, O	Alkenes with kOH $< 7.5 \times 10^4$$ppm^{-1}$ min^{-1}	30
COC_8	OLE_3+OH, O_3, NO_3, O	Monoterpenes, e.g. α-pinene and β-pinene	762

[a]Aerosol yields are in $\mu g\,m^{-3}$ ppm^{-1} of aerosol mass.

The gas-phase chemical mechanisms need to be compared with available gas chamber experiments that have been performed at several laboratories [35, 50–52, 53, 54].

The semi-volatile products (COCs) can partition from the gas to the aerosol phase. The formation of SOA from a particular VOC is often described in terms of the fractional mass yield, Y, which relates how much PM is produced when a certain amount of parent gaseous VOC is oxidised:

$$Y = \frac{M_0}{\Delta \mathrm{VOC}}$$

where M_0 ($\mu g\, m^{-3}$) is the mass concentration of SOA produced from the reaction of ΔVOC ($\mu g\, m^{-3}$). In the original SAPRC-97 mechanism the yields are constants (see Table 4.1, last column). However, several smog chamber experiments have demonstrated that the yield, Y, depends on both the ambient temperature and the concentration of organic matter present [19, 32, 35, 55]. Odum *et al.* [32] introduced the concept of multicomponent gas–particle partitioning to describe the trend of increasing Y values with increasing M_0. The following partition function has been proposed:

$$Y = M_0 \sum_i \frac{\alpha_i K_i}{1 + K_i M_0} \tag{4.1}$$

Here α_i is the individual formation yield for oxidation product i, and K_i is the corresponding absorption equilibrium constant. The above framework has been used in analysing yield data for laboratory smog chambers for a variety of parent VOC compounds. Yield data are generally not fit well by assuming a single product. However, in a two-product model ($i = 2$), assuming that two hypothetical average product compounds are formed, the representation of SOA formation is normally successful in fitting laboratory SOA yields [9]. The two-product model fits best with a non-volatile and a semi-volatile product. In the two-product model four fitting parameters are needed: Two α_i values and two K_i values are used to fit the monotonically increasing curve that goes through zero.

Most smog chamber studies are performed for high temperatures and consequently, the yields are also estimated for high temperatures. However, when the temperature is lowered, the oxidation products might shift from the gas to the SOA phase, increasing the yield significantly. Thus, the temperature is a very important parameter for SOA formation. Studies from Sheehan and Bowman [56] suggest that the temperature dependence is related to K_i and can be expressed as

$$K_i(T) = K_i \frac{T}{T_{\mathrm{ref}}} \exp\left[\frac{H}{R}\left(\frac{1}{T} - \frac{1}{T_{\mathrm{ref}}}\right)\right] \tag{4.2}$$

where T_{ref} is a reference temperature (normally ranging from 303 to 308 K), R is the ideal gas constant (0.0019872 kcal $(\mathrm{mol\,K})^{-1}$) and H is the enthalpy of vapourisation of the oxidation products (10–30 kcal mol^{-1} in most experiments) [50, 56].

4.2.6 Polycyclic aromatic hydrocarbons

PAHs are produced during the incomplete combustion of organic matter as for example wood burning, gasoline and coal fuel combustion. PAHs consist of two or more benzene rings and contain only carbon and hydrogen (Table 4.2). PAHs are suspected to be carcinogenic and mutagenic compounds and present a major constituent of atmospheric aerosols [60].

Table 4.2 Structure and properties of PAH species

Structure	Nomenclature	Molecular Weight	Melting point[a]	Boiling point[b]	Log [p (torr)] 20°C[c]	Solubility ($\mu g\,L^{-1}$)[d]
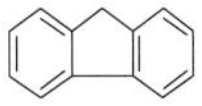	Fluorene	166.23	116	295	−2.72	31 700
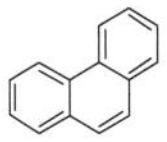	Phenanthrene	178.2	101	339	−0.3.5	1290
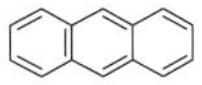	Athracene	178.2	216.2	340	−3.53	73
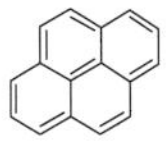	Pyrene	202.3	156	360	−4.73	135
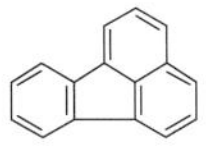	Fluorathene	202.3	111	375	−4.54	260
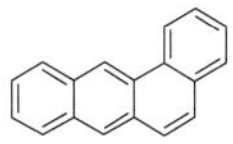	Benz(a)anthracene	228.3	160	435	−6.02	14
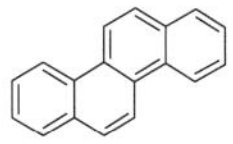	Chrysene	228.3	255	448	−6.06	2
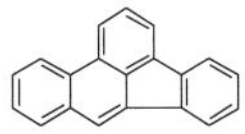	Benzo(b)fluoranthene	252.33	168	481	−5.22	14[e]
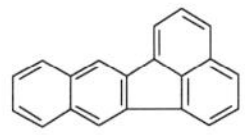	Benzo(k)fluoranthene	252.33	217	481	−7.13	4.3[e]
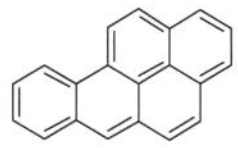	Benzo(a)pyrene	252.33	175	495	−7.33	0.05
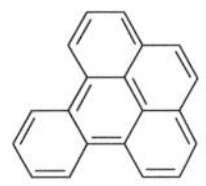	Benzo(e)pyrene	252.31	178.7	493	−7.37	3.8
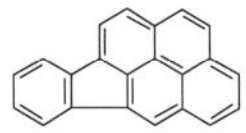	Indeno[1,2,3-cd]pyrene	276.34	163	530[e]	−10[e]	0.5[e]
	Benzo(ghi)perylene	276.34	277	525	−9.35	0.3
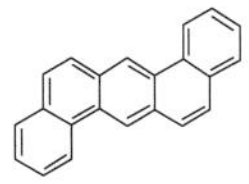	Dibenz(a,h)anthracene	278.35	267	524	−10[e]	0.5
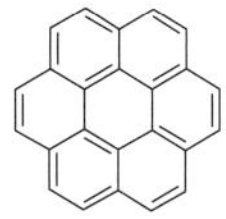	Coronene	300.36	439	590	−12.43	0.1

[a,b,d] Finlayson and Pitts[57].
[c] Pankow and Bidleman [58].
[e] ATSDR [59].

Combustion tracers such as PAHs have been used in source apportionment studies. More than 100 PAH compounds have been identified in urban areas and their composition ranges from species with two rings (such as naphthalene which is mainly in the gas phase) to species with seven and more rings (such as coronene which is mainly in the aerosol phase) [4]. Anthropogenic combustion sources contribute to PAHs from naphthalene to coronene, whereas natural biomass burning contributes mainly to alicyclic compounds.

4.2.7 Emission inventories of primary organic aerosols and gaseous precursors

POAs can originate from natural and anthropogenic sources. Direct emission of natural organic aerosols can originate from the vegetation, the soil, oceans and bioaerosols. Anthropogenic sources include biomass burning and fossil-fuel burning mainly from transport and energy production. Biomass burning is a very important source of soot particles and a mean emission factor for BC is reported by Andreae *et al.* [61] close to $0.59 \pm 0.19\ \mathrm{g\,Kg^{-1}}$ of dry matter. Furthermore, direct emissions of organic aerosols from vegetation have needlelike shape with length and width up to 200 and 30 nm, respectively.

Estimates for BC and OC emissions using energy statistics were given by Bond *et al.* [62]. Global emissions for POAs are estimated close to 9.1, 34.6 and 3.2 Tg POA y^{-1} from biofuel, vegetation fires and fuel burning, respectively, whereas the estimates for BC are 1.6, 3.3 and 3.0 Tg POA y^{-1}. Uncertainties in the above estimates due to the estimation of the emission factors have been identified in the literature [9]. Recent studies reveal the challenging task to estimate emission factors from individual sources [63].

The quantification of the emissions of gaseous precursors for the formation of organic aerosols has been focus of several studies [62]. Terpenes and reactive volatile compounds such as aromatics, terpenoid alcohols and *n*-carbonyls are two broad categories of gaseous compounds that contribute considerably in the production of secondary aerosols.

Uncertainties of the emission estimates of POAs and their gaseous precursors are quite large due to land-cover databases, emission factors and environmental conditions [9]. Future emissions of organic materials are expected to alter due to land-use changes and possible climate-change effects.

4.3 Atmospheric chemistry of organic aerosols – field studies

4.3.1 Measurements of carbonaceous aerosols

Carbonaceous aerosol sampling is mainly based on the collection of particles on a substrate. Analytical chemical techniques are used in addition for the determination of the organic mass.

Both positive and negative artefacts may occur during integrated sampling; however, there is no consensus concerning which effect prevails. Since a major part of organic aerosols consists of semi-volatile compounds that can either adsorb or desorb from the particulate phase due to the environmental conditions, this can lead to both positive and negative artefacts. Furthermore, the selection of the filter substrate is critical in the determination of the organic mass since

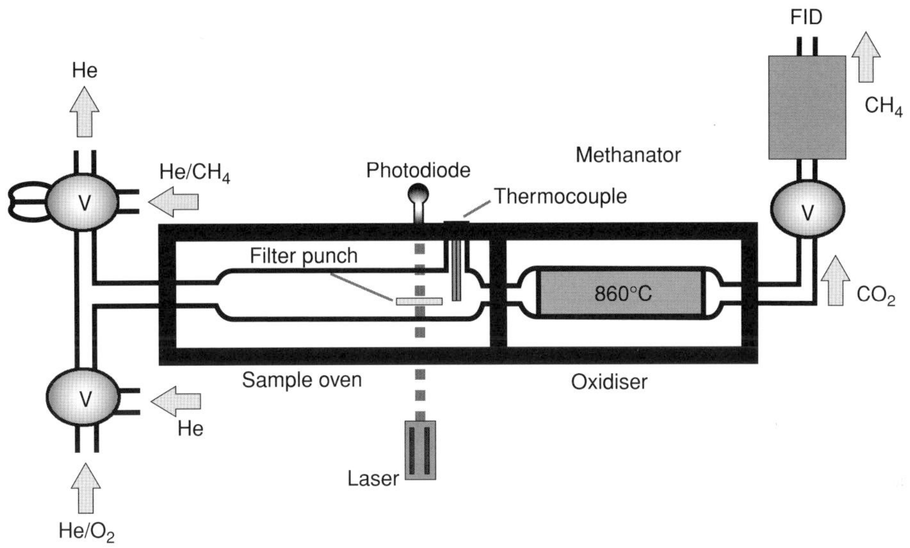

Figure 4.4 Schematic diagram of the thermal–optical instrument (V = valve). (Courtesy of Sunset Laboratory Inc.)

many substrates can absorb gas-phase organic species leading to positive artefacts. High-purity quartz is usually used as a substrate and precautions have to be taken into account during the transport and storage of the filters. The use of denuders before and/or after the quartz filter and a second backup filter greatly helps to detect or avoid artefacts. There are several approaches to correct for negative and positive artefacts [64, 65] and the QBQ approach (quartz behind quartz) (quantifying the positive artefact), described by McDow and Huntzicker [64], is used in monitoring networks.

Online techniques are also used for the determination of organic aerosols such as the measurement of BC with an aethalometer or a particle soot absorption photometer [3].

Thermal/optical analysis is another alternative methodology for the EC/OC determination in atmospheric aerosols. A detailed analysis is presented by Gelencsér [3] and Cachier [24]. The methodology used for the EC/OC analysis of atmospheric samples inside the framework of the European Monitoring and Evaluation Programme (EMEP) is presented here as an example. Figure 4.4 shows the EMEP methodology using an instrument from Sunset Laboratory Inc. A standard-sized punch is taken from the exposed filters and placed in a quartz oven. The oven is purged with helium, and a stepped temperature ramp increases the oven temperature to 870°C, thermally desorbing organic compounds and pyrolysis products into an oxidising oven. The OC is quantitatively oxidised to carbon dioxide gas. The carbon dioxide gas is mixed with hydrogen and the mixture subsequently flows through a heated nickel catalyst where it is quantitatively converted to methane. The methane is quantified using a flame ionisation detector. After cooling the oven to 600°C a second temperature ramp is initiated and the EC is oxidised off the filter by introducing a mixture of helium and oxygen into the oxidising oven. The EC is detected in the same manner as the OC [66, 67]. The use of the laser beam is needed to detect pyrolysis of organic material to char. A schematic view of the observed thermogram is shown in Figure 4.5.

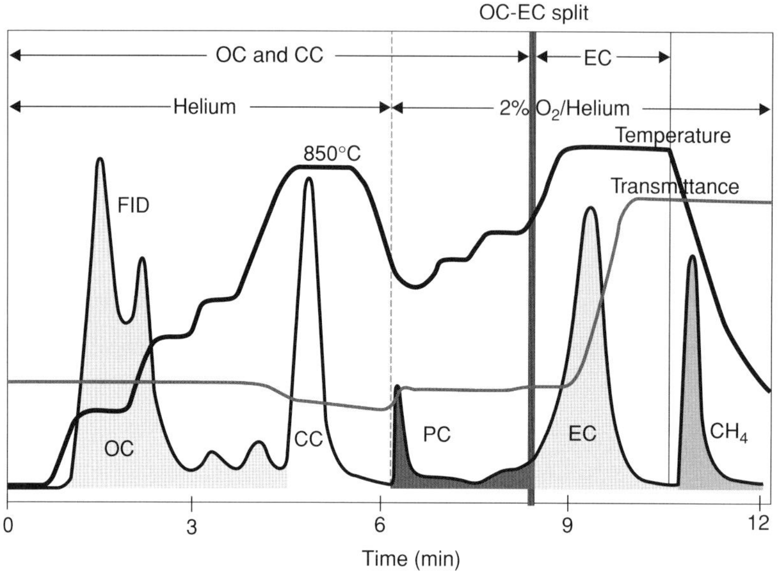

Figure 4.5 Thermogram for filter sample containing organic carbon (OC), carbonate (CC) and elemental carbon (EC). PC is pyrolytically generated carbon or 'char'. The final peak is the methane calibration peak. Carbon sources: pulverised beet pulp, rock dust (carbonate) and diesel particulate [67]. (Courtesy of Sunset Laboratory Inc.)

Furthermore, the determination of the fraction between anthropogenic and biogenic carbon in size-segregated ambient aerosols can be performed by radiocarbon analyses. ^{14}C measurement is a powerful technique, which makes it possible to distinguish between biogenic and anthropogenic carbon in ambient aerosols, as biogenic sources emit aerosols and precursors on the present $^{14}C/^{12}C$ level and anthropogenic aerosols derived only from fossil fuel, which contains only ^{12}C. Recent advances within ^{14}C measurements make it possible to perform such analysis on rather small sample quantities and even on carbonaceous subfractions, such as OC, EC, water-soluble (WSOC) and water-insoluble OC (WINSOC) [68]. Due to the oxygenated nature of SOA, the water-soluble part of the organic carbon fraction (WSOC) can be used as a proxy for SOA. By this approach, Szidat *et al.* [68] found that 76–96% of WSOC was of biogenic origin, indicating that the majority of the SOA is biogenic. However, the work of Szidat *et al.* [68] relates to the area of Zurich and the conclusions may not hold for all regions.

4.3.2 Organic compounds – source apportionment

The chemical content of organic aerosols is an important factor that affects their physicochemical properties such as their hygroscopicity and consequently their effect on the solar radiation energy reaching the atmosphere. Furthermore, the identification of specific chemical components as tracers from individual sources is of significant importance in order to quantify the anthropogenic contribution from health outcomes to climate change. Aliphatic dicarboxylated

Table 4.3 Chemical tracers identified as indicators for biomass burning as atmospheric pollution source[a]

Compound groups	Chemical tracer	Source
Monosaccharides	Levoglusan	All wood burning
Diterpenoids	Abietic, pimaric, isopimaric, sandaracopimaric acids, dihydroabietic acids, pimanthrene, retene	Gymnosperm
Triterpenoids	α-amyrin, β-amyrin, lupeol	Angiosperm
Methoxyphenols	Vanillin, vanillic acid	Gymnosperm
	Syringaldehyde, syringic acid	Angiosperm
	p-hydroxybenzaldehyde, *p*-hydroxybenzoic acid	Gramineae
Phytosterols	β-sitosterol, stigmasterol	All wood burning
	Campesterol	Gramineae

[a]Simoneit [72].

organic acids ranging from C_2 to C_{10} were identified as suitable tracers for engine emissions, whereas cholesterol was identified as a tracer for cooking activities [69].

The scientific methodology to perform source apportionment studies is focusing on the combination of receptor modelling together with the chemical quantification of source-specific tracers. For example monosaccharide anhydrides, levoglucosan (1,6-anhydro-β-D-glucopyranose), mannosan (1,6-anhydro-β-D-mannopyranose) and galactosan (1,6-anhydro-β-D-galactopyranose) have been identified as tracers of biomass burning [70, 71]. Several indicator compounds are used for identifying biomass-burning sources, as shown in Table 4.3. Furthermore, iso- and anteisoalkanes are found as tracers of cigarette smoking [73].

In addition, humic-like water-soluble substances in atmospheric aerosols have been identified in field studies (e.g. Varga *et al.* [74]). This polymeric material is refractory and resembles the chemical characteristics of humic and fulvic acids in soil and fresh water. A substantial part of WSOC is composed of HULIS [75].

4.4 Carbonaceous aerosols at urban/rural environments

The lifetime of PM in the atmosphere ranges from a few days to a few weeks and these relative long residence times result in small differences between the average total mass of $PM_{2.5}$ between

Table 4.4 Mass and chemical composition of tropospheric particulate matter[a]

		Percentage composition					
Region	Mass ($\mu g\, m^{-3}$)	Unspecified	C (elem)	C (org)	NH_4^+	NO_3^-	SO_4^{2-}
Remote (11 areas)	4.8	57	0.3	11	7	3	22
Non-urban continental (14 areas)	15	19	5	24	11	4	37
Urban (19 areas)	32	18	9	31	8	6	28

[a]Heintzenberg [76].

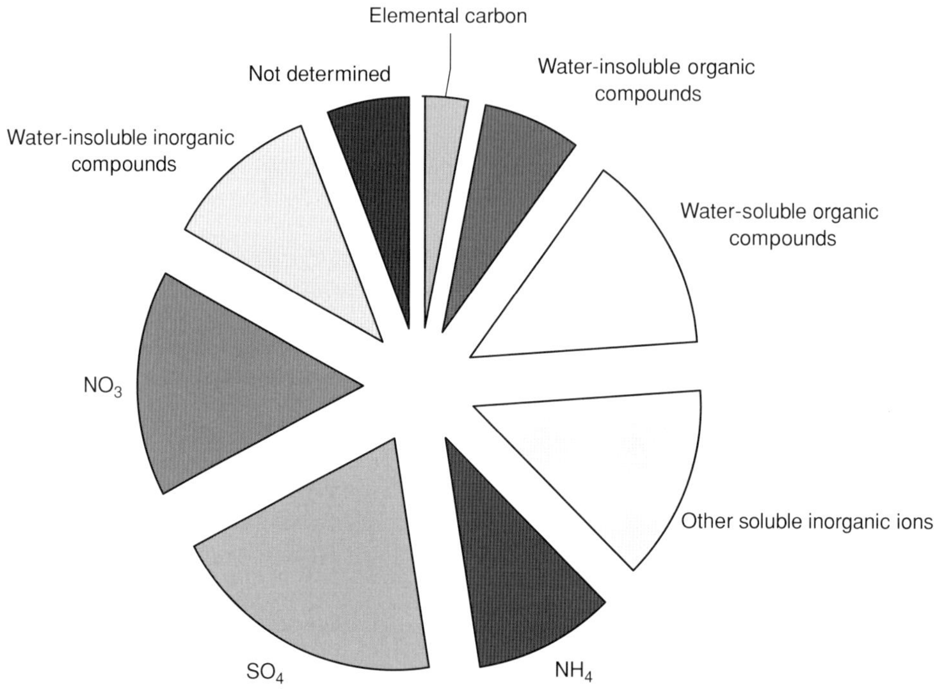

Figure 4.6 Chemical balance on element constituents of aerosol mass in Italy. (Reproduced from [75] with permission. Copyright Elsevier 1999.)

urban and non-urban continental aerosols [76]. Table 4.4 shows data on mass concentration and composition of tropospheric aerosols as summarised by Heintzenberg [76]. The data show that the average $PM_{2.5}$ mass in remote areas is about three times lower than the urban concentration, whereas the non-urban continental is two times lower. These observations show the importance of the regional component of PM that is attributed to long-range transport of pollutants.

Another example of PM chemical speciation from a study in Italy is shown in Figure 4.6. These results were taken from research by Zappoli *et al.* [75] and are derived from sampling fine particles ($d_p < 1\ \mu m$) at a site in the Po valley (highly polluted region) for a period of 2 months in 1997. Because of this, care should be taken not to regard these results as representative of a longer period or other regions in Europe. The results indicate that the composition is complex, emphasising the importance of the secondary inorganic component (SO_4^{2-}, NO_3^-, NH_4^+) and the water-soluble and water-insoluble organic components. The high relative proportion of water-soluble organic components may be a result of SOA formation. It is apparent that the chemical speciation is important for understanding the PM properties and trends in conjunction with their size distribution. These components are not sufficient to determine the full chemical mass balance. For this purpose, it is also necessary to include hydrogen, oxygen, nitrogen and other elements associated with the carbon in the organic fraction. Experience also indicates that some water is associated with the water-soluble fraction even after conditioning at 50% relative humidity. However, by multiplying the OC by a factor of 1.4, and calculating the full weight of all the other components, reasonable agreement with the PM_{10} figure can be achieved.

The organic aerosol mass can be up to 50% of the total aerosol mass at continental mid-latitudes [11], whereas at tropical forests it can reach 90% of its mass. In Europe the organic aerosol fraction can range between 20 and 50%, as indicated in a recent work by Putaud *et al.* [77]. In addition, in Europe the water-soluble fraction of the total OC ranges between 20% (in polluted areas) and 80% (at background sites) [75].

The chemical complexity of atmospheric aerosols requires a consideration of their composition and sources. In addition, because of the considerable influence of natural sources (with larger size in general than the anthropogenic emissions) there is a need for measurements of smaller particles, such as $PM_{2.5}$. This standardisation will allow a better understanding of anthropogenic influence and further implementation of control strategies [78]. New particle formation from SOA compounds is another important scientific issue and recent research studies have shown that heteromolecular nucleation including organic compounds may be a significant pathway for new particle formation in urban areas.

Measurements of PM_{10} and black smoke (BS) wintertime (winter 1993/1994) concentrations in Europe during the PEACE study [79, 80] show small differences between urban and non-urban locations. Similar winter studies for PM_{10}, $PM_{2.5}$ and BS in the Netherlands (winters of 1992/1993, 1993/1994 and 1994/1995) show the same trend in the results. PM_{10} and BS concentrations were on average 13 and 19% higher in urban areas than in non-urban areas. Non-organic secondary aerosol concentrations were on average 8% lower in urban areas, whereas $PM_{2.5}$ concentrations were similar in between urban and non-urban sites. Also, higher elemental concentrations in PM_{10} were found in urban areas for all elements except Si. Meteorological conditions seem to play a crucial role in the observations. Easterly winds resulted in smaller differences between urban and non-urban sites, showing the importance of the regional aerosol component from Eastern Europe, whereas winds from the sea led to lower average PM concentrations and larger differences between urban and non-urban locations [81].

In Europe, alongside other national and international efforts, the EMEP has been focused in the determination of OC/EC at specific sites during 2002–2003. The data set is somewhat unique, containing weekly concentrations of EC, OC and PM_{10} for 14 sites in 13 European counties for an entire year [82]. Table 4.5 summarises the EC/OC ratios and their contribution to the ambient PM_{10} levels.

The annual average of EC for the sites included in the campaign varies between 0.14 and 1.86 $\mu g\, C\, m^{-3}$. By multiplying the concentration of EC ($\mu g\ C\, m^{-3}$) by a factor 1.1, taking into account the presence of approximately 10% hydrogen as well as trace levels of other elements, EC was found to account for 1.1–5.4% of PM_{10} on an annual basis. In general the concentration of EC and the EC/TC ratio increased from summer to winter, TC being total carbon, that is EC + OC.

The annual average of OC for the sites included in the campaign varied between 1.17 and 7.90 $\mu g\, C\, m^{-3}$. In order to account for oxygen, nitrogen and hydrogen not included in the EC/OC analysis, the OC ($\mu g\ C\, m^{-3}$) concentration at the urban background sites and the 'near-city' site were multiplied by a factor 1.6. A factor of 2.0 was applied for the rural background sites and the rural site. Using these conversion factors OM (organic matter) was found to account for 12.6–44.9% of PM_{10}. The concentrations of OM increased from summer to winter except at the Scandinavian sites and at the Slovakian site where OM concentrations were higher in summer than in winter. Total carbonaceous material, EC + OM, accounted for 13.7–49.5% of the PM_{10}. However, as much as 61.6% of PM_{10} could be accounted for by carbonaceous matter at the Portuguese site during winter. The lowest fraction of EC + OM in PM_{10} was found for Ireland, not exceeding 15% in any season [82].

Table 4.5 Annual ambient concentrations of EC, OM and PM_{10} and relative contribution of EC, OM and the sum EC + OM to the PM_{10} concentration[a]

Site	Site category	EC^b ($\mu g\,m^{-3}$)	$OM^{c,d}$ ($\mu g\,m^{-3}$)	PM_{10} ($\mu g\,m^{-3}$)	EC/PM_{10}(%)	OM/PM_{10}(%)	$(EC + OM)/PM_{10}$(%)
Austria	Rural background	1.11	11.13	30.9	3.6	36.1	39.7
Czech Republic	Rural background	1.16	9.08	25.0	4.6	36.4	41.0
Finland	Rural background	0.40	4.16	11.0	3.6	37.8	41.4
Germany	Rural background	0.70	8.61	26.1	2.7	32.9	35.6
The Netherlands	Rural background	0.70	5.18	26.1	2.7	19.9	22.6
Ireland	Rural background	0.21	2.39	19.0	1.1	12.6	13.7
Portugal	Rural background	0.87	8.20	19.4	4.5	42.2	46.7
Norway	Rural background	0.16	2.33	7.4	2.1	31.4	33.5
Slovakia	Rural background	0.89	8.64	19.2	4.6	44.9	49.5
Sweden	Rural background	0.31	4.24	10.6	3.0	39.9	42.8
UK	Rural	0.60	3.06	14.3	4.2	21.3	25.5
Belgium	Urban background	1.98	6.59	37.0	5.4	17.8	23.2
Italy (S.P.C.)	Urban background	1.58	9.46	41.0	3.9	23.1	26.9
Italy (JRC)	Near-city	2.04	12.64	42.0	4.9	30.1	35.0

[a] Sampling period 07.01.2002–07.01.2003 (adopted from [82]).
[b] To account for hydrogen and trace levels of other elements, concentrations of EC ($\mu g\,C\,m^{-3}$) have been multiplied by a factor of 1.1 for all sites.
[c] To convert OC ($\mu g\,C\,m^{-3}$) to OM ($\mu g\,m^{-3}$), concentrations of OC ($\mu g\,C\,m^{-3}$) have been multiplied by a factor of 2.0 for all the rural background sites and the rural site.
[d] To convert OC ($\mu g\,C\,m^{-3}$) to OM ($\mu g\,m^{-3}$), concentrations of OC ($\mu g\,C\,m^{-3}$) have been multiplied by a factor of 1.6 for all the urban background sites and the 'near-city' site.

Table 4.6 Concentrations of WSOC and WINSOC in PM_{10}, and relative contribution of WSOC and WINSOC to OC in PM_{10}

Site	Number of samples	Samples winter: summer	WSOC ($\mu g\,C\,m^{-3}$)	WINSOC ($\mu g\,C\,m^{-3}$)	WSOC/OC (%)	WINSOC/OC (%)
AT02	$n = 5$	3:2	2.9	3.5	41	59
BE02	$n = 4$	2:2	2.2	3.2	41	59
CZ03	$n = 6$	6:0	1.2	2.5	33	67
DE02	$n = 11$	6:5	1.4	1.7	46	54
GB46	$n = 5$	5:0	0.3	0.7	31	69
IE31	$n = 4$	2:2	0.1	0.6	17	83
IT04	$n = 5$	3:2	3.0	4.9	38	62
IT08	$n = 5$	3:2	3.6	3.7	49	51
NL09	$n = 3$	2:1	1.5	1.5	49	51
NO01	$n = 4$	4:0	0.8	0.7	52	48
PT01	$n = 11$	6:5	1.5	1.1	57	43
SE12	$n = 4$	2:2	0.8	1.2	40	60
SK04	$n = 4$	2:2	1.9	4.3	30	70

Adopted from [81].
AT02, Illmitz, Austria; BE02, Ghent, Belgium; CZ03, Kosetice, Czech Republic; DE02, Langenbrugge, Germany; GB46, Penicuik, Great Britain; IE31, Mace Head, Ireland; IT04, Ispra, Italy; IT08, San Pietro Capofiume, Italy; NL09, Kollumerwaard, The Netherlands; NO01, Birkenes, Norway; PT01, Braganca, Portugal; SE12, Aspvreten, Sweden; SK04, Stara Lesna, Slovakia.

The OC fraction was additionally separated according to its solubility in water. The resulting subfractions provided by the water extraction are WINSOC and WSOC. Water extracts a significant amount of organic matter from the aerosol phase ranging between 11 and 95% of the OC fraction [83]. There is considerable interest in the water solubility of OC due to the scientific findings that WSOC influences the ambient aerosols ability to act as CCN [84].

The concentrations of WSOC and WINSOC at the EMEP sites studied are shown in Table 4.6. The lowest mean concentration of both WINSOC ($0.6\ \mu g\,C\,m^{-3}$) and WSOC ($0.1\ \mu g\,C\,m^{-3}$) was reported for the Irish site. The highest concentration for WINSOC was reported for the Italian site San Pietro Capofiume ($4.9\ \mu g\,C\,m^{-3}$), whereas the highest concentration of WSOC was reported for the Italian site Ispra ($3.6\ \mu g\,C\,m^{-3}$).

Furthermore, a chemical mass balance of fine particles (diameter smaller than 1.5 μm) collected at three European sites has been reported [75]. The sites were located in southern Sweden (S), in a rural area in the Great Hungarian Plain (H) and in the polluted Po Valley, northern Italy (I). The total fine aerosol mass ranged from a value of $38\ \mu g\,m^{-3}$(I) in the most polluted area to around $6\ \mu g\,m^{-3}$(S) in the background site, although the rural station in Hungary showed a value close to $23\ \mu g\,m^{-3}$(H). Speciation of the aerosol components with reference to water solubility of the different species is presented in Table 4.7. The results show a very high water solubility and higher fraction of WSOC in the background site compared to the rural and polluted sites [75].

4.5 Summary

The organic fraction of the ambient PM is a complex mixture of thousands of different organic compounds and a very small portion of it is molecular characterised (around 10%). Organic

Table 4.7 Speciation (in % units) of different aerosol classes of components

Site	EC	WINSOC	WSOC	OSII	NH_4^+	SO_4^-	NO_3^-	WISIC	ND
S	1	7	36	11	4	20	4	15	—
H	2	13	14	2	10	32	8	n.a.	18
I	3	7	14	14	10	20	16	11	6

From Zappoli *et al.* [75].
EC, elemental carbon; WINSOC, water-insoluble organics; WSOC, water-soluble organics; OSII, other soluble inorganic ions; WISIC, water-insoluble inorganics; ND, non-detected.

compounds that have been characterised include, among others, *n*-alkanes, *n*-alkanoic acids and polycyclic aromatic compounds. Due to difficulty in measuring organic compounds our current knowledge about organic matter is limited and incomplete but recent research findings contribute to an improvement of the situation. Primary emission sources for OC include combustion processes, geological (fossil fuels) and biogenic sources. A significant part of secondary aerosol particles in the atmosphere is composed of organic matter produced from oxidation of organic compounds. Biogenic VOC is the main source for SOA. Besides, anthropogenic VOC is significant source to SOA formation in urban areas.

There is still an open question whether the aerosol yields observed in laboratory experiments are directly transferable to atmospheric conditions. The translation of smog chamber results to the real atmosphere requires thorough inspection and further understanding of the SOA chemical mechanisms. The atmosphere is very complex with a wide variety of ambient conditions. Still we know little about the aerosol yield dependence on VOC/NOx ratio, the influence of UV light and aerosol yields past the first generation of oxidation products. In spite of large uncertainty the temperature-dependent partition coefficients approach gives an improved description of SOA formation at true atmospheric conditions.

During measurements of organic aerosols positive and negative artefacts may occur. The thermal/optical approach for the EC/OC determination requires a careful treatment of the filters and the sampling procedure.

Modelling organic aerosol transport and chemistry in mesoscale regional/global models accepts several simplifications and parameterisations. Future progress in the modelling area requires further understanding of the atmospheric organic chemistry. In addition, emission inventories for POAs and gaseous precursors contain several uncertainties, even though our understanding of the recent years is significantly improved.

Measurements of organic aerosols in Europe are based on research programmes, national and international network activities. The EMEP network has started to provide valuable information on the geographical/temporal variability of rural organic aerosol concentrations in Europe.

References

1 US-EPA, Air quality criteria for particulate matter. US-EPA/600/P-99/002aD, U.S. Environmental Protection Agency 2003.

2 Pope, C.A., Thun, M.J., Namboodiri, M.M., Dockery, D.W., Evans, J.S., Speizer, F.E. and Heath, D.W, Jr. Particulate air pollution as a predictor of mortality in a prospective study of U. S. adults. *American Journal of Respiratory Critical Care Medicine* 1995; **151**: 669–674.

3 Gelencsér, A. *Carbonaceous Aerosol, Volume Atmospheric and Oceanographic Science Library Series*, Vol. 30. Springer ISBN 1-4020-2886-5, 2004.
4 Seinfeld, J.H. and Pandis, S.N. *Atmospheric Chemistry and Physics: From Air Pollution to Climate Change*. John Wiley, New York: 1998.
5 Molnár, A., Mèszàros, E., Hansson, H.C., Karlsson, H., Gelencsèr, A., Kiss, G.Y. and Krivàcsy, Z. The importance of organic and elemental carbon in the fine atmospheric aerosol particles. *Atmospheric Environment* 1999; **33**: 2745–2750.
6 Turpin, B.J., Saxena, P. and Andrews, E. Measuring and simulating particulate organics in the atmosphere: problems and prospects. *Atmospheric Environment* 2000; **34**: 2983–3013.
7 Liousse, C., Cachier, H. and Jennings, S.G. Optical and thermal measurements of black carbon aerosol content in different environments – variation of the specific attenuation cross – section, Sigma. *Atmospheric Environment* 1993; **27**: 1203–1211.
8 Penner, J.E. and Novakov, T. Carbonaceous particles in the atmosphere: a historical perspective to the fifth international conference on carbonaceous particles in the atmosphere. *Journal of Geophysical Research – Atmospheres* 1996; **01** (D14): 19373–19378.
9 Kanakidou, M., Seinfeld, J.H., Pandis, S.N., Barnes, I., Dentener, F.J., Facchini, M.C., Van Dingenen, R., Ervens, B., Nenes, A., Nielsen, C.J., Swietlicki, E., Putaud, J.P., Balkanski, Y., Fuzzi, S., Horth, J., Moortgat, G.K., Winterhalter, R., Myhre, C.E.L., Tsigaridis, K., Vignati, E. Stephanou, G. and Wilson, J. Organic aerosol and global climate modelling: a review. *Atmospheric Chemistry and Physics* 2005; **5**: 1053–1123.
10 Jacobson, M.C., Hansson, H.-C., Noone, K.J. and Charlson, R.J. Organic atmospheric aerosols: review and state of the science. *Review of Geophysics* 2000; **38**: 267–294.
11 Saxena, P. and Hildemann, L.M. Water-soluble organics in atmospheric particles: a critical review of the literature and application of thermodynamics to identify candidate compounds. *Journal of Atmospheric Chemistry* 1996; **24**(1): 57–109.
12 Intergovernmental Panel on Climate Change (IPCC). *Climate Change: The Scientific Basis*. Cambridge University Press, UK: 2001.
13 Liousse, C., Penner, J.E., Chunag, C., Walton, J.J., Eddleman, H. and Cachier, H. A global three-dimensional model study of carbonaceous aerosols. *Journal of Geophysical Research* 1996; **101/D14**: 19411–19432.
14 Sun, J. and Ariya, A.P. Atmospheric organic and bio-aerosols as cloud condensation nuclei (CCN): a review. *Atmospheric Environment* 2006; **40**(5): 795–820.
15 Twohy, C.H., Anderson, J.R. and Crozier, P.A. Nitrogenated organic aerosols as cloud condensation nuclei. *Geophysical Research Letters* 2005; **32**(19): L19805.
16 Chughtai, A.R., Kim, J.M. and Smith, D.M. The effect of temperature and humidity on the reaction of ozone with combustion soot: implications for reactivity near the tropopause. *Journal of Atmospheric Chemistry* 2003; **45**: 231–243.
17 Pandis, S.N., Harley, R.A., Cass, G.R. and Seinfeld, J.H. Secondary organic aerosol formation and transport. *Atmospheric Environment* 1992; **26A**: 2269–2282.
18 Pankow, J.F. An absorption model of gas/particle partitioning of organic compounds in the atmosphere. *Atmospheric Environment* 1994; **26A**: 185–188.
19 Hoffmann, T., Odum, J.R., Bowman, F., Collins, D., Klockow, D., Flagan, R.C. and Seinfeld, J.H. Formation of organic aerosols from the oxidation of biogenic hydrocarbons. *Journal of Atmospheric Chemistry* 1997; **26**: 189–222.
20 Kavouras, I.G., Mihalopoulos, N. and Stephanou, E.G. Formation of atmospheric particles from organic acids produced by forests. *Nature* 1998; **395**: 683–686.
21 Anderson-Sköld, Y. and Simpson, D. Secondary organic aerosol formation in northern Europe: a model study. *Journal of Geophysical Research* 2001; **106**: 7357–7374.
22 Tsigaridis, K. and Kanakidou, M. Global modelling of secondary organic aerosol in the troposphere: a sensitivity analysis. *Atmospheric Chemistry and Physics* 2003; **3**: 1849–1869.

23 Graber, E.R. and Rudich Y. Atmospheric HULIS: how humic-like are they? A comprehensive and critical review. *Atmospheric Chemistry and Physics* 2006; **6**: 729–753.
24 Cachier, H. Carbonaceous combustion aerosols. In: R.M. Harrison and R. van Grieken, eds. *Atmospheric Particles.* John Wiley, New York: 1998.
25 Wentzel, M., Gorzawski, H., Naumann, K.H., Saathoff, H. and Weinbruch, S. Transmission electron microscopical and aerosol dynamical characterization of soot aerosols. *Journal of Aerosol Science* 2003; **34**: 1347–1370.
26 Martins, J.V., Artaxo, P., Liousse, C., Reid, J.S., Hobbs, P.V. and Kaufman, Y.J. Effects of black carbon content, particle size, and mixing on light absorption by aerosols from biomass burning in Brazil. *Journal of Geophysical Research* 1998; **103(D24)**: 32041–32050.
27 Akhter, M.S., Chughtai, A.R. and Smith, D.M. The structure of hexane soot, I: extraction studies. *Applied Spectroscopy* 1985; **39**: 154–167.
28 Hamilton, R.S. and Mansfield, T.A. Airborne particulate elemental carbon – its sources, transport and contribution to dark smoke and soiling. *Atmospheric Environment* 1991; **25**: 715–723.
29 Cass, G.R., Boone, P.M. and Macias, E.S. Emissions and air quality relationships for atmospheric carbon particles in Los Angeles. In: G.T. Wolff and R.L. Klimisch, eds. *Particulate Carbon: Atmospheric Life Cycle.* Plenum Press, New York: 1982; pp. 207–243.
30 Grosjean, D. In situ organic aerosol formation during a smog episode: estimated production and chemical functionality. *Atmospheric Environment* 1992; **26**: 953–963.
31 Turpin, B.J. and Huntzicker, J.J. Identification of secondary organic aerosol episodes and quantitation of primary and secondary organic aerosol concentrations during SCAQS. *Atmospheric Environment* 1995; **29**: 3527–3544.
32 Odum, J.R., Hoffmann, T., Bowman, F., Collins, D., Flagan, R.C. and Seinfeld, J.H. Gas/particle partitioning and secondary organic aerosol yield. *Environmental Science and Technology* 1996; **30**: 2580–2585.
33 Lazaridis, M. Gas-particle partitioning of organic compounds in the atmosphere. *Journal of Aerosol Science* 1999; **30**(9): 1165–1170.
34 Limbeck, A., Kraxner, Y. and Puxbaum, H. Gas to particle distribution of low molecular weight dicarboxylic acids at two different sites in central Europe (Austria). *Journal of Aerosol Science* 2005; **36**(8): 991–1005.
35 Griffin, R.J., Cocker, D.R., III, Flagan, R.C. and Seinfeld, J.H. Organic aerosol formation from oxidation of biogenic hydrocarbons. *Journal of Geophysical Research* 1999; **104**: 3555–3567.
36 Seigneur, C., Pai, P., Louis, J.-F., Hopke, P. and Grosjean, D. Review of air quality models for particulate matter. Prepared for the American Petroleum Institute. API (Report No. CP015-97-1b), Washington, D.C.: 1997.
37 Johnson, D., Utembe, S.R., Jenkin, M.E., Derwent, R.G., Hayman, G.D., Alfarra, M.R., Coe, H. and McFiggans, G. Simulating regional scale secondary organic aerosol formation during the TORCH 2003 campaign in the southern UK. *Atmospheric Chemistry and Physics* 2006; **6**: 403–418.
38 Pun, B.K., Seigneur, C., Vijayaraghavan, K., Wu, S.Y., Chen, S.Y., Knipping, E.M. and Kumar, N. Modeling regional haze in the BRAVO study using CMAQ-MADRID: 1. Model evaluation. *Journal of Geophysical Research – Atmospheres* 2006; **111(D6)**: D06302.
39 O'Neill, S.M., Lamb, B.K., Chen, J., Claiborn, C., Finn, D., Otterson, S., Figueroa, C., Bowman, C., Boyer, M., Wilson, R., Arnold, J., Aalbers, S., Stocum, J., Swab, C., Stoll, M., Dubois, M. and Anderson, M. Modeling ozone and aerosol formation and transport in the pacific north-west with the community multi-scale air quality (CMAQ) modeling system. *Environmental Science and Technology* 2006; **40**(4): 1286–1299.
40 Matthias-Maser, S. and Jaenicke, R. The size distribution of primary biological aerosol particles with radii >0.2 mu m in an urban rural influenced region. *Atmospheric Research* 1995: **39**: 279–286.
41 Matthias-Maser, S. Primary biological aerosol particles: their significance, sources, sampling methods and size distribution in the atmosphere. In: R.M. Harrison and R.E. van Grieken, eds. *Atmospheric Particles.* John Wiley, Chichester: 1998; pp. 349–368.

42 Tunved, P., Hansson, H.C., Kerminen, V.M., Strom, J., Dal Maso, M., Lihavainen, H., Viisanen, Y., Aalto, P.P., Komppula, M. and Kulmala, M. High natural aerosol loading over boreal forests. *Science* 2006; **312**(5771): 261–263.

43 Baltensperger, U., Kalberer, M., Dommen, J., Paulsen, D., Alfarra, M.R., Coe, H., Fisseha, R., Gascho, A, Gysel, M., Nyeki, S., Sax, M., Steinbacher, M., Prevot, A.S.H., Sjoren, S., Weingartner, E. and Zenobi, R. Secondary organic aerosols from anthropogenic and biogenic precursors. *Faraday Discussions* 2005; **130**: 265–278.

44 Staudt, M., Bertin, N., Frenzel, B. and Seufert, G. Seasonal variation in amount and composition of monoterpenes emitted by young Pinus pinea trees – implications for emission modeling. *Journal of Atmospheric Chemistry* 2000; **35**: 77–99.

45 Odum, J.R., Jungkamp, T., Griffin, R., Flagan, R. and Seinfeld, J. The atmospheric aerosol-formation potential of whole gasoline vapor. *Science* 1997; **276**: 96–99.

46 Atkinson, R. Gas-phase tropospheric chemistry of volatile organic compounds: 1 alkanes and alkenes. *Journal of Physical Chemistry Reference Data* 1997; **26**: 215–290.

47 Gery, M.W., Whitten, G.Z. and Killus, J.P. Development and testing of the CMB-IV for urban and regional modelling. Report prepared by Systems Applications Inc., San Rafael, CA EPA Contract No. 68-02-4136, 1988.

48 Griffin, R.J., Cocker, D.R., III and Seinfeld, J.H. Estimate of global atmospheric aerosol from oxidation of biogenic hydrocarbons. *Geophysical Research Letters* 1999; **26**(17): 2721–2724.

49 Svendby, T.M. and Lazaridis, M. Modelling of secondary biogenic aerosols. In: The first international WeBIOPATR Workshop, Belgrade, 20–22 May 2007. Book of Extended Abstracts. VINCA Institute of Nuclear Sciences, Belgrade: 2007; pp. 64–66.

50 Stanier, C.O. and Pandis, S.N. Secondary organic aerosols: laboratory results for gas-aerosol partitioning and its dependence on temperature and relative humidity. In: *Aerosol Conference in Taiwan*, September 2002.

51 Winterhalter, R., Van Dingenen, R., Larsen, B.R., Jensen, N.R. and Hjort, J. LC-MC analysis of aerosol particles from the oxodation of a-pinene by ozone and OH-radicals. *Atmospheric Chemistry and Physics Discussions* 2003; **3**: 1–39.

52 Cocker, D.R., Flagan, R.C. and Seinfeld, J.H. State-of-the-art chamber facility for studying atmospheric aerosol chemistry. *Environmental Science and Technology* 2001; **35**: 2594–2601.

53 Khamaganov, V.G. and Hites, R.A. Rate constants for the gas-phase reaction of ozone with isoprene, α-pinene, β-pinene, and limonene as a function of temperature. *Journal of Physical Chemistry* 2001; **105**: 815–822.

54 Wängberg, I., Barnes, I. and Becker, K.H. Product and mechanistic study of the reaction of NO_3 radicals with α-pinene. *Environmental Science and Technology* 1997; **31**: 2130–2135.

55 Barthelmie, R.J. and Pryor, S.C. A model mechanism to describe oxidation of monoterpenes leading to secondary organic aerosol, 1 α-pinene and β-pinene. *Journal of Geophysical Research* 1999; **104**: 23657–23699.

56 Sheehan, P.E. and Bowman, F.M. Estimated effects of temperature on secondary organic aerosol concentration. *Environmental Science and Technology* 2001; **35**: 2129–2135.

57 Finlayson, B.J. and Pitts, J.N. *Chemistry of Upper and Lower Atmosphere.* Academic Press, San Diego: 2000.

58 Pankow, J.F. and Bidleman T.F. Effects of temperature, TSP and per cent non-exchangeable material in determining the gas-particle partitioning of organic compounds. *Atmospheric Environment* 1991; **25A**: 997–1008.

59 ATSDR (Agency for Toxic Substances and Disease Registry). Toxicological Profile for Polycyclic Aromatic Hydrocarbons. Acenaphthene, Acenaphthylene, Anthracene, Benzo(a)anthracene, Benzo(a)pyrene, Benzo(b)fluoranthene, Benzo(g,i,h)perylene, Benzo(k)fluoranthene, Chrysene, Dibenzo(a,h)anthracene, Fluoranthene, Fluorene, Indeno(1,2,3-c,d)pyrene, Phenanthrene, Pyrene. Prepared by Clement International Corporation, under Contract No. 205-88-0608. ATSDR/TP-90-20 1990.

60 Smith, D.J.T. and Harrison, R.M. Polycyclic aromatic hydrocarbons in atmospheric particles. In: R.M. Harrison and R. van Grieken, eds. *Atmospheric Particles*. John Wiley, New York: 1998.

61 Andreae, M.O., Browell, E.V., Garstang, M., Gregory, G.L., Harris, R.C., Hill, G.F., Jacob, D.J., Pereira, M.C., Sachse, G.W., Setzer, A.W., Dias, P.L.S., Talbot, R.W., Torres, A.L. and Wofsy, S.C. Biomass-burning emissions and associated haze layers over Amazonia. *Journal of Geophysical Research* 1998; **93**: 1509–1527.

62 Bond, T.C., Streets, D.G., Yarber, K.F., Nelson, S.M., Woo, J.H. and Klimont, Z. A technology-based global inventory of black and organic carbon emissions from combustion. *Journal of Geophysical Research* 2004; **109**: D14203, doi: 10.1029/2003JD003697.

63 Rose, D., Wehner, B., Ketzel, M., Engler, C., Voigtländer, J., Tuch, T. and Wiedensohler, A. Atmospheric number size distributions of soot particles and estimation of emission factors. *Atmospheric Chemistry and Physics* 2006; **6**: 1021–1031.

64 McDow, S.R. and Huntzicker, J.J. Vapor adsorption artifact in the sampling of organic aerosol: face velocity effects. *Atmospheric Environment* 1990; **24A**: 2563–2571.

65 Mader, B.T., Flagan, R.C. and Seinfeld, J.H. Sampling atmospheric carbonaceous aerosols using a particle trap impactor/denuder sampler. *Environmental Science and Technology* 2001; **35**: 4857–4867.

66 Birch, M.E. and Cary, R.A. Elemental carbon-based method for monitoring occupational exposures to particulate diesel exhaust. *Aerosol Science and Technology* 1996; **23**: 221–241.

67 NIOSH. *NIOSH Manual of Analytical Methods (NMAM)*, 4th edn. Method 5040. Cincinnati, National Institute for Occupational Safety and Health: 2003.

68 Szidat, S., Jenk, H.W., Gäggler, T.M, Synal, H.-A., Fisseha, R., Baltensperger, U., Kalberer, M., Samburova, V., Reimann, S., Kasper-Giebl, A. and Hajdas, I. Radiocarbon (14C)-deduced biogenic and anthropogenic contributions to organic carbon (OC) of urban aerosols from Zürich, Switzerland. *Atmospheric Environment* 2004; **38**: 4035–4044.

69 Rogge, W.F., Hildemann, L.M., Mazurek, M.A., Cass, G.R. and Simoneit, B.R.T. Sources of fine organic aerosol. 1: Charbroilers and meat cooking operations. *Environmental Science and Technology* 1991; **25**: 1112–1125.

70 Simoneit, B.R.T. Biomass burning – a review of organic tracers for smoke from incomplete combustion. *Applied Geochemistry* 2002; **17**: 129–162.

71 Dye, C. and Yttri, K.E. Determination of monosaccharide anhydrides in atmospheric aerosols by use of high-performance liquid chromatography combined with high-resolution mass spectrometry. *Analytical Chemistry A* 2005; **77**(6): 1853–1858.

72 Simoneit, B.R.T. A review of biomarker compounds as source indicators and tracers for air pollution. *ESPR – Environmental Science and Pollution Research* 1999; **6/3**: 159–169.

73 Rogge, W.F., Hildemann, L.M., Mazurek, M.A., Cass, G.R. and Simoneit, B.R.T. Sources of fine organic aerosol. 6: Cigarette smoke in the urban atmosphere. *Environmental Science and Technology* 1994; **28**: 1375–1388.

74 Varga, B., Kiss, G., Ganszky, I., Gelencser, A. and Krivacsy, Z. Isolation of water-soluble organic matter from atmospheric aerosol. *Talanta* 2001; **55**: 561–572.

75 Zappoli, S., Andracchio, A., Fuzzi, S., Facchini, M.C., Gelencser, A., Kiss, G., Krivacsy, Z., Molnar, A., Meszaros, E., Hansson, H.C., Rosman, K. and Zebuhr, Y. Inorganic, organic and macromolecular components of fine aerosol in different areas of Europe in relation to their water solubility. *Atmospheric Environment* 1999; **33**(17): 2733–2743.

76 Heintzenberg, J. Fine particles in the global troposphere – a review. *Tellus* 1989; **41B**: 149–160.

77 Putaud, J.P., Raes, F., Van Dingenen, R., Bruggemann, E., Facchini, M.C., Decesari, S., Fuzzi, S., Gehrig, R., Huglin, C., Laj, P., Lorbeer, G., Maenhaut, W., Mihalopoulos, N., Mulller, K., Querol, X., Rodriguez, S., Schneider, J., Spindler, G., ten Brink, H., Torseth, K. and Wiedensohler, A. European aerosol phenomenology-2: chemical characteristics of particulate matter at kerbside, urban, rural and background sites in Europe.*Atmospheric Environment* 2004; **38**: 2579–2595.

78 Position Paper on PM (Second). CAFE Working Group on Particles, 2004.

79 Hoek, G., Forsberg, B., Borowska, M., Hlawiczka, S., Vaskovi, E., Welinder, H., Branis, M., Benes, I., Kotesovec, F., Hagen, L.E., Cyrys, J., Jantunen, M., Roemer, W. and Brunekreef, B. Wintertime PM_{10} and black smoke concentrations across Europe: results from the PEACE Study. *Atmospheric Environment* 1997; **31**: 3609–3622.

80 Hoek, G., Brunekreef, B., Goldbohm, S., Fischer, P. and van den Brandt, P.A. Association between mortality and indicators of traffic-related air pollution in the Netherlands: a cohort study. *Lancet* 2002; **360**: 1203–1209.

81 European Monitoring and Evaluation Programme (EMEP). *Measurements of Particulate Matter: Status Report 2005.* EMEP/CCC-Report 5/2005.

82 European Monitoring and Evaluation Programme (EMEP). *Measurements of Particulate Matter: Status Report 2004.* EMEP/CCC-Report 5/2004.

83 Mader, B.T., Yu, J.Z., Xu, J.H., Li, Q.F., Wu, W.S., Flagan, R.C. and Seinfeld, J.H. Molecular composition of the water-soluble fraction of atmospheric carbonaceous aerosols collected during ACE-Asia. *Journal of Geophysical Research* 23 March 2004; **109(D6)**: D06206.

84 Novakov, T. and Corrigan, C.E. Cloud condensation nucleus activity of the organic component of biomass smoke particles. *Geophysical Research Letters* 1996; **23**: 2141–2144.

Chapter 5
Metals in Aerosols

Irena Grgić

5.1 Introduction

Atmospheric aerosols have important role in the biogeochemistry and transportation of metals in the atmosphere. On a mass basis, so-called trace metals represent a relatively small proportion of the atmospheric aerosol (generally less than 1%). Among them are transition metals (TM) (e.g. vanadium, chromium, manganese, iron, cobalt, nickel, copper, etc.), which have several oxidation states, and thus can participate in many important atmospheric redox reactions. Field measurements provide strong evidence that TM are common components in aerosol particles as well as in atmospheric liquid water [1–5].

The concentrations of trace metals in atmospheric aerosols are a function of their sources. Natural emissions of trace metals result from different processes acting on crustal minerals (e.g. erosion, surface winds and volcanic eruptions), as well as from natural burning and from the oceans. On a global scale, resuspended surface dusts make a large contribution to the total natural emission of trace metals to the atmosphere (e.g. more than 50% of Cr, Mn and V, and more than 20% of Cu, Mo, Ni, Pb, Sb and Zn), whereas volcanism presumably generates about 20% of Cd, Hg, As, Cr, Cu, Ni, Pb and Sb [6]. Sea-salt aerosols generated by spray and wave action may contribute to about 10% of total trace metals emissions. Biomass combustion can contribute to emissions of Cu, Pb and Zn [7]. The predominant anthropogenic sources are due to high-temperature processes, biomass burning, fossil-fuel combustion, industrial activity, incineration, etc. Anthropogenic high-temperature processes result in the release of volatile metals as vapours forming particles by condensation or gas-to-particle reactions. Combustion of fossil fuels is the most important anthropogenic source of atmospheric Be, Co, Hg, Mo, Ni, Sb, Se, Sn and V; it also contributes to emissions of As, Cr, Cu, Mn and Zn [6]. Industrial metallurgical processes produce the largest emissions of As, Cd, Cu, Ni and Zn. Exhaust emissions from gasoline- and diesel-fuelled vehicles contribute to atmospheric Pb, Fe, Cu, Zn, Ni and Cd, while tyre-rubber abrasion to Zn [8]. In the last decade, the use of new automobile catalytic converters also contributes to the emissions of Pt, Pd and Rh [9, 10]. The composition of aerosols depends on the occurrence of metals in combustion processes and their volatility as well as on their amount produced by decomposition of continental and oceanic surfaces. Thus, the size distributions of different metals depend on the balance of different sources.

The source of trace metals in atmospheric liquid water is the dissolution of aerosol particles incorporated in water droplets. The solubility of metal aerosols is influenced by many factors, for example pH of the aqueous phase [11], crustal enrichment factor (EF_{crust}) of metal [12], aerosol type and size [12, 13], photoreduction [14], presence of organic [15] and carbonaceous components [16], etc. Dissolved trace metals, especially TM such as iron, manganese and copper, are involved in different chemical processes, both in aqueous aerosols and in the atmospheric liquid phase. Many studies have pointed out that transition metal ions (TMI) and their complexes can participate in a variety of reactions, including the oxidation of sulphur(IV) and organic compounds and the catalytic oxidation of sulphur(IV) [17–19]. TMI can efficiently react with many oxidising and reducing components, and therefore they influence the free-radical budget in the atmospheric liquid phase [20].

Atmospheric concentrations of many trace metals have been significantly affected by man's activities, which have drastically changed the biogeochemical cycles and balance of some metals. For example, between 1850 and 1990, the anthropogenic emissions of Cu, Zn and Pb exponentially increased due to the increase in mine production of metals [21]. Elevated concentrations of atmospheric aerosol particles have been associated with adverse effects on human health [22, 23]. Particle size and shape are key factors that control the extent of penetration of particles into the human respiratory tract. In addition, the potential health effects depend on many other factors, such as chemical and physical characteristics of aerosols, the amount of toxic substances, their solubility in biological fluids, etc. [24–26]. The exact chemical and physical characteristics connected with health effects as well as the toxicity mechanisms are still uncertain [27]. Nevertheless, it was found that PM_{10} toxicity can be related to soluble components (especially soluble TM) [28], which are an important factor in lung inflammation, most likely because of their bioavailability [29].

5.2 Physico-chemical characteristics of metals in aerosols

5.2.1 Size distribution

Environmental effects of atmospheric aerosol particles depend on their particle size and chemical composition (as a function of size and time) [30]. Knowledge of the size distribution of atmospheric particles, which also consist of trace metals, is important in understanding the effects of particles on human health; it gives information on the sources and transformation processes during transport in the atmosphere and is required for estimating the dry deposition of metals to the earth's surface. In the last few years, many studies on the size distribution of metals in aerosol particles in very diverse environments have been undertaken [31–41].

The size distributions of trace metals in atmospheric aerosols were measured using impactors (a ten-stage micro-orifice uniform deposit impactor (MOUDI) and an isokinetic impactor system) at three background sites in central England and southern Scotland by Allen *et al.* [32]. The elements were determined by inductively coupled plasma mass spectrometry analysis after digestion in nitric acid, which allowed a recovery of $>80\%$. For Scotland, the size distributions were typically trimodal, while for central England more variable (Figure 5.1). On the basis of characteristic size distributions they identified metals (Cd, Sn, Pb and Se) whose mass resided mainly within the accumulation mode, metals (Ni, Zn, Cu, Co, Mn and Hg) which were distributed between fine, intermediate and coarse modes and metals (Fe, Sr and Ba) which

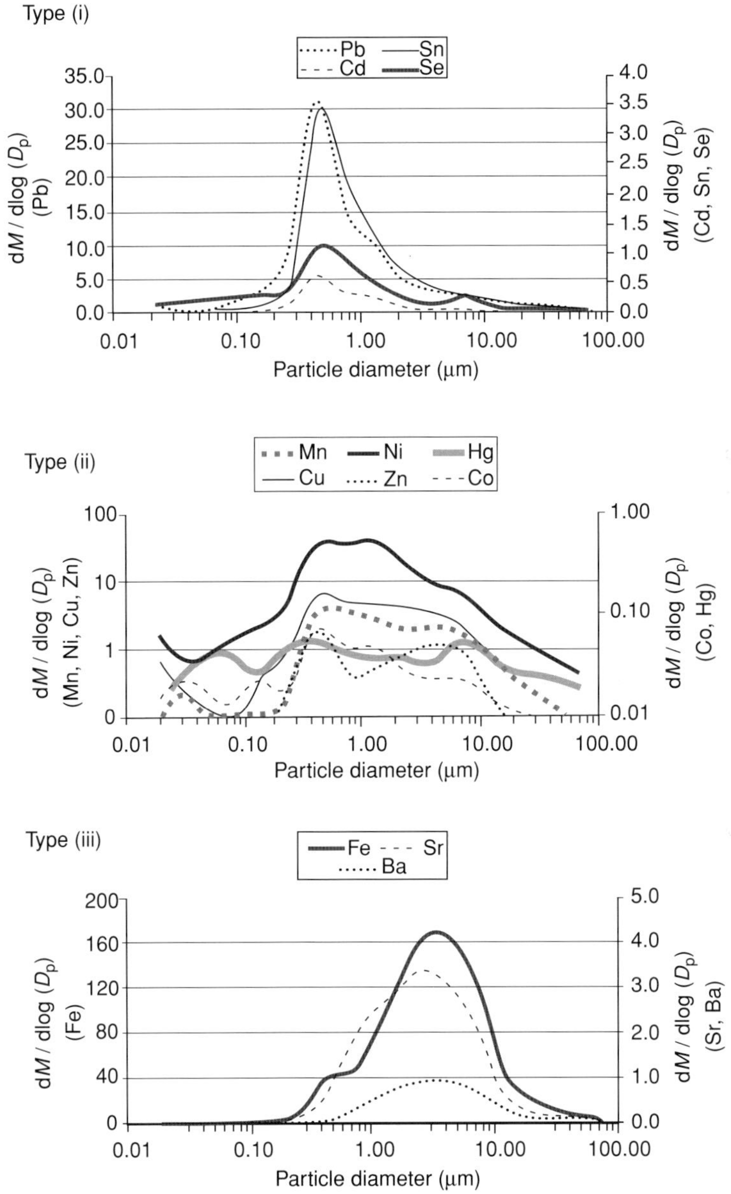

Figure 5.1 Trace metal size distributions (central England). (Reprinted from [32]. Copyright 2001, with permission from Elsevier.)

were mainly found within coarse particles (Figure 5.2). It is believed that the measured size distributions are a result of a combination of different processes, that is local anthropogenic and natural sources, long-range transport and resuspension.

In addition to the size distribution, Singh *et al.* [33] also studied diurnal characteristics of particle-bound metals in two different areas that are representative of distinct air pollution

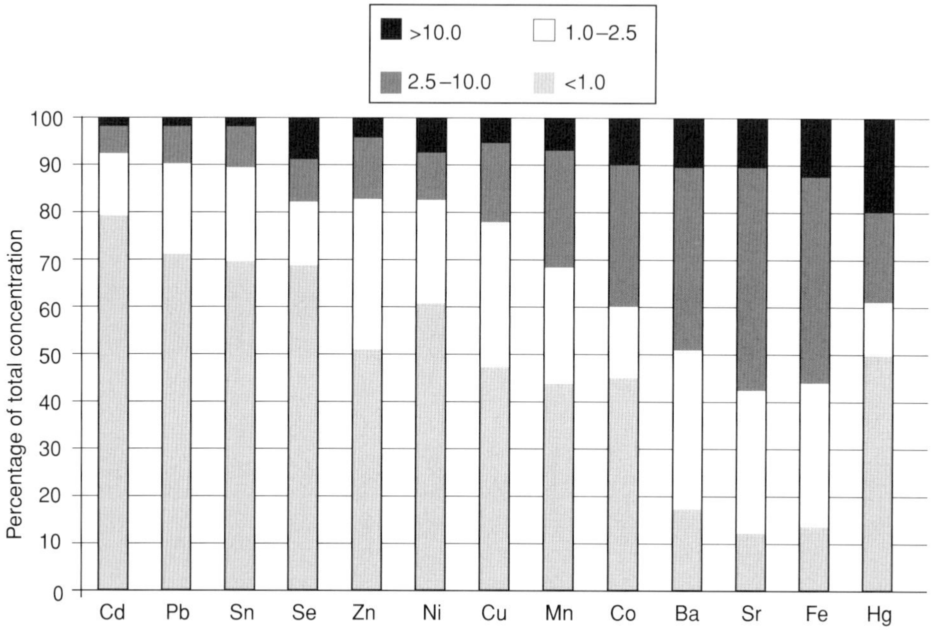

Figure 5.2 Distribution of individual metals according to size fraction (central England). (Reprinted from [32]. Copyright 2001, with permission from Elsevier.)

regimes in the Los Angeles Basin. Size-segregated particle samples were collected by means of five-stage MOUDI impactors on teflon filters. The amount of trace elements was determined using X-ray fluorescence. An example of average diurnal trends of the most predominant metals in the coarse and fine modes of aerosols is presented in Figure 5.3. It is obvious that diurnal patterns of metals in the coarse and fine modes are different and can be explained by different sources of coarse and fine particles and formation mechanisms as well as the meteorological conditions. So, in this particular case, the concentrations of metals in the coarse mode (mainly resuspended by wind) were higher during the time of day at which the wind speed was high (wind from the West, a maximum velocity at around 3 p.m.). In contrast, the concentrations of metals in the fine mode reached maximum values during early morning and the nighttime, mainly due to the traffic, but also due to the decrease in the wind speed and depression of mixing depth.

Due to increased size resolution of modern cascade impactors and improved programmes for obtaining aerosol size distribution [36], there are now possibilities for determining the overall shape of the mass size distributions in detail (i.e. fine structure of the size distribution). The fine structure of particulate matter (PM) and elemental mass size distributions with aerodynamic diameter (AD) in the range 50 nm–10 μm was investigated, in downtown Budapest, in 2002 by Salma *et al.* [38]. For measuring the PM mass size distribution a MOUDI with 11 impaction stages and a backup filter stage was used, while for measuring elemental mass size distributions a small-deposit-area low-pressure impactor with 12 impaction stages [39] was used. The samples were analysed by particle-induced X-ray emission analysis. For the typical crustal elements (Al, Si, Ca, Ti and Fe) they identified two modes, that is a major coarse one and an intermediate one at about 1-μm AD, which represents about 4% of the elemental mass

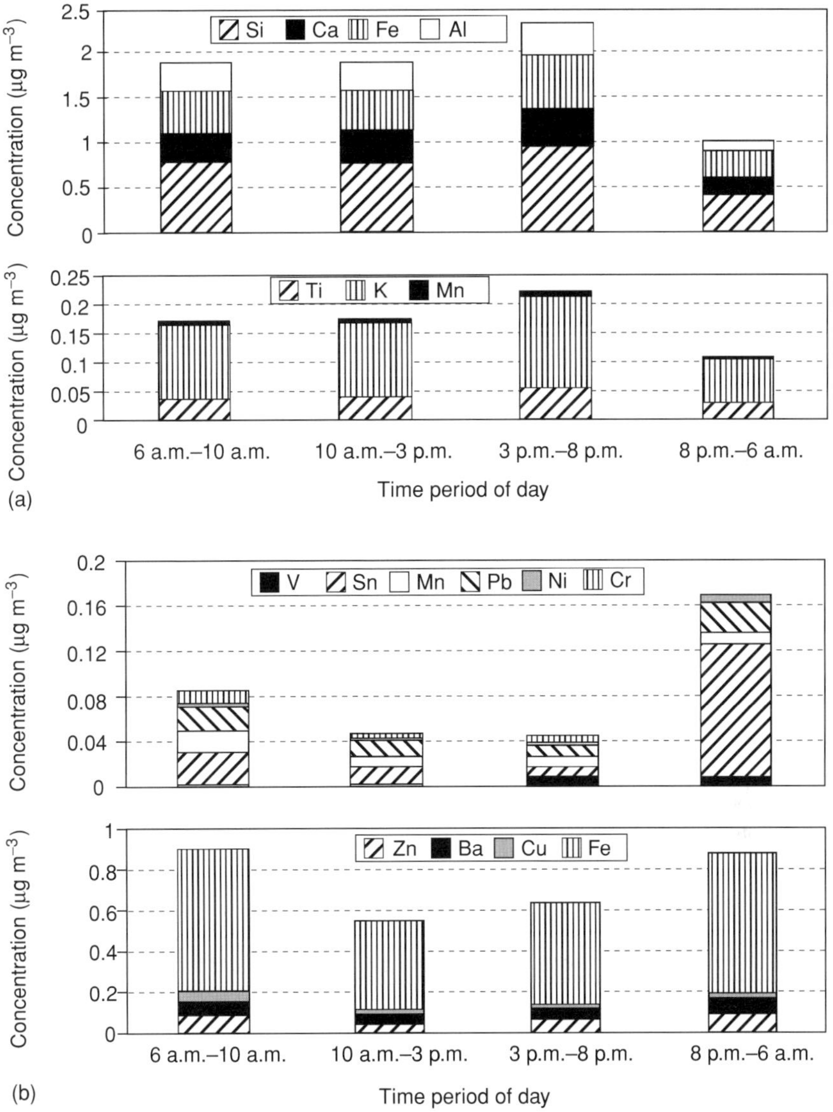

Figure 5.3 Metal concentrations as a function of time period of day in (a) coarse and (b) fine particles mode at Downey, Los Angeles Basin. (Reprinted from [33]. Copyright 2002, with permission from Elsevier.)

(Figure 5.4). The predominant source of the coarse particles is most likely road dust, while the intermediate particles may originate also from other sources (besides road dust also fly ash from coal burning). For the typical anthropogenic elements (Cl, Zn, Na, Mn, Ni, Cu, Pb, S and Br) the distributions were generally trimodal, with a coarse mode and two submicrometer modes instead of a single accumulation mode. The different relative intensities of submicrometer modes for the anthropogenic elements found in this study indicate the existence of several sources.

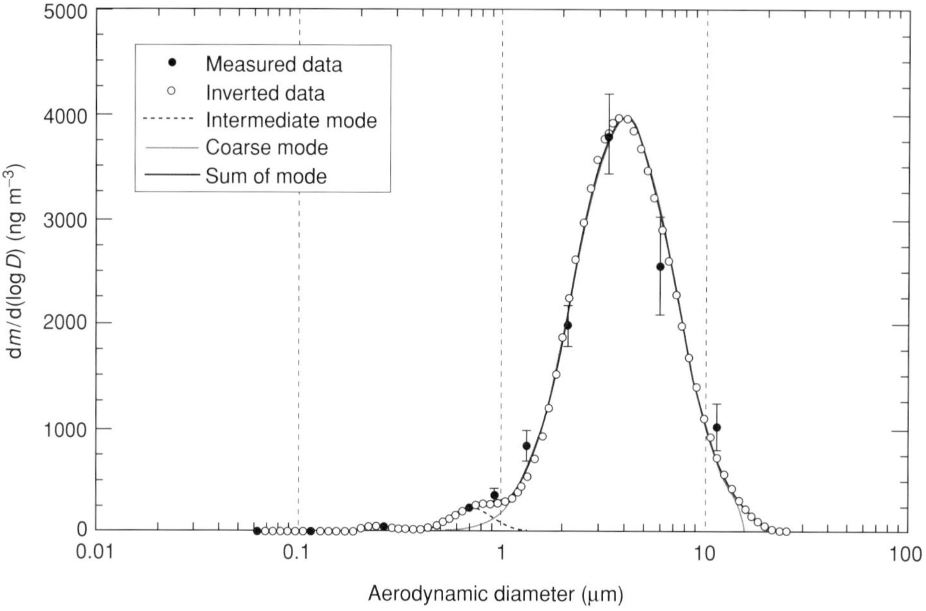

Figure 5.4 Mass size distribution of Fe in downtown Budapest for the daytime period on 29 April 2002. The error bars indicate ±1 standard deviation. The atmospheric concentrations in the intermediate and coarse modes were 51 and 2400 ng m^{-3}, respectively, and the Gads (geometric mean aerodynamic diameters) for the two modes were 0.73 μm (GSD – geometric standard deviation: 1.3) and 3.9 μm (GSD: 1.7), respectively. (Reprinted from [38]. Copyright 2005, with permission from Elsevier.)

5.2.2 Chemical speciation

Chemical speciation defines the chemical form of metal, in which it is present in a certain environment. So, speciation of metals in the atmosphere can indicate the distribution of metals between the aerosol and atmospheric water, the distribution between different metal complexes formed by different ligands present in the aqueous phase or the partition between different oxidation states of a certain metal. Anyhow, chemical speciation of atmospheric metals needs a clear definition as well as agreement on terms and procedures used for the determination of a certain metal species.

5.2.2.1 Dissolved/particulate distribution

Speciation of metals in aerosol particles is usually different than that in the aqueous phase. The information on the chemical speciation of metals in aerosols is very important, because it determines the mobility of metals once the aerosol gets into contact with water. So, the determination of soluble fraction of metals in different leaching agents provides information on the nature of the species in which metals are bound, their origin, their potential bioavailability and toxicity. The most useful procedure for the estimation of different chemical forms of trace metals in aerosol particles according to their bonding properties is a sequential leaching [42–47]. Using selective reagents, trace metals can be separated into a soluble fraction

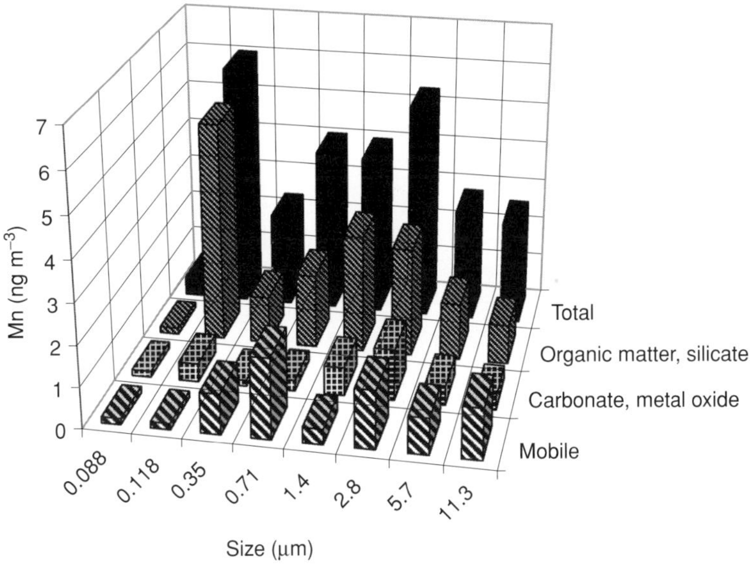

Figure 5.5 Distribution of Mn in three fractions (using a sequential method) as a function of particle size. Aerosol samples were collected in Veszprém, Hungary. (Reproduced from [44] by permission of The Royal Society of Chemistry.)

(environmentally mobile), metal carbonates and oxides and metals bound to organic matter and silicates (environmentally immobile) [42].

The fractionation of aerosols according to particle size and chemical bonding was studied by Hlavay *et al.* [44]. Aerosol samples were collected in Veszprém, Hungary, using a cascade impactor (Berner type) with eight collecting stages. A three-stage sequential leaching procedure was applied to the aerosol samples to establish the distribution of metals among three size fractions. Figure 5.5 is an example of the distribution of Mn in three fractions as a function of particle size. The size distribution is bimodal (with higher concentrations in the coarse fraction) as a result of anthropogenic and natural sources. It is also evident that Mn is concentrated in the stable fraction (i.e. bound to organics and silicates) and thus with no direct influence on the environment. The important result of this study is that the loosely bound and thus highly soluble fraction of metals (like Pb and Cr and especially Cd) corresponds to the fine particles, which are characteristic of anthropogenic sources.

The chemical speciation of trace metals in the finest particles of the impactor system (<0.61 μm) collected in 12 areas of the city of Seville (Spain) was studied by Espinoza and co-workers [47], using a new speciation scheme. This scheme is specific for urban aerosols and the experimental conditions are closer to the conditions of deposition and dissolution of trace metals into the lung. It is consisted of four fractions: (1) soluble and exchangeable metals; (2) carbonates, oxides and reducible metals; (3) bound to organic matter, oxidisable and sulphidic metals; and (4) residual metals. The important finding of this study is that trace metals prevailing in fine urban aerosols are toxic, like V, Ni, Pb and Cd. In addition, Ni and V compounds are very soluble; that is, about 50% of V and nearly 40% of Ni was determined in the soluble and exchangeable fraction (Figure 5.6).

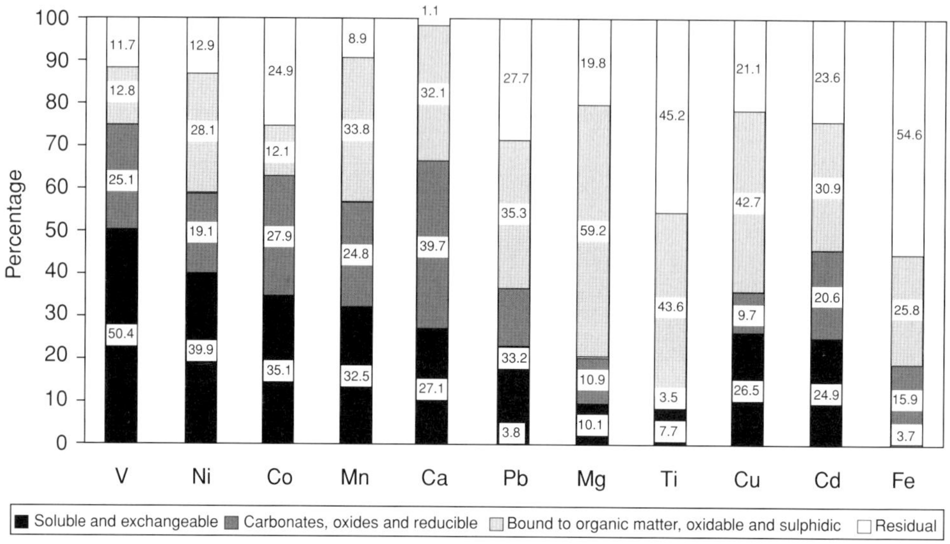

Figure 5.6 Chemical speciation of different metals by applying the sequential extraction scheme as percentage of the elemental concentration in fine particles. (Reprinted from [47]. Copyright 2002, with permission from Elsevier.)

5.2.2.2 *Redox speciation*

Knowledge of the chemical speciation in a sense of partition between different oxidation states of TM is crucial in understanding of chemical changes that occur in atmospheric liquid water. The reactivity of dissolved TM compounds (especially compounds of iron, manganese and copper) towards different pollutants and photo-oxidants depends on the oxidation state and upon the chemical form in which the metals are present in the aqueous phase (e.g. as cations or complexes with different ligands).

In previous studies of the solubility of atmospheric iron, it has generally been assumed that dissolved iron was present predominantly as Fe(III) [48, 49]. Nevertheless, it was shown that Fe(II) is an important constituent of atmospheric liquid water. Measurements of iron oxidation states in clouds, fog and rainwater demonstrate that the relative concentrations of Fe(II) and Fe(III) can vary considerably between different events ([3, 50–52]; see also [53] and references therein). The distribution of iron between the two oxidation states is a complex function of different factors, such as sunlight intensity, the concentration of oxidants (e.g. HO_2/O_2^-, H_2O_2, O_3, OH and O_2), reductants (e.g. HO_2/O_2^-, Cu(I) and S(IV)) and complexing agents (e.g. organic ligands) [4, 50, 52–56]. Behra and Sigg [50] found that a large fraction (20–90%) of dissolved iron in fog water was present as Fe(II), which increased with exposure to light as well as with decreasing pH. Similarly, the percentage of Fe(II) was correlated with the solar irradiation during the most of the cloud events during the Great Dun Fell campaign [3]. Deutsch *et al.* [57] evaluated the Fe(II)/Fe(III) partition in cloud water samples during two field campaigns on the top of the Kleiner Feldberg, Germany, and also found a correlation between the concentration of Fe(II) in the liquid phase and the sunlight intensity for most of the samples (Figure 5.7). To investigate the iron chemistry in aqueous-phase measurements of iron oxidation states in fog

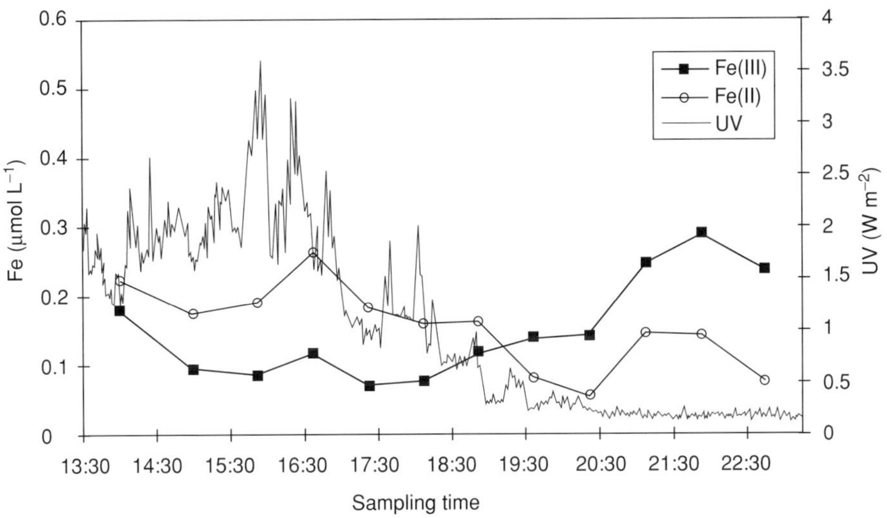

Figure 5.7 Concentrations of Fe(II) and Fe(III) and intensity of the solar irradiation during the cloud event at 28 April 1997. (Reprinted from [57] with kind permission of Springer Science and Business Media.)

and cloud water, samples have been combined with a kinetic model [4]. Results, which predicted that Fe(II) would be the predominant oxidation state during daylight and Fe(III) during nighttime conditions, were in agreement with field measurements. In addition, recent modelling studies show strong dependence of the Fe(II)/Fe(III) ratio on photolysis reactions [58, 59]. So, it appears that during the daylight the photochemical reduction of Fe(III) is the most important source of Fe(II) species in atmospheric water, with concentration levels dependent on pH.

Chemical speciation of iron in aerosol particles has also been determined. Although the experiments were performed under different leaching conditions, they clearly show that dissolved iron can and does exist as Fe(II) [14, 60–63]. It has been reported that at low pH, Fe(II) contributed to about 7–12% of the total soluble iron and only 1% of the total iron in marine aerosols [14], 8.4% of the total iron in urban aerosols [63] and also a small fraction (about 1%) of the total iron in aerosols of mainly crustal origin [62, 63]. Nevertheless, it was suggested that besides the photochemical reduction of Fe(III), and to a lesser extent thermal reduction of Fe(III), direct dissolution of Fe(II) compounds from aerosol particles can also contribute to the amount of Fe(II) in the atmospheric aqueous phase [62]. So, the Fe(II)/Fe(III) concentration ratio in solution depends on iron speciation in aerosol particles [64], on processes which control iron solubility [65] as well as on the reactivity of dissolved iron species towards different reductants and oxidants [53].

Like iron, the copper chemistry in the atmospheric aqueous phase is also very complex. Unfortunately, the information on the oxidation state of copper (Cu(I)/Cu(II)) in atmospheric waters is rather scarce. Although copper in its reduced form is not thermodynamically stable in the atmospheric liquid phase because of the presence of different oxidants, there are evidences that dissolved copper can exist as Cu(I) [4, 66]. Redox speciation of copper in rainwater samples was studied by Kieber *et al.* [66]. They determined relatively high concentrations of dissolved Cu(I) (on average about 25% of the total Cu) with an average ratio, Cu(II)/Cu(I), of about 2.3.

A large abundance of Cu(I) closely resembled the proportion of other redox-active trace metals found in rain (i.e. 24% of dissolved chromium found as Cr(III) and 16% of dissolved iron as Fe(II)). It was suggested that the conditions in rainwater would favour reduced metals if the H_2O_2/O_2 couple controls the redox conditions with H_2O_2 as a reductant [56]. Even if the O_2/H_2O couple controls the redox conditions, substantial amounts of reduced metals can exist because of photoreduction of metals complexed with different organic ligands [67]. It was found that concentrations of strong Cu-complexing ligands were consistently lower than dissolved copper concentrations, indicating that some proportion of dissolved copper is likely to be complexed [66]. A portion of these ligands may be Cu(I)-complexing ligands.

By using a kinetic model for fog and cloud water, Siefert *et al.* [4] found that Cu(II) was the predominant oxidation state during both the day and the nighttime. During the daytime the Cu(I) concentration increased, whilst during the nighttime it decreased rapidly due to the oxidation with H_2O_2 and O_2. On the other hand, modelling studies by Ervers *et al.* [58] and Deguillaume *et al.* [59] showed no diurnal variation in the oxidation state of copper with most of the simulated copper in Cu(II) oxidation state due to the oxidation of Cu(I) with O_2 and Fe(III). The inconsistency between model results and field measurements clearly shows that further research is needed to understand the complicated copper speciation.

Concerning manganese species in the atmospheric aqueous phase, very limited information is available. Deutsch *et al.* [68] determined the total manganese concentrations as well as soluble fraction and oxidation state of soluble manganese in rain and snow samples. They found that manganese exists entirely as soluble Mn(II). Their investigations also showed that Mn(III) is unstable under atmospheric conditions. In the absence of ligands, Mn(III) immediately disproportionates. At pH values of rainwater (pH: 4–5) it is stable only in the presence of diphosphate ($P_2O_7^{4-}$), whose presence in rainwater is unlikely. The existence of Mn(IV) in the aqueous phase is also uncertain since its concentration measured in atmospheric waters was usually below limit of detection [4, 69]. Modelling studies have also predicted that Mn(II) is the prevailing oxidation state in an aqueous solution [59, 70].

5.2.2.3 Chemical forms

As stressed above, the reactivity of dissolved TM depends not only on the oxidation state but also on the chemical form in which the metals are present in atmospheric liquid water. Iron in its reduced form is primarily present as Fe^{2+} cation, while Fe(III) is able to coordinate different inorganic as well as organic ligands, producing complexes of diverse stability and solubility. However, in the absence of other ligands, Fe(III) in aqueous solutions is found in different forms due to the hydrolysis processes. The hydrolytic speciation of Fe(III) in aqueous solutions can be represented by the following equilibriums:

$$Fe^{3+} + H_2O \leftrightarrow Fe(OH)^{2+} + H^+$$
$$Fe(OH)^{2+} + H_2O \leftrightarrow Fe(OH)_2^+ + H^+$$
$$Fe(OH)_2^+ + H_2O \leftrightarrow Fe(OH)_3(s) + H^+$$
$$2Fe(OH)^{2+} \leftrightarrow Fe_2(OH)_2^{4+}$$

For simplicity, coordinated water molecules are not included in the chemical formulae (e.g. Fe^{3+} corresponds to $Fe(H_2O)_6^{3+}$). The relative amount of different forms is strongly pH dependent [71]. At low pH (i.e. below 3), Fe(III) is mainly in the form of Fe^{3+}, while with

Table 5.1 Concentrations of soluble Fe(III) species calculated from appropriate equilibriums with the solid phase [72] in an aqueous solution for pH in the range 3–5.

Fe(III) species	pH = 3.0	pH = 4.0	pH = 5.0
$[Fe^{3+}]$	3×10^{-5} M	3×10^{-8} M	3×10^{-11} M
$[Fe(OH)^{2+}]$	3×10^{-5} M	3×10^{-7} M	3×10^{-9} M
$[Fe(OH)_2^+]$	1.2×10^{-5} M	1.3×10^{-6} M	1.2×10^{-7} M
$[Fe_2(OH)_2^{4+}]$	8×10^{-7} M	7.9×10^{-11} M	7.9×10^{-15} M
$[Fe(OH)_4^-]$	10^{-16} M	10^{-15} M	10^{-14} M

increasing pH, besides the formation of insoluble $Fe(OH)_3(s)$, complex reactions between Fe^{3+} and OH^- also take place, leading to the formation of hydroxo complexes containing one or more Fe^{3+} ions (mono- and polyhydroxo complexes). The concentration of these species is usually higher than the concentration of Fe^{3+}, and therefore this leads to a considerable increase in the solubility of the solid phase. In equilibrium the activities of these species are determined by the presence of solid $Fe(OH)_3(s)$. From appropriate equilibriums with the solid phase [71, 72],

$$Fe(OH)_3(s) + 3H^+ \leftrightarrow Fe^{3+} + 3H_2O$$
$$Fe(OH)_3(s) + 2H^+ \leftrightarrow Fe(OH)^{2+} + 2H_2O \text{ etc.}$$

the concentration of soluble Fe(III) species for pH range 3–5, which is typical for atmospheric liquid water, can be calculated (Table 5.1). The amount of all soluble Fe(III) species at pH 3.0 is about 7×10^{-5} M, at pH 4.0 about 1.6×10^{-6} M and at pH 5.0 about 1×10^{-7} M. It is evident that in pH range 3–5, Fe^{3+}, $Fe(OH)^{2+}$ and $Fe(OH)_2^+$ coexist in an aqueous solution, with concentrations of the same order of magnitude at pH 3.0, while at pH 4.0 and 5.0 $Fe(OH)_2^+$ is the main species.

Deguillaume *et al.* [53] performed calculations for the partitioning of these three main monomeric Fe(III) species as a function of pH at an ionic strength of zero and temperature of 298 K. As shown in Table 5.1 they also found that at pH 4.0, $Fe(OH)_2^+$ was the prevailing species. On the other hand, Faust and Hoigné [73] determined different equilibrium speciation of Fe(III) species at an ionic strength of 0.03 M and temperature of 293 K. At pH 4.0 they found that more than 90% of Fe(III) was present as $Fe(OH)^{2+}$. The reason for the different speciation is in the modification of the equilibrium constant values due to the influence of ionic strength [17].

In atmospheric liquid water, the kinetics of the Fe(II)/Fe(III) redox cycle is further complicated because of the presence of different inorganic and, especially, organic ligands. Low-molecular-weight carboxylic acids have been shown to be one of the dominant classes of organic compounds found in a variety of phases in the troposphere [74]. As they have low molecular weights and high polarity, they are water soluble. Among them, oxalic acid is a very common constituent of atmospheric waters [75] and is usually observed at concentrations of the same order of magnitude as dissolved iron. Oxalate has a strong affinity for Fe(III) [76]:

$$Fe^{3+} + C_2O_4^{2-} \leftrightarrow FeC_2O_4^+ \qquad K_1 = 2.5 \times 10^9\ M^{-1}$$
$$FeC_2O_4^+ + C_2O_4^{2-} \leftrightarrow Fe(C_2O_4)_2^- \qquad K_2 = 6.3 \times 10^6\ M^{-1}$$
$$Fe(C_2O_4)_2^- + C_2O_4^{2-} \leftrightarrow Fe(C_2O_4)_3^{3-} \qquad K_3 = 3.8 \times 10^4\ M^{-1}$$

Thus, its presence has a significant effect on the speciation of Fe(III) in the aqueous phase. Unfortunately, there are no field measurements on the partitioning of different Fe(III) complexes in atmospheric waters. However, model calculations indicate that the Fe(III)–oxalato complexes can be the predominant species of Fe(III) in cloud, fog and rainwaters for pH in the range 3–5 [76–78]. Erel *et al.* [52] calculated that for an atmospheric water sample with appreciable concentration of oxalate, more than 97% of the Fe(III) is complexed by oxalate, while 95% of Fe(II) is in the Fe^{2+}form. Thermodynamic-speciation models by Siefert *et al.* [4] have revealed two different cases of Fe(III) speciation in fog and cloud water. Fe(III) was found to exist either as $Fe(OH)_2^+$ or as $Fe(C_2O_4)_2^-$, but an unidentified strong chelating ligand with Fe(III) was also suggested. Model calculations and correlation analysis of Willey *et al.* [55] indicate that Fe(III) occurs as both oxalato complexes and $Fe(OH)_2^+$ with large variations depending on the composition and pH of rainwater. In summer rain, Fe(III) occurs mainly complexed with oxalate (as $Fe(C_2O_4)_2^-$, with lower concentrations of $FeC_2O_4^+$ and $Fe(C_2O_4)_3^{3-}$). In winter rain, Fe(III)–oxalato complexes and $Fe(OH)_2^+$ are almost equal, while in marine rain, $Fe(OH)_2^+$ is the main species.

Potential organic ligands, which can form strong complexes with Fe(III), are also polycarboxylic acids, such as malonic and citric acid [79]. It has also been suggested that humic substances are also potential complexing agents for dissolved Fe(III) if they are present in sufficient concentrations [55].

The information on the chemistry of manganese and copper complexes in atmospheric waters is very scarce. However, some studies have observed metal–organic complexing ligands in rainwater [80–82]. The results of Spokes *et al.* [80] showed that more than 99.9% of copper in the precipitation samples measured was complexed. They assumed that ligands were organic and, as they occurred at concentrations higher than total copper, complexation should be organic in character. In addition, the study of Nimmo and Fones [82] confirmed the existence of rainwater- and seawater-soluble metal (Cu, Pb, Cd, Ni and Co) complexing organic ligands associated with aerosol material. Nevertheless, further research, which should be focused on the chemical composition of the soluble organic ligands, their influence on the metal solubility as well as on the metal reactivity, is needed.

5.3 Reactivity of transition metals

Dissolved TM, especially iron, manganese and copper, capable of undergoing one-electron redox processes play an important role in many atmospheric processes. Recently, Deguillaume *et al.* [53] presented a detailed overview on the reactivity of TM in the atmospheric liquid phase. Thus, only some important interactions of TM are presented with the emphasis on the reactivity of TM connected with their oxidation state and chemical form.

5.3.1 Photochemical/chemical cycling

On the basis of a detailed kinetic model Graedel *et al.* [83] first attempted to resolve the complex redox chemistry of TM in atmospheric waters. They suggested that iron, as well as copper and manganese, could play an important role in the oxidation of aqueous SO_2 (i.e. S(IV) species) and the production of OH radicals. Many studies [58, 59, 70, 73, 83, 84] showed that photolysis of Fe(III) complexes can be an important source of Fe^{2+} as well as OH radicals in the atmospheric liquid phase. As shown above, in the absence of other ligands, $Fe(OH)^{2+}$ predominates at low pH and is also the most photoreactive among the Fe(III)–hydroxo species [53]. The photochemical reduction of $Fe(OH)^{2+}$ provides an important source of Fe^{2+} and OH due to the high reaction rate [85]. However, laboratory studies showed that the significance of this reaction depends on many factors, such as the concentration of dissolved iron, the solution pH, the intensity of solar radiation and concentration of complexing ligands.

Some complexes of Fe(III) with organic ligands (e.g. with low chain carboxylic acids) can undergo photochemical redox reactions [14, 15, 52, 76, 77, 79, 86–88]. The photoredox behaviour has been found for some simple Fe(III) complexes with carboxylic acids (e.g. oxalic, citric, malonic, glyoxalic and maleic). In the mechanism of the photolysis of Fe(III) complexes, shown below, $h\nu$ is the energy of a photon, L is an organic ligand and $L^{\bullet}$ is an organic radical.

$$\mathrm{Fe(III) - L} + h\nu \rightarrow \mathrm{Fe(II) - L^{\bullet}}$$
$$\mathrm{Fe(II) - L^{\bullet}} \rightarrow \mathrm{Fe(II) + L^{\bullet}}$$
$$\mathrm{L^{\bullet} + O_2 \rightarrow O_2^{-}} + \text{oxidised L}$$
$$\mathrm{2\,O_2^{-} + 2H^{+} \rightarrow H_2O_2 + O_2}$$

Photochemical reactions lead to the reduction of Fe(III) to Fe(II) and formation of an organic radical. In the presence of O_2, the organic radical can reduce O_2; this leads to the production of O_2^-/HO_2 (superoxide ion and its conjugate acid), which further disproportionates to H_2O_2 and O_2 [76]. Fe(III)–oxalato complexes are especially sensitive to sunlight. Laboratory studies of the photolysis of iron(III)–oxalato complexes under the conditions typical for acidified atmospheric waters have shown the depletion of oxalic acid and formation of H_2O_2 [76, 79, 86]. In addition, photochemical redox reactions of Fe(III) complexed with citric and malonic acid can be a potential important source of Fe(II), O_2^-/HO_2, H_2O_2 and OH [79].

Under ambient conditions Fe(II) can be readily reoxidised by H_2O_2 (Fenton reaction), O_2 and HO_2/O_2^-. The reoxidation rates are strongly dependent on solution pH, and can also be influenced by the presence of inorganic and organic compounds.

5.3.2 Interactions of transition metals with S(IV) species

Numerous laboratory studies have shown that the oxidation of S(IV) species (i.e. HSO_3^-/SO_3^{2-}) by dissolved oxygen (so-called autoxidation) under non-photochemical conditions is catalysed by TMI (for a detailed overview, see [17]). The catalytic effect of iron is reasonably well understood [89–92]. S(IV) oxidation by O_2 in the presence of iron proceeds via a free-radical-chain mechanism initiated by the interaction of Fe(III) with S(IV) (Table 5.2). An example of the redox cycling of Fe(II)/Fe(III) during the catalysed S(IV) autoxidation, which was followed online by a continuous flow analyser with a spectrophotometric detection, can be seen in Figure 5.8 [93]. It is evident that Fe(III) ions are rapidly reduced by HSO_3^- (reactions 1 and 2) and then reoxidised by SO_5^-, HSO_5^- and/or by SO_4^- (reactions 6, 7 and 8). After HSO_3^- oxidation is completed, iron present in the solution remains primarily as Fe(III).

As stressed above, the distribution of Fe(III) species depends on the pH of the solution. Furthermore, the presence of S(IV) significantly influences the speciation and iron solubility at higher pH. The results of Novič *et al.* [91] indicated that at pH ≥ 4 the formation of polymerised hydroxo–Fe(III) species occurs rapidly. The catalytic activity of Fe(III) at higher initial pH

Table 5.2 Proposed reaction mechanism for the iron-catalysed S(IV) autoxidation[a]

Reaction	Number
$FeOH^{2+} + HSO_3^- = [FeOHSO_3H]^+$	(1)
$[FeOHSO_3H]^+ \rightarrow Fe^{2+} + SO_3^- + H_2O$	(2)
$SO_3^- + O_2 \rightarrow SO_5^-$	(3)
$SO_5^- + HSO_3^- \rightarrow SO_3^- + HSO_5^-$	(4)
$SO_5^- + HSO_3^- \rightarrow SO_4^- + HSO_4^-$	(5)
$Fe^{2+} + SO_5^- (+H_2O) \rightarrow FeOH^{2+} + HSO_5^-$	(6)
$Fe^{2+} + HSO_5^- \rightarrow FeOH^{2+} + SO_4^-$	(7)
$Fe^{2+} + SO_4^- (+H_2O) \rightarrow FeOH^{2+} + SO_4^{2-} + H^+$	(8)
$HSO_5^- + HSO_3^- \rightarrow 2SO_4^{2-} + H^+$	(9)
$SO_3^- + SO_3^- \rightarrow S_2O_6^{2-}$	(10)
$SO_5^- + SO_5^- \rightarrow S_2O_6^{2-} + 2O_2$	(11)

[a]Condensed from the mechanism presented by Brandt and van Eldik [17] and Warneck and Ziajka [89].
(1) Formation of Fe(III)–sulphito complex.
(2) Redox decomposition (rate-determining step).
(3) In the presence of O_2, the redox cycle is controlled by the formation of SO_5^- radical.
(4, 5) SO_5^- may oxidise either sulphite ion or the reduced metal ion (6).
(7, 8) Oxidation of the reduced metal ion by HSO_5^- and by SO_4^-.
(9, 10, 11) Product formation.

decreases with longer hydrolysis time. The effect of initial pH is first of all due to the catalytic activity of hydroxo–Fe(III) species. Monomeric $Fe(OH)^{2+}$ is much more reactive than Fe^{3+} [94], or dimeric and polymeric hydroxo–Fe(III) species. Hence, pH is a very important parameter that controls the concentration of Fe(III) ions and thus the catalytic cycle of Fe(II)/Fe(III) in aqueous solutions of S(IV) and consequently the reaction rate of S(IV) oxidation by oxygen.

It was also shown that the reaction rate is not considerably affected by the oxidation state of iron at the start of the reaction, because the equilibrium between Fe(II) and Fe(III) is established a few minutes after the start of the reaction [91, 95, 96]. Nonetheless, if Fe(II) is used as a catalyst, the conditions have to allow its oxidation in Fe(III), which is the catalytically active species.

Manganese is potentially capable of replacing iron as a catalyst in the aqueous-phase oxidation of S(IV) species by dissolved oxygen. As shown above, the most abundant oxidation state in nature is Mn(II), but the effective catalyst for S(IV) oxidation is Mn(III) [18]. Catalysis by manganese has been the subject of many experimental and theoretical studies [18, 92, 97–102], but discrepancies still remain, particularly for catalysis above pH 4. Due to the highly complex nature of this reaction, even a minor change in the experimental conditions can result in a change of the dominant path of the course of reaction. Nevertheless, it has been suggested that Mn(II) reacts with S(IV) in a similar way than Fe(II). Thus far no experimental evidence exists for initiation of the radical chain mechanism of S(IV) oxidation catalysed by Mn(II). The prevailing opinion is that in the absence of Mn(III) the chain reaction could be initiated via the oxidation of Mn(II) by molecular oxygen [103], by traces of Fe(III) or by an interaction between HSO_3^- and traces of Fe(III) present as impurities in chemicals [18].

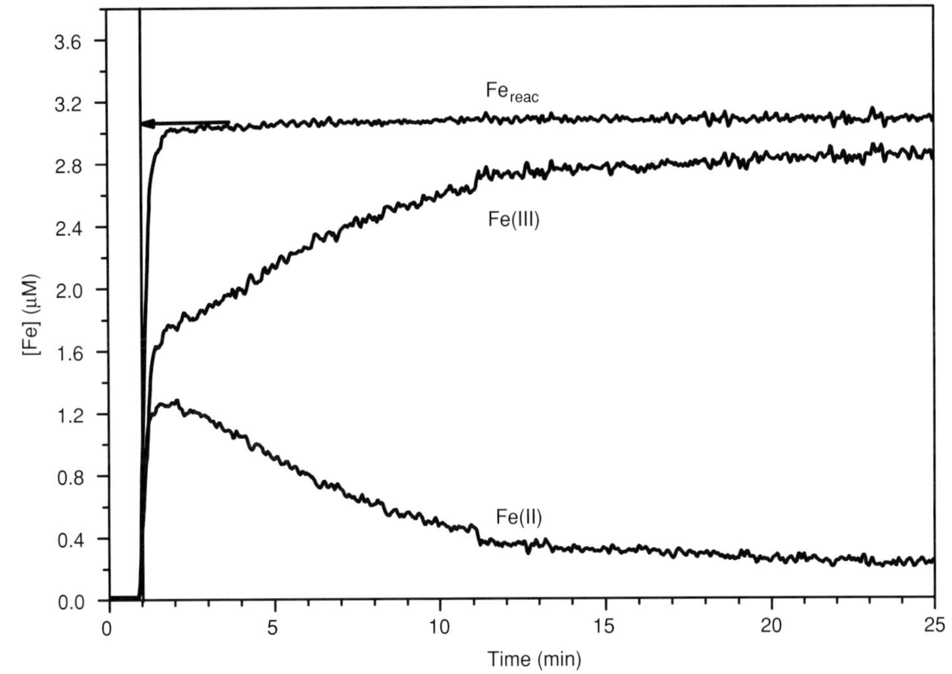

Figure 5.8 Redox cycling of Fe(II)/Fe(III) during the iron-catalysed S(IV) autoxidation. Initial conditions: total [Fe] = 3.6 μM, [Fe(III)] ≥3.5 μM, [HSO_3^-] = 0.2 mM, pH = 3.7; arrows pointing to the left Fe$_{reac}$ designate the concentration of total soluble iron. (Reprinted from [93] with kind permission of Springer Science and Business Media.)

Although the concentrations of copper in atmospheric waters are generally one or more orders of magnitude lower than the concentrations of iron [77], it has been suggested that it can significantly influence the interactions of S(IV) with other TMI as well as the free-radical chemistry. Since copper reacts efficiently with HO_2/O_2^- radicals, it can influence the redox cycling of iron [19, 77]. Even at concentrations more than a hundred times lower than concentrations of iron, copper consumes most of the HO_2/O_2^- and cycles between Cu(II) and Cu(I). Cu^+ reacts with $FeOH^{2+}$, where Fe(II) and Cu(II) are formed.

Further, the presence of two or more catalytically active TMI in atmospheric water may exhibit a significant synergistic effect on S(IV) autoxidation; that is, the reaction rate of S(IV) oxidation catalysed with a mixture of TMI is significantly higher than that in the presence of the individual metal ions. The synergistic effect has been shown by a combination of different TMI, such as Fe(III)/Mn(II), Fe(II)/Mn(II), Fe(III)/Cu(II), Mn(II)/Cu(II) and Co(II)/Mn(II) [104–106].

5.3.2.1 Influence of organic compounds on the oxidation of S(IV)

The presence of different organic compounds in the atmospheric aqueous phase may influence the rate of S(IV) oxidation [90]. A complex chain reaction of S(IV) oxidation catalysed by TMI involves various sulphur-oxy radical anions (Table 5.2), which are very sensitive to organic

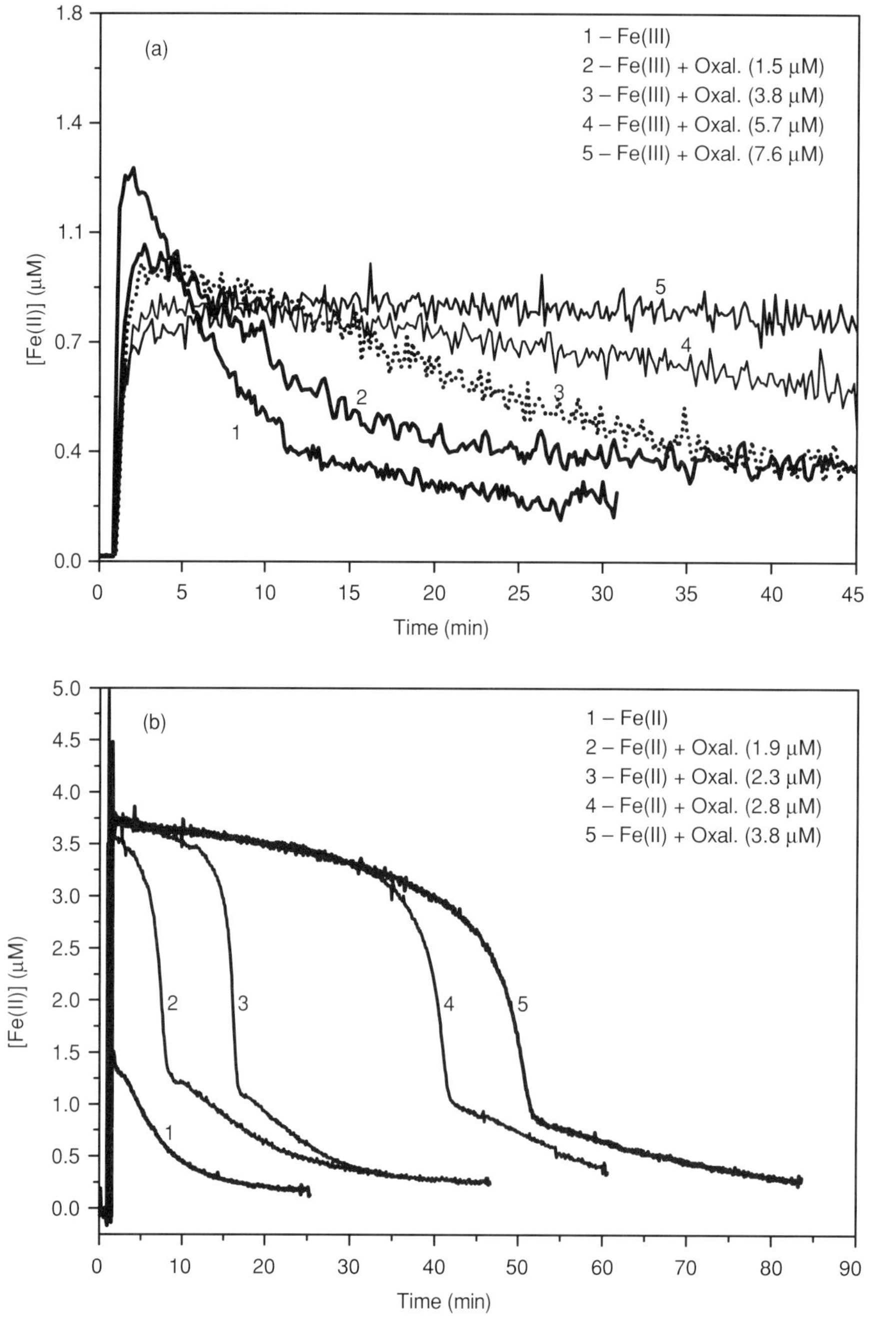

Figure 5.9 Comparison of Fe(II) concentrations during the iron-catalysed S(IV) autoxidation in the presence of different concentrations of oxalate. Initial conditions: $[HSO_3^-] = 0.2$ mM, total $[Fe] = 3.6$ μM; (a) $[Fe(III)] \geq 3.5$ μM, pH = 3.7; (b) $[Fe(II)] \geq 3.4$ μM, pH = 3.8. (Reprinted from [93, 96] with kind permission of Springer Science and Business Media.)

substances. These substances may act as inhibitors of S(IV) oxidation processes, distorting the expected pathways of the reaction, as well as the yield and distribution of reaction products. They may react with sulphur-oxy radicals [107] and/or TMI [93, 96, 108]. Up to now, the following organic compounds were found as effective inhibitors of S(IV) oxidation: phenols [109], alcohols [110], benzene [95], toluene, naphthalene and paraffin oil [103], terpenes [107] and carboxylic acids [93, 96, 108, 111].

Oxalate has a strong inhibiting effect on the iron-catalysed S(IV) oxidation [93, 96, 108, 112]. The formation of stable complexes with Fe(III) (see Section 5.2.2.3) affects the redox cycling of Fe(II)/Fe(III) and thus the reaction rate. By online measurements of the redox cycling of Fe(II)/Fe(III) (pH = 2.8, 3.7 and 4.5) during the reaction course, Grgić *et al.* [93] demonstrated that the presence of oxalate increases the concentration of dissolved Fe(III), while on the other hand it prevents the reduction of Fe(III). The formation of Fe(III)–oxalato complexes competes with the formation of Fe(III)–sulphito complexes, that is the first step of a radical chain mechanism (Table 5.2, reaction 1). Therefore, the concentration of Fe(II) formed from catalytically active Fe(III) species after the reduction with HSO_3^- decreases with increasing concentration of oxalate (Figure 5.9a). It was also shown that the effect of oxalate on the redox cycling of iron and thus on the reaction rate differs according to whether iron is initially present as Fe(III) or Fe(II) [96] (Figure 5.9). In the case of Fe(II) (Figure 5.9b), an induction period appears during which Fe(III) is formed and immediately complexed with oxalate. Consequently, the equilibrium between the Fe(III)–oxalato complexes and free Fe(III) depends upon the concentration of oxalate. When the equilibrium is reached, free Fe(III) initiates the reaction and the concentration of Fe(II) sharply decreases to the value that corresponds to the concentration of Fe(II) formed after the reduction of catalytically active Fe(III). Carboxylates like acetate and formate, which are not strong complex formers, have also been shown to have an inhibiting effect on the iron-catalysed S(IV) oxidation at pH higher than 4 [93, 111]. However their effect is much lower in comparison to the effect of oxalate.

The influence of carboxylic acids on the oxidation of S(IV) catalysed by Mn(II) under conditions representative for tropospheric liquid water was also investigated [113]. Among the monocarboxylic acids (formic, glycolic, lactic and acetic) formic acid was found to have the strongest inhibiting effect. The results suggest that the most probable reason for the inhibition is the interaction between sulphate radicals and carboxylic acids, that is scavenging of sulphate radicals. Oxalic acid has also been shown to slow down the Mn(II)-catalysed oxidation of S(IV); however, its influence is much lower than the influence of monocarboxylic acids as well as its influence on iron catalytic reaction.

References

1 Schroeder, W.H., Dobson, M., Kane, D.M. and Johnson, N.D. Toxic trace elements associated with airborne particulate matter: a review. *Journal of the Air Pollution Control Association* 1987; **37**: 1267–1285.

2 Deutch, F., Hoffmann, P. and Ortner, H.M. Analytical characterization of manganese in rainwater and snow samples. *Journal of Analytical Chemistry* 1997; **357**: 105–111.

3 Sedlak, D.L., Hoigné, J., David, M.M., Colvile, R.N., Seyffer, E., Acker, K., Wiepercht, W., Lind, J.A. and Fuzzi, S. The cloudwater chemistry of iron and copper at Great Dunn Fell, U.K. *Atmospheric Environment* 1997; **31**: 2515–2526.

4 Siefert, R.L., Johansen, A.M. and Hoffmann, R. Measurements of trace metal (Fe, Cu, Mn, Cr) oxidation states in fog and stratus clouds. *Journal of Air & Waste Management Association* 1998; **48**: 128–143.

5 Weber, S., Hoffmann, P., Ensling, J., Dedik, A.N., Weinbruch, S., Miehe, G., Gütlich, P. and Ortner, H.M. Characterization of iron compounds from urban and rural aerosol sources. *Journal of Aerosol Science* 2000; **31**: 987–997.

6 Pacyna, J.M. Source inventories for atmospheric trace metals. In: R.M. Harrison and R.E. van Grieken, eds. *Atmospheric Particles, IUPAC Series on Analytical and Physical Chemistry of Environmental Systems*, Vol. 5. John Wiley, Chichester, UK: 1998; pp. 385–423.

7 Mirage, J.O. A global assessment of natural sources of atmospheric trace metals. *Nature* 1989; **338**: 47–49.

8 Pacyna, J.M. Emission factors of atmospheric elements. In: J.O. Nriagu and C.I. Davidson, eds. *Toxic Metals in the Atmosphere*. John Wiley, New York: 1986; pp. 33–52.

9 Moldovan, M., Gomez, M.M. and Palacios, M.A. Determination of platinum, rhodium and palladium in car exhaust fumes. *Journal of Analytical Atomic Spectrometry* 1999; **14**: 1163–1169.

10 Limbeck, A., Rendl, J., Heimburger, G., Kranabetter, A. and Puxbaum, H. Seasonal variation of palladium, elemental carbon and aerosol mass concentrations in airborne particulate matter. *Atmospheric Environment* 2004; **38**: 1979–1987.

11 Desboeufs, K.V., Losno, R., Vimeux, F. and Cholbi, S. The pH-dependent dissolution of wind-transported Saharan dust. *Journal of Geophysical Research* 1999; **104**: 21287–21299.

12 Chester, R., Murphy, K.J.T., Lin, F.J., Berry, A.S., Bradshaw, G.A. and Corcoran, P.A. Factors controlling the solubilities of trace metals from non-remote aerosols deposited to the sea surface by the 'dry' deposition mode. *Marine Chemistry* 1993; **42**: 107–126.

13 Chester, R., Bradshaw, G.F. and Corcoran, P.A. Trace-metal chemistry of the north sea particulate aerosol-concentrations, sources and sea-water fates. *Atmospheric Environment* 1994; **28**: 2873–2883.

14 Zhu, X., Prospero, J.M., Savoie, D.L., Millero, F.J., Zika, R.G. and Saltzman, E.S. Photoreduction of Iron(III) in marine mineral aerosol solutions. *Journal of Geophysical Research* 1993; **98**: 9039–9046.

15 Zuo, Y. Kinetics of photochemical/chemical cycling of iron coupled with organic substances in cloud and fog droplets. *Geochimica et Cosmochimica Acta* 1995; **59**: 3123–3130.

16 Desboeufs, K.V., Sofikitis, A., Losno, R., Colin, J.L. and Ausset, P. Dissolution and solubility of trace metals from natural and anthropogenic aerosol particulate matter. *Chemosphere* 2005; **58**: 195–203.

17 Brandt, Ch. and van Eldik, R. Transition metal-catalyzed oxidation of sulfur(IV)-oxides: atmospheric-relevant processes and mechanisms. *Chemical Reviews* 1995; **95**: 119–190.

18 Berglund, J., Fronaeus, S. and Elding, L.I. Kinetics and mechanism for manganese-catalyzed oxidation of sulfur(IV) by oxygen in aqueous solution. *Inorganic Chemistry* 1993; **32**: 4527–4538.

19 Sedlak, D.L. and Hoigné, J. Oxidation of S(IV) in atmospheric water by photooxidants and iron in the presence of copper. *Environment Science & Technology* 1994; **28**: 1898–1906.

20 Losno, R. Trace metals acting as catalysts in a marine cloud: a box model study. *Physics and Chemistry of the Earth* 1999; **24**: 281–286.

21 Nriagu, J.O. History of global metal pollution. *Science* 1996; **272**: 223–224.

22 Schwartz, J., Dockery, D.W. and Neas, L.M. Is daily mortality associated specifically with fine particles? *Journal of the Air & Waste Management Association* 1996; **46**: 927–939.

23 Katsouyanni, K., Touloumi, G., Samoli, E., Gryparis, A., le Tertre, A., Monopolis, Y., Rossi, G., Zmirou, D., Ballester, F., Boumghar, A., Anderson, H.R., Wojtyniak, B., Braunstein, R., Pekkanen, J., Schindler, C. and Schwartz, J. Confounding and effect modification in the short-term effects of ambient particles on total mortality: results from 29 European cities within the APHEA2 project. *Epidemiology* 2001; **12**: 521–531.

24 Bérubé, K.A., Jone, T.P., Williamson, B.J., Winters, C., Morgan, A.J. and Richards, R.J. Physicochemical characterisation of diesel exhaust particles: factors for assessing biological activity. *Atmospheric Environment* 1999; **33**: 1599–1614.

25 Kodavanti, U.P., Hauser, R., Christiani, D.C., Meng, Z.H., McGee, J., Ledbetter, A., Richards, J. and Costa, D.L. Pulmonary responses to oil fly ash particles in the rat differ by virtue of their specific soluble metals. *Toxicological Sciences* 1998; **43**: 204–212.
26 Laden, F., Neas, L.M., Dockery, D.W. and Schwartz, J. Association of fine particulate matter from different sources with daily mortality in six U.S. cities. *Environmental Health Perspectives* 2000; **108**: 941–947.
27 Paoletti, L., De Barardis, B. and Diociaiuti, M. Physico-chemical characterization of the inhalable particulate matter (PM10) in an urban area: an analysis of the seasonal trend. *Science of the Total Environment* 2002; **292**: 265–275.
28 Voutsa, D. and Samara, C. Labile and bioaccessible fractions of heavy metals in the airborne particulate matter from urban and industrial areas. *Atmospheric Environment* 2002; **36**: 3583–3590.
29 Dreher, K.L, Jaskot, R.H., Lehmann, J.R., Richards, J.H., McGee, J.K., Ghio, A.J. and Costa, D.L. Soluble transition metals mediate residual oil fly ash induce acute lung injury. *Journal of Toxicology and Environmental Health* 1997; **50**: 285–305.
30 Mészáros, E. *Fundamentals of Atmospheric Aerosol Chemistry*. Académiai Kiadó, Budapest: 1999.
31 Espinoza, A.J.F., Rodrigez, M.T., de la Rosa, F.J.B. and Sanchez, J.C.J. Size distribution of metals in urban aerosols in Seville (Spain). *Atmospheric Environment* 2000; **35**: 2595–2601.
32 Allen, A.G., Nemitz, E., Shi, J.P., Harrison, R.M. and Greenwood, J.C. Size distribution of trace metals in atmospheric aerosols in the United Kingdom. *Atmospheric Environment* 2001; **35**: 4581–4591.
33 Singh, M., Jaques, P.A. and Sioutas, C. Size distribution and diurnal characteristics of particle-bound metals in source and receptor sites of the Los Angeles basin. *Atmospheric Environment* 2002; **36**: 1675–1689.
34 Salma, I., Maenhaut, W. and Záray, G. Comparative study of elemental mass size distributions in urban atmospheric aerosol. *Journal of Aerosol Science* 2002; **33**: 339–356.
35 Maenhaut, W., Scwartz, J., Cafmeyer, J. and Annegarn, H.J. Study of elemental mass size distributions at Skukuza, South Africa, during the SAFARI 2000 dry season campaign. *Nuclear Instruments and Methods* 2002; **189B**: 254–258.
36 Talukdar, S.S. and Swihart, M.T. An improved data inversion program for obtaining aerosol size distribution from scanning differential mobility analyzer data. *Aerosol Science and Technology* 2003; **37**: 145–161.
37 Pakkanen, T.A., Kerminen, V.-M., Loukkola, K., Hillamo, R.E., Aarnio, P., Koskentalo, T. and Maenhaut, W. Size distribution of mass and chemical components in street-level and rooftop PM1 particles in Helsinki. *Atmospheric Environment* 2003; **37**: 1673–1690.
38 Salma, I., Ocskay, R., Raes, N. and Maenhaut, W. Fine structure of mass size distributions in an urban environment. *Atmospheric Environment* 2005; **39**: 5363–5374.
39 Maenhaut, W., Hillamo, R., Mäkelä, T., Jaffrezo, J.L., Bergin, M.H. and Davidson, C.I. A new cascade impactor for aerosol sampling with subsequent PIXE analysis. *Nuclear Instruments and Methods* 1996; **109/110B**: 482–487.
40 Maenhaut, W., Raes, N., Chi, X., Cafmeyer, J., Wang, W. and Salma, I. Chemical composition and mass closure for fine and coarse aerosols at kerbside in Budapest, Hungary, in Spring 2002. *X-Ray Spectrometry* 2005; **34**: 290–296.
41 Heal, M.R., Hibbs, L.R., Agius, R.M. and Beverland, I.J. Total and water-soluble trace metal content of urban background PM_{10}, $PM_{2.5}$ and black smoke in Edinburgh, UK. *Atmospheric Environment* 2005; **39**: 1417–1430.
42 Chester, R., Lin, F.J. and Murphy, K.J.T. A three stage sequential leaching scheme for the characterization of the sources and environmental mobility of trace metals in the marine aerosol. *Environmental Technology Letters* 1989; **10**: 887–900.
43 Hlavay, J., Polyák, K., Bódog, A. and Molnár, Á. Mészáros, distribution of trace elements in filter-collected aerosol samples. *Fresenius Journal of Analytical Chemistry* 1996; **354**: 227–232.

44 Hlavay, J., Polyák, K., Molnár, Á. and Mészáros, E. Determination of the distribution of elements as a function of particle size in aerosol samples by sequential leaching. *Analysts* 1998; **123**: 859–863.
45 Bikkes, M., Polyák, K. and Hlavay, J. Fractionation of elements by particle size and chemical bonding from aerosols followed by ETAAS determination. *Journal of Analytical Atomic Spectrometry* 2001; **16**: 74–81.
46 Fernández, A.J., Ternero, M., Barragán, F.J. and Jiménez, J.C. An approach to characterization of sources of urban airborne particles through heavy metal speciation. *Chemosphere – Global Change Science* 2000; **2**: 123–136.
47 Fernández Espinoza, A.J., Ternero Rodriguez, M., Barragán de la Rosa, F.J. and Jimenez Sánchez, J.C. A chemical speciation of trace metals for fine urban particles. *Atmospheric Environment* 2002; **36**: 773–780.
48 Moore, R.M., Milley, J.E. and Chatt, A. The potential for biological mobilization of trace elements from aeolian dust in the ocean and its importance in the case of iron. *Oceanologica Acta* 1984; **7**: 221–228.
49 Zhuang, G.S., Duce, R.A. and Kester, D.R. The solubility of atmospheric iron in surface seawater of the open ocean. *Journal of Geophysical Research* 1990; **95**: 16207–16216.
50 Behra, P. and Sigg, L. Evidence for redox cycling of iron in atmospheric water droplets. *Nature* 1990; **344**: 419–421.
51 Kotronarou, A. and Sigg, L. SO_2 oxidation in atmospheric water: role of Fe(II) and effect of ligands. *Environmental Science & Technology* 1993; **27**: 2725–2735.
52 Erel, Y., Pehkonen, S.O. and Hoffmann, M.R. Redox chemistry of iron in fog and stratus clouds. *Journal of Geophysical Research* 1993; **98**: 18423–18434.
53 Deguillaume, L., Leriche, M., Desboeufs, K., Mailhot, G., George, C. and Chaumerliac, N. Transition metals in atmospheric liquid phases: sources, reactivity and sensitive parameters. *Chemical Reviews* 2005; **105**: 3388–3431.
54 Kieber, R.J., Peake, B., Willey, J.D. and Jacobs, B. Iron speciation and hydrogen peroxide concentrations in New Zealand rainwater. *Atmospheric Environment* 2001; **35**: 6041–6048.
55 Willey, J.D., Kieber, R.J., Williams, K.H., Crozier, J.S., Skrabal, S.A. and Avery, G.B., Jr. Temporal variability of iron speciation in coastal rainwater. *Journal of Atmospheric Chemistry* 2000; **37**: 185–205.
56 Kieber, R.J., Williams, K., Willey, J.D., Skrabal, S. and Avery, G.B., Jr. Iron speciation in coastal rainwater: concentration and deposition to seawater. *Marine Chemistry* 2001; **73**: 83–95.
57 Deutsch, F., Hoffmann, P. and Ortner, H.M. Field experimental investigations on the Fe(II)- and Fe(III)-content in cloudwater samples. *Journal of Atmospheric Chemistry* 2001; **40**: 87–105.
58 Ervers, B., George, C., Williams, J.E., Buxton, G.V., Salmon, G.A., Bydder, M., Wilkinson, F., Dentener, F., Mirabel, P., Wolke, R. and Herrmann, H. CAPRAM2.4 (MODAC Mechanism): an extended and condensed tropospheric aqueous phase mechanism and its application. *Journal of Geophysical Research* 2003; **108**: 4426, doi: 10.1029/2002JD002202.
59 Deguillaume, L., Leriche, M., Monod, A. and Chaumerliac, N. The role of transition metal ions on HO_x radicals in clouds: a numerical evaluation of its impact on multiphase chemistry. *Atmospheric Chemistry and Physics* 2004; **4**: 95–110.
60 Dedik, A.N. and Hoffmann, P. Chemical characterization of iron in atmospheric aerosols. *Atmospheric Environment* 1992; **26A**: 2545–2548.
61 Zhuang, G., Yi, Z., Duce, R.A. and Brown, P.R. Link between iron and sulphur cycles suggested by detection of Fe(II) in remote marine aerosols. *Nature* 1992; **355**: 537–539.
62 Zhu, X.R., Prospero, J.M. and Millero, F.J. Diel variability of soluble Fe(II) and soluble total fe in North African dust in the trade winds at Barbados. *Journal of Geophysical Research* 1997; **102**: 21297–21305.
63 Spokes, L.J. and Jickells, T.D. Factors controlling the solubility of aerosol trace metals in the atmosphere and on mixing into seawater. *Aquatic Geochemistry* 1996; **1**: 355–374.

64 Weber, S., Hoffmann, P., Ensling, J., Dedik, A.N., Weinbruch, S., Miehe, G., Gütlich, P. and Ortner, H.M. Characterization of iron compounds from urban and rural aerosol sources. *Journal of Aerosol Science* 2000; **31**: 987–997.

65 Jickells, T.D. and Spokes, L.J. Atmospheric iron inputs to the oceans. In: D.R. Turner and K.A. Hunter, eds. *The Biogeochemistry of Iron in Seawater*. John Wiley, New York: 2001; pp. 85–121.

66 Kieber, R.J., Skrabal, S.A., Smith, C. and Willey, J.D. Redox speciation of copper in rainwater: temporal variability and atmospheric deposition. *Environmental Science & Technology* 2004; **38**: 3587–3594.

67 Kieber, R.J., Hardison, D.R., Whitehead, R.F. and Willey, J.D. Fotochemical production of Fe(II) in rainwater. *Environmental Science & Technology* 2003; **37**: 4610–4616.

68 Deutsch, F., Hoffmann, P. and Ortner, H.M. Analytical characterization of manganese in rainwater and snow samples. *Fresenius Journal of Analytical Chemistry* 1997; **357**: 105–111.

69 Pehkonen, S.O., Erel, Y., Siefert, R.L., Klewicki, K., Hoffmann, M. and Morgan, J.J. The dynamic chemistry of transition metals in the troposphere. *Israel Journal of Earth Sciences* 1994; **43**: 279–295.

70 Herrmann, H., Ervers, B., Jacobi, H.-W., Wolke, R., Nowacki, P. and Zellner, R. CAPRAM2.3: a chemical aqueous phase radical mechanism for tropospheric chemistry. *Journal of Atmospheric Chemistry* 2000; **36**: 231–284.

71 Stumm, W. and Morgan, J.J. *Aquatic Chemistry*, 3rd edn. John Wiley, New York: 1996.

72 Bolt, G.H. and Bruggenwert, M.G.M. *Soil Chemistry: A. Basic Elements*. Elsevier, Amsterdam: 1978.

73 Faust, B.C. and Hoigné, J. Photolysis of Fe(III)-hydroxy complexes as source of OH radicals in clouds, fog and rain. *Atmospheric Environment* 1990; **24**: 79–89.

74 Chebbi, A. and Carlier, P. Carboxylic acids in the troposphere, occurrence, sources, and sinks: a review. *Atmospheric Environment* 1996; **30**: 4233–4249.

75 Kawamura, K., Steiberg, S., Ng, L. and Kaplan, I.R. Wet deposition of low molecular weight mono- and di-carboxylic acids, aldehydes and inorganic species in Los Angeles. *Atmospheric Environment* 2001; **35**: 3917–3926.

76 Zuo, Y. and Hoigné, J. Formation of hydrogen peroxide and depletion of oxalic acid in atmospheric water by photolysis of iron(III)-oxalato complexes.*Environmental Science & Technology* 1992; **26**: 1014–1022.

77 Sedlak, D.L. and Hoigné, J. The role of copper and oxalate in the redox cycling of iron in atmospheric waters. *Atmospheric Environment* 1993; **27A**: 2173–2185.

78 Hoigné, J., Zuo, Y. and Nowell, L. Photochemical reactions in atmospheric waters; role of dissolved iron species. In: G. Heltz, R. Zepp and D. Crosby, eds. *Aquatic and Surface Photochemistry*. Lewis Publishers, Inc., Michigan: 1994; pp. 75–84.

79 Faust, B.C. and Zepp, R.G. Photochemistry of aqueous iron(III)-polycarboxylate complexes: roles in the chemistry of atmospheric and surface waters. *Environmental Science & Technology* 1993; **27**: 2517–2522.

80 Spokes, L.J., Campos, M.L.A.M. and Jickells, T.D. The role of organic matter in controlling copper speciation in precipitation. *Atmospheric Environment* 1996; **30**: 3959–3966.

81 Nimmo, M., Fones, G.R. and Chester, R. Atmospheric deposition: a potential source of trace metal organic complexing ligands to the marine environment. *Croatica Chemica Acta* 1998; **71**: 323–341.

82 Nimmo, M. and Fones, G.R. The potential pool of Co, Ni, Cu, Pb and Cd organic complexing ligands in coastal and urban rain waters. *Atmospheric Environment* 1997; **31**: 693–702.

83 Graedel, T., Mandich, M.L. and Weschler, C.J. Kinetic studies of atmospheric droplet chemistry. 2. Homogeneous transition metal chemistry in raindrops. *Journal of Geophysical Research* 1986; **91**: 5205–5221.

84 Leriche, M., Voisin, D., Chaumerliac, N., Monod, A. and Aumont, B. A model tropospheric multiphase chemistry: application to one cloudy event during the CIME experiment. *Atmospheric Environment* 2000; **34**: 5015–5036.

85 Warneck, P. Chemistry and photochemistry in atmospheric water drops. *Berichte der Bunsen-Gesellschaft für Physikalische Chemie* 1992; **96**: 454–460.

86 Zuo, Y. and Hoigné, J. Evidence for photochemical formation of H_2O_2 and oxidation of SO_2 in authentic fog water. *Science* 1993; **260**: 71–73.
87 Zuo, Y. and Hoigné, J. Photochemical decomposition of oxalic, glyoxalic and pyruvic acid catalysed by iron in atmospheric waters. *Atmospheric Environment* 1994; **28**: 1231–1239.
88 Franch, M.I., Ayllón J.A., Peral, J. and Domènech, X. Fe(III) photocatalyzed degradation of low chain carboxylic acids implications of the iron salt. *Applied Catalysis B: Environment* 2004; **50**: 89–99.
89 Warneck, P. and Ziajka, J. Reaction mechanism of the iron(III)-catalyzed autoxidation of bisulfite in aqueous solution: steady state description for benzene as radical scavenger. *Berichte der Bunsen-Gesellschaft für Physikalische Chemie* 1995; **99**: 59–65.
90 Warneck, P., Mirabel, P., Salmon, G.A., van Eldik, R., Vinckier, C., Wannowius, K.J. and Zetzsch, C. Transport and chemical transformation of pollutants in the troposphere. In: P. Warneck, ed. *Heterogeneous and Liquid Phase Processes*, Vol. 2. Springer-Verlag, Berlin: 1996; pp. 7–74.
91 Novič, M., Grgić, I., Poje, M. and Hudnik, V. Iron-catalyzed oxidation of S(IV) species by oxygen in aqueous solution: influence of pH on the redox cycling of iron. *Atmospheric Environment* 1996; **30**: 4191–4196.
92 Grgić, I. and Berčič, G. A simple kinetic model for autoxidation of S(IV) oxides catalyzed by iron and/or manganese ions. *Journal of Atmospheric Chemistry* 2001; **39**: 155–170.
93 Grgić, I., Dovžan, A., Berčič, G. and Hudnik, V. The effect of atmospheric organic compounds on the Fe-catalyzed S(IV) autoxidation in aqueous solution. *Journal of Atmospheric Chemistry* 1998; **29**: 315–337.
94 Fábián, I. and Gordon, G. Kinetics and mechanism of the complex formation of the chlorite ion and iron(III) in aqueous solution. *Inorganic Chemistry* 1991; **30**: 3994–3999.
95 Ziajka, J., Beer, F. and Warneck, P. Iron-catalysed oxidation of bisulphite aqueous solution: evidence for a free radical chain mechanism. *Atmospheric Environment* 1994; **28**: 2549–2552.
96 Grgić, I., Poznič, M. and Bizjak, M. S(IV) autoxidation in atmospheric liquid water: the role of Fe(II) and the effect of oxalate. *Journal of Atmospheric Chemistry* 1999; **33**: 89–102.
97 Coichev, N. and van Eldik, R. A mechanistic study of the sulfite-induced autoxidation of Mn(II) in aqueous azide medium. *Inorganica Chimica Acta* 1991; **185**:69–73.
98 Grgić, I., Hudnik, V., Bizjak M. and Levec, J. Aqueous S(IV) oxidation – I. Catalytic effects of some metal ions. *Atmospheric Environment* 1991; **25A**: 1591–1597.
99 Connick, R.E. and Zhang, Y.X. Kinetics and mechanism of the oxidation of HSO_3^- by O_2: the manganese(II) catalysed reaction. *Inorganic Chemistry* 1996; **35**: 4613–4621.
100 Fronaeus, S., Berglund, J. and Elding, L.I. Iron-manganese redox processes and synergism in the mechanism for manganese-catalyzed autoxidation of hydrogen sulfite. *Inorganic Chemistry* 1998; **37**: 4939–4944.
101 Ermakov, A.N. and Purmal, A.P. Catalysis of HSO_3^-/SO_3^{2-} oxidation by manganese ions. *Kinetics and Catalysis* 2002; **43**: 273–284.
102 Podkrajšek, B., Berčič, G., Turšič, J. and Grgić, I. Aqueous oxidation of sulfur(IV) catalyzed by manganese(II): a generalized simple kinetic model. *Journal of Atmospheric Chemistry* 2004; **47**: 287–303.
103 Pasiuk-Bronikowska, W., Bronikowski, T. and Ulejczyk, M. Solubilization of organics in water coupled with sulphite autoxidation. *Water Research* 1997; **31**: 1767–1775.
104 Martin, L.R. and Good, T.W. Catalyzed oxidation of sulfur dioxide in solution: the iron-manganese synergism. *Atmospheric Environment* 1991; **25**: 2395–2399.
105 Grgić, I., Hudnik, V., Bizjak M. and Levec, J. Aqueous S(IV) oxidation – II. Synergistic effects of some metal ions. *Atmospheric Environment* 1992; **26**: 571–577.
106 Coichev, N., Bal Reddy, K. and van Eldik, R. The synergistic effect of manganese(II) in the sulfite-induced autoxidation of metal ions and complexes in aqueous solution. *Atmospheric Environment* 1992; **26**: 2295–2300.

107 Pasiuk-Bronikowska, W., Bronikowski, T. and Ulejczyk, M. Inhibition of the S(IV) autoxidation in the atmosphere by secondary terpenic compounds. *Journal of Atmospheric Chemistry* 2003; **44**: 97–111.

108 Wolf, A., Deutsch, F., Hoffmann, P. and Ortner, H.M. The influence of oxalate on Fe-catalyzed S(IV) oxidation by oxygen in aqueous solution. *Journal of Atmospheric Chemistry* 2000; **37**: 125–135.

109 Altwicker, E.R. Oxidation and oxidation/inhibition of sulfur dioxide. In: J.R. Pfafflin and E.N. Ziegler, eds. *Advances in Environmental Science and Engineering*, Vol. 3. Gordon and Breach Science Publishers, New York: 1980; pp. 80–91.

110 Ziajka, J. and Pasiuk-Bronikowska, W. Autoxidation of sulphur dioxide in the presence of alcohols under conditions related to the tropospheric aqueous phase. *Atmospheric Environment* 2003; **37**: 3913–3922.

111 Martin, L.R., Hill, M.W., Hill, A.F. and Good, T.W. The iron catalyzed oxidation of sulfur(IV) in aqueous solution: differing effects of organics at high and low pH. *Journal of Geophysical Research* 1991; **96**: 3085–3097.

112 Zou, Y. and Zhang, J. Effects of oxalate on Fe-catalyzed photooxidation of dissolved sulfur dioxide in atmospheric water. *Atmospheric Environment* 2005; **39**: 27–37.

113 Podkrajšek, B., Grgić, I., Turšič, J. and Berčič, G. Influence of atmospheric carboxylic acids on catalytic oxidation of sulfur(IV). *Journal of Atmospheric Chemistry* 2006; **54**: 103–120.

Chapter 6
Thermodynamics of Aqueous Systems

David Topping

6.1 Introduction

It is necessary that the underpinning processes governing the impacts that aerosol particles have on the surrounding environment are understood to permit development of appropriate prognostic tools. Crucially, this includes our ability to interrelate the chemical and physical characteristics of atmospheric aerosols, through both analytical and theoretical methods, the two ideally working in tandem. Particulate matter in the atmosphere may exist as a completely solid, completely liquid or a combined solid/liquid system and predicting the phase state in relation to the associated chemical composition requires the use of thermodynamic models. However, this broad classification masks a hierarchy of models and techniques with greatly varying complexity and range of applicability. This can be dictated, for example, by the need for computational efficiency or simply a lack of experimental data from which important model parameters can be derived. To cover all of the branching areas of thermodynamics in detail is far beyond the scope of this chapter. Indeed, for an in-depth discussion of generalised thermodynamics, both macroscopic and microscopic, the reader is referred to sources in the literature [1, 2]. Rather, the following body of text is aimed at providing a relatively robust overview of ideas pertinent to analysing the phase characteristics of atmospheric aerosol whilst focusing specifically on the aerosol–water vapour interactions, thus aqueous systems, and equilibrium thermodynamics.

6.2 Equilibrium

Calculations of composition and phase state commonly assume thermodynamic equilibrium, thus neglecting any kinetic effects. This is valid so long as processes that lead towards equilibrium are quicker than those that lead away. There are instances where this assumption may introduce errors. For example, various studies have focused on the kinetic effects of organic compounds on the water content of mixed inorganic/organic particles, both through the possible presence of surface films but also through chemical effects [3–9]. Inference of influences from surfactant material in atmospheric particles has been carried out for some time now (see [10] and references therein). It is important however to realise that thermodynamic equilibrium is the end point of all kinetic processes and so calculating the extent of a reaction often requires a knowledge of the thermodynamics involved [11]. Even when equilibrium applies, calculating the phase state can be a highly complex problem. Indeed, this may be conceptually obvious since the atmospheric aerosol can be a complex mixture of both inorganic and organic compounds. Currently, there is a dearth of information from which important model parameters can be derived regarding the interaction between these two separate fractions. This lack of data applies to both bulk- and surface-phase phenomena. Conversely, theoretically it can be difficult to build a 'complete' model, combining both the inorganic and organic fractions [12], or indeed the bulk and surface phases [13]. Both issues are addressed to varying extents in the following chapter. Firstly, however, it is now pertinent to provide a brief overview of basic thermodynamic principles before focus is given to more specific areas. For systems at constant temperature and pressure, the equilibrium composition of an aerosol system corresponds to that which minimises the Gibbs free energy (G) (see [1]). Hence the criterion for spontaneity is

$$\mathrm{d}G < 0 \tag{6.1}$$

Or at equilibrium,

$$\mathrm{d}G = 0 \tag{6.2}$$

Hence, one has to find the set of variables that satisfies the condition given above, and the problem becomes that of a mathematical minimisation procedure. A differential relationship can be derived [1] to give

$$\mathrm{d}G = -S\,\mathrm{d}T + V\,\mathrm{d}p + \sum_{i=1}^{k} \mu_i\,\mathrm{d}n_i \tag{6.3}$$

where V is the volume and the variable μ_i, known as the chemical potential, is defined as

$$\mu_i = \left(\frac{\partial G}{\partial n_i}\right)_{T,p,nj} \tag{6.4}$$

The chemical potential has an important feature similar to pressure and temperature. Whilst a temperature difference between two bodies determines the tendency of heat to pass from one body to the other, and pressure determines the difference of bodily movement, a difference in the chemical potential can be viewed as the cause for chemical reaction or for mass transfer from one phase to another [14]. Finally, at constant temperature and pressure, we can write

$$\mathrm{d}G = \sum_{i=1}^{k} \mu_i\,\mathrm{d}n_i \tag{6.5}$$

This important result shows that for a single reaction, the sum of the chemical potentials is zero, as defined by the stoichiometry of the reaction. If there are multiple reactions taking place, then at equilibrium, the above condition applies to all of these. For example, for solid–liquid equilibrium take the dissolution of sodium chloride in water into its respective ions:

$$\mu_{\mathrm{NH_4NO_3(s)}} = \mu_{\mathrm{NH_4^+(aq)}} + \mu_{\mathrm{NO_3^-(aq)}} \tag{6.6}$$

Similarly, for gas–liquid equilibrium consider the partitioning of HNO_3 between the gas and aqueous phase.

$$\mu_{\mathrm{HNO_3(g)}} = \mu_{\mathrm{H^+(aq)}} + \mu_{\mathrm{NO_3^-(aq)}} \tag{6.7}$$

We can now simply formulate the problem quantitatively as

$$G = \sum_i \mu_i n_i \tag{6.8}$$

The Gibbs free energy (G) is a function of temperature, pressure and the number of moles of each component being considered. Applying Equation (6.8) to individual reactions and correlating with experimental data one can derive equilibrium constants, as defined by Equation (6.9):

$$K = \prod_{i=1}^{k} a_i^{wi} \tag{6.9}$$

where K is the equilibrium constant for the reaction, a_i the 'activity' of specie 'i', discussed shortly, and wi the stoichiometric coefficient of specie 'i' in the reaction. If an equilibrium module were to rely on the use of equilibrium constants (see Jacobson [15]) then the appropriate experimental data would have to be available for species involved in the appropriate reactions. For organic or mixed inorganic/organic systems, this may not be available. Indeed, often more direct approaches to solving Equation (6.2) are required [16–20]. Another equation, known as the Gibbs–Duhem relationship, is given as [1]

$$\sum_i n_i \, \mathrm{d}\mu_i = 0 \tag{6.10}$$

It shows us that the chemical potentials of a mixture cannot change independently. This is another way to highlight the fact that component properties in mixtures interact and influence each other.

6.2.1 The chemical potential

In order to utilise the above formulations we obviously need a way of calculating the chemical potential 'μ_i'. To do this, the general formulation given below is used:

$$\mu_i = \mu_i^0 + \mathrm{RT}\,\ln(a_i) \tag{6.11}$$

where μ_i^0 is the standard-state chemical potential (function of temperature only (kJ mol^{-1})), R the ideal gas constant (0.0083145 kJ mol^{-1}), T the temperature (K) and 'a_i' represents the *activity* of the substance. The chemical potential has a value that is independent of the choice of concentration scale and standard state. The standard-state chemical potential, the activity and the activity coefficient, however, all have values that do depend on the choice of concentration scale and standard state. This is discussed in more detail in Section (6.2.2). The temperature dependence of the standard-state chemical potential is calculated from Equation (6.12) [16].

$$\mu_i^0(T) = T\left[\frac{\Delta G_\mathrm{f}^0}{T_0} + \Delta H\left(\frac{1}{T} - \frac{1}{T_0}\right) + c_\mathrm{p}\left(\ln\frac{T_0}{T} - \frac{T_0}{T} + 1\right)\right] \tag{6.12}$$

Table 6.1 Thermodynamic parameters and reported deliquescence points for common inorganic compounds of atmospheric significance

Species	ΔG_f° (kJ mol^{-1})	ΔH_f° (kJ mol^{-1})	C_p° (J mol^{-1})	Adj. $\Delta G_f^{\circ a}$ (kJ mol^{-1})	DRH (%)
NH_4NO_3(s)	−183.87	−365.56	139.3	−184.4116215	62
$(NH_4)_2SO_4$(s)	−901.67	−1180.85	187.49	−902.9684592	80
NH_4HSO_4(s)	−823	−1026.96	127.5	−822.7440844	40
$(NH_4)_3H(SO_4)_2$(s)	−1730	−2207	315	−1731.801894	69
Na_2SO_4(s)	−1270.16	−1387.08	128.2	−1268.1116	84
$NaHSO_4$(s)	−992.8	−1125.5	85	−1005.081095	52
NaCl(s)	−384.14	−411.15	50.5	−384.0132158	75
$NaNO_3$(s)	−367	−461.75	92.88	−367.0149944	74
NH_4Cl(s)	−202.87	−314.43	84.1	−205.1639246	80
NH_3(g)	−16.45	−46.11	35.06		
HNO_3(g)	−74.72	−135.06	53.35		
HCl(g)	−95.3	−92.31	29.13		
H_2O(g)	−228.57	−241.82	33.58		
H_2O(l)	−237.13	−285.83	75.29	−237.16619243942[b]	
NH_4(aq)	−79.31	−132.51	79.9		
Na^+(aq)	−261.91	−240.12	46.4		
NO_3^-(aq)	−111.25	−207.36	−86.6		
SO_4^{2-}(aq)	−744.53	−909.27	−293		
Cl^-(aq)	−131.23	−167.16	−136.4		
HSO_4^-(aq)	−755.91	−887.34	−84		

All values are reproduced from Ansari and Pandis [16] unless otherwise noted.

[a] All values were derived by using the chemical potentials of the ionic components at saturation as discussed by Topping *et al.* [19, 21]

[b] Value derived for the minimal difference of water activity from the known RH as discussed by Topping *et al.* [19, 21]

where ΔG_f^0 is the free energy of formation (kJ mol^{-1}), ΔH the standard heat of formation (kJ mol^{-1}) and c_p the heat capacity (kJ mol^{-1}), all at T_0, a reference temperature (298.15 K). Such data for common inorganic compounds are tabulated in Table 6.1. Unfortunately, pure component data for relevant organic compounds are not readily available in the literature. This includes recorded energies of formation for both solid and aqueous phases. In this instance, one has to rely on predictive schemes. However, as noted by Marrero and Gani [22], many of these are unable to distinguish between isomers due to an oversimplification of the molecular structure. Similarly, properties of large complex and polyfunctional substances of interest in biochemical and environmental studies cannot be determined with high levels of accuracy with these methods due to the relatively small data set from which functional group parameters can be derived. Indeed, this seems to be an issue with other group contribution revisions, regarding modelling of activity coefficients. In the light of this, Marrero and Gani [22] developed a third-order group contribution method for predicting pure component properties, such as the Gibbs free energy of formation. The first level (used solely in first-order groups) is intended to deal with simple and monofunctional compounds. The second involves groups that permit a better description of proximity effects and differentiation amongst isomers, whilst the third provides more structural information about molecular fragments of compounds that are inadequately described by the first two.

6.2.2 Activity

The 'ideal' solution is defined as one for which the chemical potential of every component is related to its mole fraction by the relation

$$\mu_i = \mu_i^0(T) + \text{RT} \ln(X_i) \quad (6.13)$$

Commonly attributed properties to ideal solutions are all deducible from this relation [1]. For example, for a single ideal gas one would replace the mole fraction by the partial pressure and the ideal gas law could be easily defined (e.g. [14]). In dealing with real solutions, the method usually adopted is to find the magnitude of a pure number, which, when multiplied by the mole fraction of the particular species, makes applicable a relationship of the above form. Thus an activity coefficient is defined, such that:

$$\mu_i = \mu_i^0(T) + \text{RT} \ln(f_i X_i) \quad (6.14)$$

Here μ_i^0 is a function of temperature and pressure only, but f_i may be a function of these variables together with the mole fractions of all substances in solution [1]. When relationship (6.14) is applied to a single ideal gas and it is assumed that that this is in equilibrium with an ideal solution, then one can easily derive Raoult's law (see [14]):

$$P_i = P_i^0 X_i \quad (6.15)$$

where P_i is the partial pressure of component 'i' over a solution, P_i^0 the saturation pressure of component 'i' and X_i the mole fraction of component 'i' in solution. This simple formulation shows that the equilibrium pressure of a component over a solution is directly proportional to its mole fraction in solution. However, empirical studies of vapour pressure curves of binary liquid mixtures show that this rule actually represents limiting behaviour and intermediate compositions actually deviate positively or negatively from Raoult's law. Another feature is that for the other component, the equilibrium pressure has a finite slope as its mole fraction in solution approaches zero. This is represented by Henry's law:

$$P_i = H_i X_i \quad (6.16)$$

where H_i is the Henry's law constant for component 'i'. Deviations from ideality are measured in relation to excess functions (e.g. excess Gibbs free energy). For example, the excess Gibbs free energy is the difference between that observed and that predicted for an ideal solution. It is important to consider how the different definitions of standard state and activity relate to each other and the above discussions. Because all solvents obey Raoult's law increasingly closely as the concentration of solute approaches zero, the activity of the solvent 'A' approaches the mole fraction as X_A tends to 1 [23]. Thus, a convenient way of expressing this convergence is to define the activity coefficient by $f_A \rightarrow 1$ as $X_A \rightarrow 1$. The chemical potential of the solvent is then

$$\mu_A = \mu_A^0 + \text{RT} \ln(f_A X_A) \quad (6.17)$$

where the standard state is taken as the pure liquid. When defining the activity coefficient for solutes, two definitions of standard states are often used. Normally, the reference state infinitely dilute with respect to the solvent, for example water, is adopted for solute specie 'B' [24]. Indeed, the problem with defining activity coefficients and standard states for solutes is that they approach ideal-dilute (Henry's law) behaviour as X_B tends to 0, not as X_B tends to 1 (corresponding to the pure solute) [23]. Following Clegg and Pitzer [24], the symbol f_B^* is used and hence f_B^* tends to 1 as X_B tends to 0. Thus, the standard-state chemical potential must change accordingly. The activity coefficients based on the two standard states are interconvertible as follows [24, 25]:

$$f_i^* = \frac{f_i}{f_i^\infty} \tag{6.18}$$

where f_i^∞ is the activity coefficient at infinite dilution. This can normally be found by setting the mole fraction of specie 'i' to zero in the appropriate models. The selection of standard state is entirely arbitrary, and compositions are often expressed as molalities in place of mole fractions. In this case the chemical potential of the solute has its standard value when the molality of 'i' is equal to 1. The mole fraction activity coefficient f_B^* (infinite dilution reference state on a mole fraction scale) is related to the corresponding molal value γ_B by Equation (6.19) [24]:

$$f_B^* = \gamma_B \left(1 + (M_A/1000) \sum_i m_i\right) \tag{6.19}$$

where M_A is the molar mass of the solvent (grams), m_i the molal concentration of solute species 'i' and the summation is over all solute species. For systems in which water is the only solvent, this reduces to

$$f_B x_{water} = \gamma_B \tag{6.20}$$

Thus, the new expression for μ_A would become

$$\mu_A = \mu_A^\theta + RT\ \ln(\gamma_A m_A) \tag{6.21}$$

where μ_A^θ is the new standard state according to the new scale. Thus, non-ideal solutions approach ideality when the concentrations of all components but one approach zero. In that case, the solutes satisfy Henry's law where the solvent satisfies Raoult's law. An aqueous aerosol solution droplet approaches ideality as it becomes increasingly dilute; that is, the composition approaches that of pure water. It is also important to recognise that the activity coefficient is a parameterisation of very complicated thermodynamic behaviour and it is the modelling of such behaviour that provides a great challenge in attempting to represent real behaviour in the aerosol particle. For solids, the activity is simply 1, and for gases it is defined as the partial pressure divided by 1 atmosphere. Indeed, the chemical potentials for solids and gases can be defined, respectively, as

$$\mu_A(s) = \mu_A^\beta \tag{6.22}$$

$$\mu_A(g) = \mu_A^\alpha + RT\,\mathrm{Ln}(a_A) = \mu_A^\alpha + RT\,\mathrm{Ln}\left(\frac{p_A}{p^\alpha}\right) \tag{6.23}$$

where μ_A^β is the standard-state chemical potential of the solid A, p_A the partial pressure of the gas A, p^α the standard pressure (1 atm) and μ_A^α the standard-state chemical potential of the gas. The thermodynamics of a solution mixture however depends on the intermolecular forces that operate between molecules that are dependent on the nature of the solvent(s) and solutes. It is much more difficult to consider a mixture than a pure substance, as one must consider interactions between dissimilar molecules [2]. In describing the thermodynamics of a system, Gmehling [26] identifies two main approaches, namely the excess Gibbs energy models and the equation of state approach. For more information on these two different approaches the reader is referred to the literature (e.g. [26, 27] and references therein). We are interested with activity coefficient models and thus the g^E approach. These models first appeared in the pioneering work of Margules in 1890 and Van Laar in 1910 where they identified the idea of liquid-phase non-ideality [27]. There are numerous theoretical arguments, often providing quantitative representations, reviewed in some textbooks [2, 28]. In Sections 6.6–6.8 inorganic, organic and mixed inorganic/organic activity coefficient models are discussed with direct focus given to aqueous systems.

6.3 Mixed solvent systems – a brief outline

As stated by Clegg and Seinfeld [29], water and solute activities and gas–liquid–solid equilibrium of atmospheric aerosols containing dissolved organic compounds are currently modelled in a relatively simple way. This is partly because of lack of information regarding the properties and reactions of the compounds that are present, and partly because of the lack of a suitable general method of representing the thermodynamic properties of aqueous solutions containing both ions and uncharged solutes to very high concentration [29]. Whilst the concentration of organic compounds may be such that water can be considered as the only solvent, there may be cases where organic concentrations are similar to or larger than water itself whereby the solution may have to be considered as a mixed solvent system. In extractive distillation processes involving mixed solvents, salts are often added to enhance separation. This can cause the formation of immiscible liquid phases in equilibrium with one another. Depending on the relative amounts of solvents and salt(s) initially present, the two phases can have very different compositions, with each typically being dominated by a single solvent [12]. Organics exhibiting low aqueous solubility are often treated as insoluble solids in various model calculations. The behaviour in the atmosphere however is such that organic matter is perhaps expected to exist predominantly in the form of a liquid or amorphous phase [30, 31]. Hence, the organic matter could constitute an absorptive medium in which gaseous species and inorganic ions may dissolve [30], even at low relative humidity (RH). This phase separation would of course depend on the type and quantity of species present. Hybrid models in which the particle is assumed to exist as two phases have been proposed [32]. For example, hydrophobic substances would not partition into an aqueous phase, rather into a water-insoluble organic fraction consisting of a wider range of hydrophobic substances. Studies neglecting and employing such assumptions have modelled gas/particulate partitioning of gas-phase organic oxidation products with varying success [32–36]. For example, in the theory of Pankow [37], partitioning between the gas phase and the condensed organic phase is described with an equilibrium partitioning coefficient:

$$K_p = \frac{7.501 \times 10^{-9}\,RT}{MW_{om}\xi p_L^0} \tag{6.24}$$

where R is the ideal gas constant (8.314 J K^{-1} mol^{-1}), T the temperature (K), MW_{om} the mean molecular weight of the absorbing particulate organic matter (g mol^{-1}), ξ the activity coefficient of the species in the condensed organic phase and p_L^0 its (probably subcooled) liquid vapour pressure (Torr). However, the value of ξ is often assumed to be unity for oxidation products, as it is assumed that the aerosol particles comprise mixtures of similar types of molecules [38].

Recently, Johnson *et al.* [39] found they had to scale all partitioning coefficients by 500 to match with ambient data. This was rationalised in terms of association reactions occurring in the organic aerosol phase, but also put down to a possible contribution from aqueous uptake of water-soluble organics not accounted for in the model. The assumption that the aerosol particle consists of these two phases is essentially a result of the fact that thermodynamic treatments of homogeneous organic/water/electrolyte systems are not yet sufficiently well developed [40]. As stated by Clegg *et al.* [12], the general problem of predicting activities of all components in organic/ion/water solutions for all possible compositions would require a complex model and an unrealistically large amount of data for compounds present and their mixtures. For the remaining part of the discussion, focus is given to aqueous systems. Indeed, water vapour continually interacts with the atmospheric aerosol particles and it is this interaction that plays a large role in determining its influence on the environment in which it resides.

6.4 Aqueous systems

An aerosol particle may consist of an aqueous phase at high RH, one or more solid phases at low RH and both aqueous and solid phases at intermediate RH [41]. This phase variability is highly dependent on the aerosol composition, ambient conditions and in some cases the size of the particle. At low RH, where equilibrium thermodynamics can dictate that the aerosol of an assumed composition remains in an anhydrous state, adsorption may occur at the surface. Adsorption behaviour is a complex process, leading to the formation of monolayers and often multilayers (for more information, see e.g. [42–44]). As an anhydrous particle is subject to increasing RH, there may be a transition whereby it starts to take up water through absorption, that is the deliquescence point. Predicting the deliquescence point of an aerosol particle usually involves identifying the RH at which the dissolved salt droplet is at a lower Gibbs free energy than the solid particle. However, this neglects the influence from the energetics of the surface–air interface, the possible influences discussed briefly in Section 6.10. After deliquescence, water uptake is determined by condensation and maintenance of thermodynamic equilibrium. As the RH increases further, a maximum supersaturation of water vapour is attained where water molecules continue to condense on the droplet in an unstable state, unlike the regime determined by lower supersaturations. At this point, the aerosol particle is said to have been 'activated' and it starts growing rapidly. Köhler [45] was the first to theoretically describe this behaviour, the resulting environment curve analysed in Section 6.10.2. Water may exist in 'bound' or 'unbound' states, although 'unbound' water represents that mostly associated with aerosol particles [14]. Bound water, or water of hydration, is water chemically bound to the salt. 'Unbound' water and its concentration are governed by thermodynamic equilibrium. This may occur due to changes in ambient RH or to condensation and evaporation of solutes.

6.4.1 Water content in the regime subsaturated with respect to water vapour

The varying nature of primary emissions and processes to which particles are subject to are manifest in measurements of aerosol hygroscopicity in the real atmosphere, of which there have been a number of field studies [46–52, 53]. Such field studies consist of selecting monodisperse ambient aerosol distributions, which are 'dried', and then subject to varying humidity in the hygroscopic tandem differential mobility analyser (H-TDMA) instrument. For details regarding the operation of the H-TDMA the reader is referred to Cubison [8]. Briefly, this instrument samples an aerosol particulate population of a given dry size, subjects it to a varying RH and observes the associated change in size. Results are given as the growth factor, or the ratio of the 'wet' diameter usually at an RH of 90% to a 'dry' reference state, usually taken as 10% RH. The most common result concerned with these studies is the existence of certain hygroscopicity 'modes'. When ambient aerosols are subject to a given RH, at a given dry size using the H-TDMA, the hygroscopic characteristics are usually subdivided into four categories [54]. These are usually classified as (1) hydrophobic particles that do not grow at any RH, (2) 'less' hygroscopic particles that have growth factors smaller than pure salts but do show some response to increasing RH, (3) more hygroscopic particles that have growth factors similar to common inorganic salts (e.g. $(NH_4)_2SO_4$) and those with higher growth factors commonly attributed to sea-salt particles (NaCl) [55]. However, depending on the environment, the actual GF corresponding to each of these classes (particularly the 'less' and 'more' hygroscopic fractions) can show large variation [56]. Cocker *et al.* [57] found that the use of only four classifications was insufficient to characterise 50- and 150-nm particles sampled from urban air. Table 6.2 shows a compilation of observed growth factors from a variety of location types. Kanakidou *et al.* [56] provide a more extensive list though split into only two fractions (<1.2 and >1.2), Table 6.2 serving only as a rough compilation of varying hygroscopicity. Similarly, McFiggans *et al.* [53] recently provided an in-depth review of such measurements. The relative dominance of these modes is indicative of the mixing state of the particle, which when coupled with appropriate chemical and meteorological information may be used to highlight the relative importance of primary and secondary production sources for a given location. For example, the hygroscopicity of both inorganic and organic primary emissions can have characteristic signatures. However, as the aerosol resides in the atmosphere it is subject to processes that drive the particle towards a state of internal mixing. Thus for an aged aerosol population, one may report only one hygroscopic mode. Kanakidou *et al.* [56] analysed an extensive set of ambient hygroscopicity data and drew two broad observations:

- *Under traffic-dominated conditions (street canyon and urban background), a significant fraction of the 'less' hygroscopic particles is quasi-permanently observed. This fraction decreases with increasing particle size for a given urban site.* This is most likely linked with the chemical nature of particles produced in urban areas at the smaller sizes.
- *The time fraction of occurrence of these less hygroscopic particles decreases progressively when moving from the urban background to continental and remote marine locations. Simultaneously, a population of more hygroscopic particles is always present.*

For example, measurements in Minneapolis [62], Los Angeles [46] and the Po Valley, Italy [50], showed that particles of a given size separated into more and less hygroscopic fractions when humidified to 90% indicating an external mixture [63]. In urban areas often the TDMA cannot detect any significant growth for the 'less' hygroscopic particles, whereas only hygroscopic properties were found in Japan and in the Arctic [63]. Relating growth factors, that is the directly observable measure of water affinity, to the associated composition of course relies on theoretical frameworks.

In the subsaturated humid environment, the water content is invariably calculated using thermodynamic equilibrium alone. Considering the thermodynamic equilibrium of water in the particulate and vapour phase, neglecting curvature (see Section 6.10), it can be shown that

Table 6.2 Measured hygroscopic growth factors of aerosol particles with varying dry sizes in a range of different environments

Dry diameter (nm)	Location (aerosol type)	Less hygroscopic mode growth factor	More hygroscopic mode growth factor	Fraction in more hygroscopic mode (%)	RH (%)
20	Bresso, Italy (urban)[f]	<1. 02 (0.02)	None	—	90
50	San Pietro Capofiume, Italy (industrial)[a]	1.1 (0.05)	1.49 (0.12)	37 (21)	85
50	Los Angeles, USA (urban)[b]	1.03 (0.03)	1.14 (0.05)	—	90
50	Bresso, Italy (urban)[f]	1.03 (0.04)	1.21 (0.09)	49 (22)	90
50	Laboratory (fresh gasoline soot)[c]	1.0 (0.01)	None	—	90
50	Laboratory (fresh gasoline soot)[a]	0.99 (0.01)	None	—	90
100	San Pietro Capofiume, Italy (industrial)[a]	1.11 (0.06)	1.46 (0.12)	41 (24)	85
100	Munich, Germany (urban)[e]	1.0 (0.02)	1.3 (0.04)	28 (11)	85
100	Bresso, Italy (urban)[f]	1.02 (0.05)	1.25 (0.1)	62 (19)	90
100	Laboratory (fresh gasoline soot)[c]	0.99 (0.01)	None	—	90
100	Laboratory (fresh gasoline soot)[d]	0.99 (0.01)	None	—	90
200	San Pietro Capofiume, Italy (industrial)[a]	1.09 (0.09)	1.44 (0.07)	53 (13)	85
200	Los Angeles, USA (urban)[c]	1.02 (0.02)	1.23 (0.08)	—	90
200	Bresso, Italy (urban)[f]	1.02 (0.02)	1.28 (0.06)	72 (15)	90

[a] Svenningson *et al.* [50].
[b] Zhang *et al.* [47].
[c] Weingartner *et al.* [60].
[d] Weingartner *et al.* [61].
[e] Ferron *et al.* [145].
[f] Baltensperger *et al.* [109].

in the subsaturated humid regime the water activity is equivalent to the ambient saturation ratio of water vapour [14]:

$$a_w = \frac{RH}{100\%} \tag{6.25}$$

Measurements of water activity versus concentration can be provided by analysis of bulk solutions or suspended particles. Whilst bulk measurements can be very accurate, they have concentration limitations that can be circumvented by using suspended particles. Suspended particles are less likely to crystallise, thus permitting the measurement of supersaturated solutions that are of particular use in atmospheric applications. Techniques that utilise these principles include the electrodynamic balance (EDB) (e.g. [64]) and the hygroscopic tandem differential mobility analyser (H-TDMA) (e.g. [8]). Each has its own advantage. With the H-TDMA, one focuses on submicron aerosol particles, thus being able to probe the influence of surface tension. With the EDB, however, one can suspend the particle for hours, thus largely bypassing the influence of equilibration time effects. Here however the particles are normally supermicron, the two techniques therefore deemed essentially complimentary for our purposes. It should be noted however that for model validation purposes, on a pure equilibrium basis, the EDB is more suited since there are fewer issues such as assumed state of the dry particle and equilibration time effects can be circumvented. Whilst the H-TDMA can be employed in the laboratory and the field, the EDB is purely a laboratory instrument.

There are numerous options available for calculating the water content associated with atmospheric aerosols in the subsaturated humid regime ranging from simple parameterisations, which are numerically relatively easy to implement, to more complex thermodynamic treatments. These simple parameterisations usually refer to the use of mixing rules that employ, in one way or another, data from binary systems to describe mixtures. Indeed, for a basic calculation of water content in an aqueous aerosol particle, neglecting any solid formation or gas–aerosol equilibration, one can employ a variation of the well-known Zdanovski–Robinson–Stokes (ZSR) mixing rule [65]. Based on the simple assumption that the water content of mixtures can be described by a summation of the contributions from individual components, it has been shown that the ZSR relationship has been able to predict the water content of supersaturated aqueous aerosols [66–68]. The generalised form presented in most publications (e.g. [69]) is given by (6.26).

$$W = \sum \frac{M_i}{m_{io}(a_w)} \tag{6.26}$$

where W is the total water content, M_i the mass number of moles of the solute present and $m_{io}(a_w)$ the molality at a given RH (a_w = RH/100%). Data only for single-solute solutions are required to predict the water content of a mixture of solutes. Data for water activity as a function of binary electrolyte molality are available in the literature [69–73]. However, as noted by Chan *et al.* [74], there are two major challenges to mixing rules: (1) performance at high concentrations and (2) performance in the presence of strong electrolytes.

In principle, the ZSR relationship can be applied to any number of systems, including organic mixtures, as only water activity data from single electrolytes is required. However, it is important to note that the traditional ZSR relationship is based on the assumption of semi-ideality in that interactions between solutes are neglected. Thus, the water content associated with a complex mixture is assumed to be the sum of binary solutions that act independently of each other.

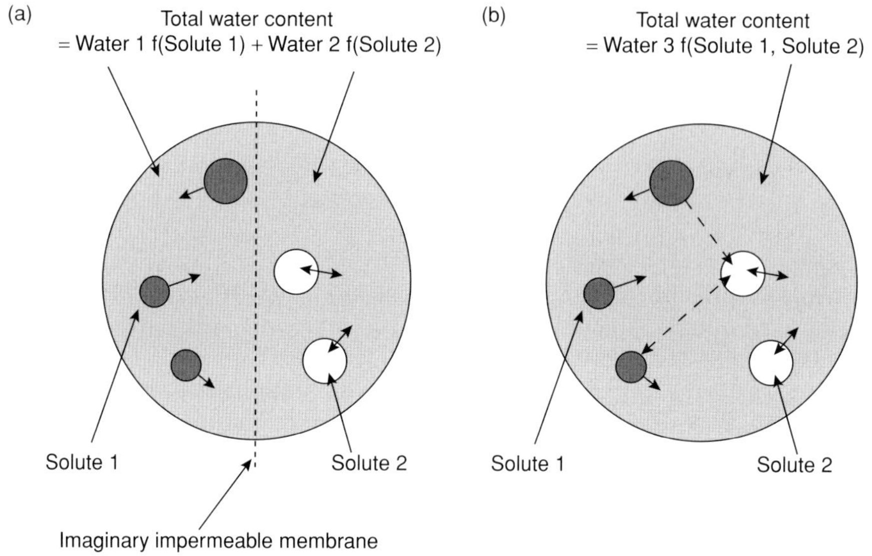

Figure 6.1 (a) Simple representation of the theory behind semi-ideality for a ternary aqueous system. Here the total water content is calculated by adding the individual contents from binary mixtures where the interactions between different solutes are neglected. (b) No ideality assumed. Here the interactions between different solutes are also taken into account. The solid arrows in the circle represent interactions between solute–solvent, whereas the dashed arrows represent interactions between the respective solutes.

This is shown schematically in Figure 6.1. Similarly, if said binary data are not available then this precludes its use in predictive models for systems where experimental data are going to be lacking. This is likely to be the case for analysing the water-soluble organic fraction of atmospheric aerosols using suggested representative species given from studies such as that provided by Decesari *et al.* [75]. Indeed, though water activity data for inorganic compounds are relatively abundant (e.g. [73]), corresponding data for water-soluble organic compounds are scarce. This is partly because these compounds were previously considered unimportant for the hygroscopic and cloud-forming properties of aerosol particles and also because organic compounds are more difficult to measure than inorganic compounds, especially the low-molecular-weight organics due to their high volatility [56].

Attempts have been made to improve the traditional ZSR relationship and account in some manner for deviations from semi-ideality (e.g. [73, 76]). A different approach involves attempting to explicitly deal with the mutual influence of solutes. Such methods are discussed for both inorganic and organic aqueous solutions in the following sections.

6.5 General aqueous thermodynamic equilibrium models for atmospheric aerosols

The fundamental relationships defined earlier generally present two ways of searching for the chemical equilibrium of the aerosol. The first utilises equilibrium constants defined for a set of reactions, which the user assumes to occur in the aerosol and the gaseous phase, housed in varying levels of iteration. The second centres on a search for the global Gibbs free energy minimum. For an equation-based approach, provided that the equilibrium relations represent the actual behaviour of the aerosol system, the solution should correspond to the global minimum. However, as discussed by Ansari and Pandis [16], if not properly accounted for, the model will return a local minimum. The authors also note that conditions suggesting multistage behaviour and quantifying the degree of deliquescence depression are not known a priori. As such, modifying an equation-based algorithm to predict multistage behaviour and deliquescence depression is difficult [16]. With the direct minimisation approach one can treat water as a free variable, thus allowing the use of potentially more accurate activity coefficient models for calculating the water activity of the solution, rather than using simpler mixing rule parameterisations. Most of the attempts at modelling the equilibrium behaviour of aqueous aerosols in the subsaturated humid regime have been driven by the antiquated view that only inorganic species contribute to the observed water affinity. Alongside the use of simple empirical parameterisations, there have been numerous inorganic aerosol aqueous thermodynamic modules developed. These include EQUIL [77], KEQUIL [78], MARS [79], SEQUILIB [80], SCAPE and SCAPE 2 [81–84], MARS-A [85], EQUISOLV and EQUISOLV II [15, 69, 86], AIM and AIM 2 [17, 41, 87, 88], ISSORROPIA [89, 90], GFEMN [16], ADDEM [19, 21], EQSAM [91–93] and UHAERO [20]. For a comparative review of a selection of the above models the reader is referred to Zhang *et al.* [94]. It is also interesting that the majority of the current available modules neglect the influence of curvature. However it is this aspect of the aerosol that largely separates it from more general equilibrium problems and directly influences hygroscopicity in both the sub- and supersaturated humid regime, as discussed in Section 6.10. Including the above schemes only the models of Bassett and Seinfeld [78], Ming and Russell [18] and Topping *et al.* [19, 21] treat the influence of curvature using different approaches.

6.6 Inorganic systems

6.6.1 Activity coefficients

Upon deliquescence of a crystalline inorganic aerosol the electrolyte solute dissociates into its respective ions. Calculating the activity coefficients of individual ions must be done theoretically, as ions cannot be separated in solution. For single-electrolyte solutions such as NaCl or NH_4NO_3, both ions have identical activity coefficients and empirical data may be used relatively easily. However, for unsymmetrical electrolytes and mixtures of various ions, one must choose empirical mixing rules or semi-empirical models to calculate the respective activities. Perhaps the simplest method of estimating activity coefficients (in single-electrolyte systems) is the use of empirical data. Jacobson [69] provides data for a number of species of atmospheric importance, usually in the form of polynomial fits:

$$\ln \gamma_{12b}^{o} = B_0 + B_1 m_{12}^{1/2} + B_2 m_{12} + B_3 m_{12}^{3/2} + \cdots \tag{6.27}$$

where γ_{12b}^{o} is the mean binary activity coefficient at 298.15 K and B_0, B_1, etc. are the given polynomial coefficients. The mean binary activity coefficient is related to the single-ion activity coefficients by

$$\gamma = (\gamma_+^{V+} \gamma_-^{V-})^{1/(V_+ + V_-)} \tag{6.28}$$

where γ_+ and γ_- are the activity coefficients of a cation and anion, respectively, $V+$ and $V-$ the corresponding stoichiometric coefficients. However, the range of use is restricted by the maximum molality at which the data were obtained. Empirical mixing rules are available, such as Bromley's (1973) or Kusik and Meisner's rule (1978). (The equations are not presented here, but can be found in the literature, e.g. [95].) These schemes have been used extensively in atmospheric models, and shown to accurately predict activity coefficients over a range of molalities. The disadvantage of such techniques is the lack of being able to treat the system on a more fundamental level. That is to say, the techniques of Pitzer [96] and Clegg *et al.* [24, 97], hereafter referred to as the CSB approach following Clegg and Seinfeld [29], use binary and ternary interaction parameters derived from numerous systems, whereas the above techniques rely solely on data from binary mixtures. Carslaw *et al.* [98] also notes that since these rules do not account directly for specific interactions between solution components, in a study regarding HNO_3–H_2SO_4–H_2O mixtures, the modelled activity of pure aqueous H_2SO_4 had to be adjusted in order to reproduce the measured vapour pressures. Following this, the use of these rules often results in the employment of similar mixing rules used to calculate the water content. Metzger *et al.* [91] formulated empirical equations, designed specifically to improve computational performance, for the activity coefficients of solutes as a function of RH. Although noted to be applicable to mixed solutions, and comparable to other mixing rules (see [93]), Metzger *et al.* [93] note how these equations break down for a sulphuric acid solution. Similarly, this formulation is based on semi-ideality in that interactions between different solutes are not treated explicitly. As stated by Zaveri *et al.* [99], as this scheme simply assumes that the multicomponent activity coefficient of each electrolyte is equal to its mean binary activity coefficient, which may be reasonable for dilute solutions (RH $>$ 80%), it generally tends to break down for concentrated solutions. More recently, Zaveri *et al.* [99] developed the multicomponent Taylor expansion method, based on the series expansion rule used for dilute alloy solutions. Here the authors extend this to aqueous electrolyte solutions at any concentration, using the ZSR technique for calculating the water content and the CSB model for deriving most of the important model parameters at 298.15 K.

6.6.1.1 Theoretical methods

The long-range nature of Coulombic interactions between ions means it is likely to be responsible for departures from ideality in ionic solutions and to dominate contributions to non-ideality [23]. This is the basis of the Debye–Huckel theory of ionic solutions. However, it generally cannot be used to describe the properties of real solutions above 0.1 mol kg^{-1} and the effect of complex ion-specific short-range forces that dominate in solutions at high concentrations must be described empirically [100]. One can treat non-Coulombic interactions of ions in a series expansion of individual ion concentrations analogous to the virial expansion used for studying imperfect gases [28]. In doing this one is trying to model the excess Gibbs free energy of the solution, which leads onto the development of semi-empirical models. They are semi-empirical as short-range interaction forces are dealt with through empirically determined parameters for various systems, whereas the long-range contributions remain theoretical in nature. In solution mixtures of strong electrolytes where association of ions is assumed not to occur, an ion-interaction rather than ion-pairing model is the appropriate choice [101]. The same authors note how Pizter's technique (1975) was the most powerful of these. The theory behind the Pitzer's equations, along with required interaction parameters, is given in the literature [96, 102–104]. The activity coefficient equations are based on a model for the excess Gibbs free energy that was represented by a virial expression of terms in concentration. These are similar in form to series used to describe the properties of non-ideal gases [101]. The excess energy comprises a Debye–Huckel term that represents the long-range interactions, which dominate in very dilute solutions. The short-range interactions are accounted for by virial coefficients for binary and ternary interactions. One of the successes of Pitzer's technique was the dependence of the binary interaction parameters on the ionic strength of solution, which was not accounted for in earlier techniques. Pitzer's equations have been used in existing atmospheric models to varying extents (MARS-A (binary activity coefficients) [79], SEQUILIB (binary activity coefficients) [105], SCAPE 2 (optional multicomponent model) [82, 83] and EQUISOLV II (binary activity coefficients)), along with numerous theoretical models of natural waters ([106] and references therein).

Unfortunately, as noted by Pitzer and Simonson [107], any virial expansion becomes unsatisfactory at sufficiently high concentration. Indeed, the presence of very high aqueous concentrations in aerosols at low RH prevents the application of potentially more accurate techniques such as Pitzer's method (relative to empirical mixing rules) for calculating activities. Many of the concentration limitations of the molality-based model of Pitzer are overcome by using the CSB approach [24, 97], which has the advantage over mixing rules that interactions between ions can be treated explicitly as in Pitzer's method. A comparison between the two different models is given in Figure 6.2, in which the RH corresponding to the molalities of interest are given. Pitzer and Simonson [107] developed a new model comprising an ideal term, an extended Debye–Huckel term and a Margules expansion up to the four-suffix level. The Gibbs free energy is modelled by both a short-range and long-range contribution. The short-range term is accounted for by a three-suffix Margules expansion, whereas the long-range contribution is an extended Debye–Huckel model analogous to Pitzer's model but on a mole fraction scale. A correction to the molality scale for the solute ions (see Section 6.2.2) is implicit in the model formulations. Clegg *et al.* [24, 97, 108] showed that the model was able to represent osmotic and activity coefficients various systems to high concentration.

6.6.2 Bulk inorganic contributions to particulate water content in the regime subsaturated with respect to water vapour

The hygroscopic characteristics of common inorganic components are relatively well characterised. For example, Figure 6.3 shows the uptake characteristics of five salts and one acid of atmospheric significance calculated using AIM online [24, 97], neglecting the Kelvin effect (see Section 6.10). Those compounds exhibiting finite solubility in water exhibit the phenomena of deliquescence. Identifying the RH at which the dissolved inorganic solute droplet is at a lower Gibbs free energy than the solid particle requires energies of formation for both the ionic components and the solid particle. Such information can be found in the literature and is tabulated in Table 6.1 for common inorganic species. However, both Ansari and Pandis [16] and Topping *et al.* [19] found that in order to match deliquescence behaviour to that reported experimentally, the energies of formation for each solid had to be adjusted slightly. The deliquescence relative humidities (DRHs) of various common inorganic salts are also given in Table 6.1. Similarly, Figure 6.4 gives an example of how the most stable dry state for an inorganic system varies with the relative ionic composition.

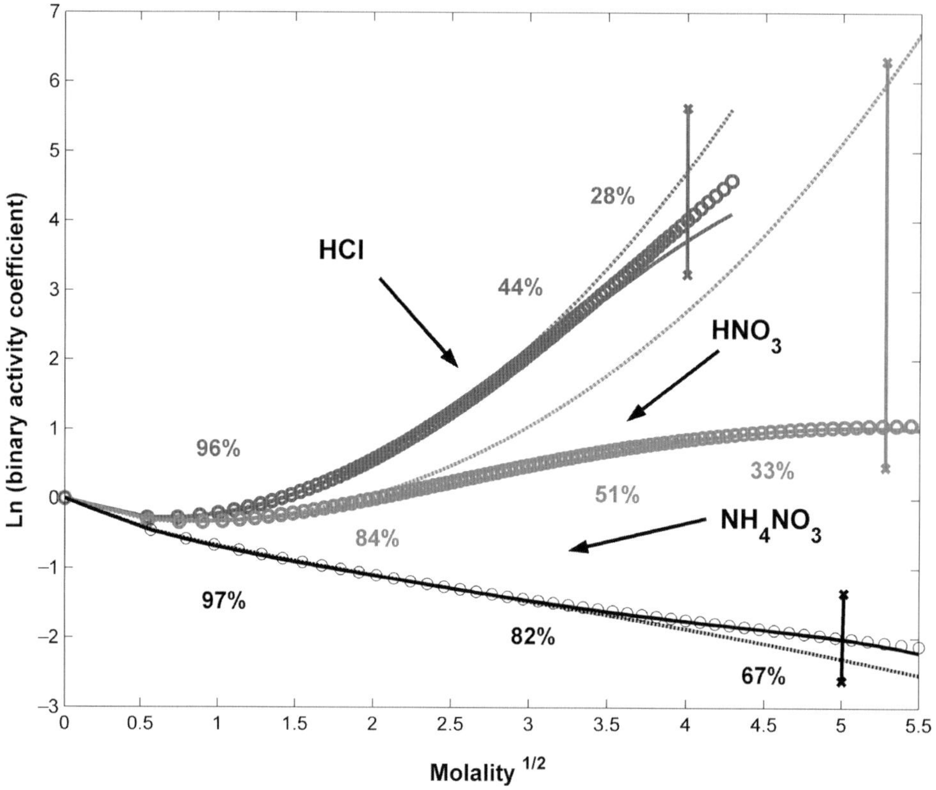

Figure 6.2 Mean binary activity coefficients calculated using Pitzer's method (dashed lines, see text), the CSB model (open circles, see text) compared with the empirical parameterisations published by Jacobson [69] (solid lines) for HCl, HNO_3 and NH_4NO_3. The vertical lines represent the maximum concentration at which the experimental data were taken. Also shown are the RH values corresponding to the molality for the given binary system.

The majority of models available do not take double or hydrated salts into account. There is contrasting evidence in the literature for their importance. Experimental evidence has shown that inorganic aerosols may exist in dry states that are composed of complex or hydrated salts. Stelson and Seinfeld [110] showed that the double salt $2NH_4NO_3.(NH_4)_2SO_4$ is present over a small RH range (63–66%) for the mixed $NH_4NO_3/(NH_4)_2SO_4$ particle. Similarly, Harrison and Sturges [111] provide evidence of the same salt. Hydrated salts are those salts to which water is chemically bound. Pilinis *et al.* [112] note that Na_2SO_4 is the only inorganic salt present in atmospheric aerosols that has bound water associated with it. Tang [73] also discusses the behaviour of sodium decahydrate ($Na_2SO_4.10H_2O$ – its stable form). It is noted that under ambient conditions, a supersaturated solution droplet at room temperature rarely crystallises to form the decahydrate. Instead, it crystallises into an anhydrous particle (no water) and obtains a lower DRH as a result. Also mixed salt particles containing Na_2SO_4 are not observed to form hydrates [73]. However, thermodynamic calculations show it may be the most stable form under certain conditions. Wexler and Clegg [11], in developing AIM

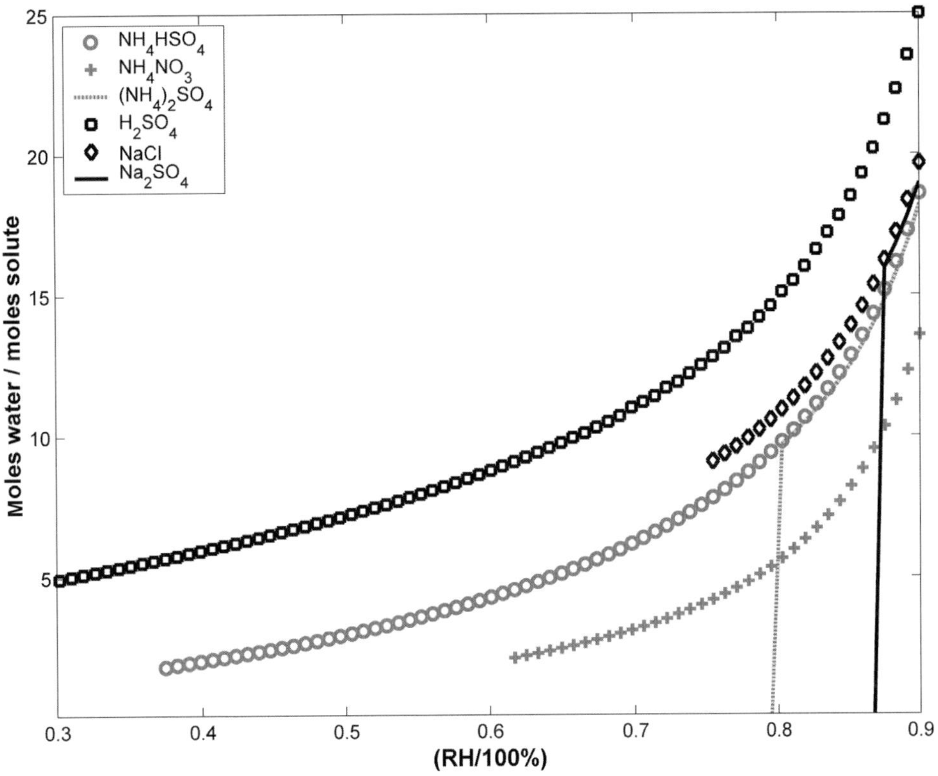

Figure 6.3 Water uptake characteristics of common inorganic aerosol components. Calculations made using AIM online (Clegg *et al.* [17, 88]).

online, analysed the importance of double salts in the atmosphere. They found that for a mixed $NH_4/SO_4/NO_3/H_2O$ system, model calculations that do not include double salt formation predict an aqueous phase to persist until lower RH. The equilibrium constants for these salts are a function of RH (water vapour pressure), and there is a dissociation pressure associated with each salt [113].

6.7 Organic systems

6.7.1 Activity coefficients

Unlike electrolytes, organic components have diverse chemical structures and possess quite different properties, both from each other and from electrolytes [18]. This necessitates activity model frameworks that are different to inorganic models. Indeed, because electrolytes and organics have different interactions in aqueous solutions, a large fraction of all activity coefficient models are applicable either for non-electrolyte solutions or for aqueous electrolyte solutions [114]. The breakthrough in the field of g^E models was achieved by Wilson (1964), who introduced the so-called local composition concept. This removed a major assumption

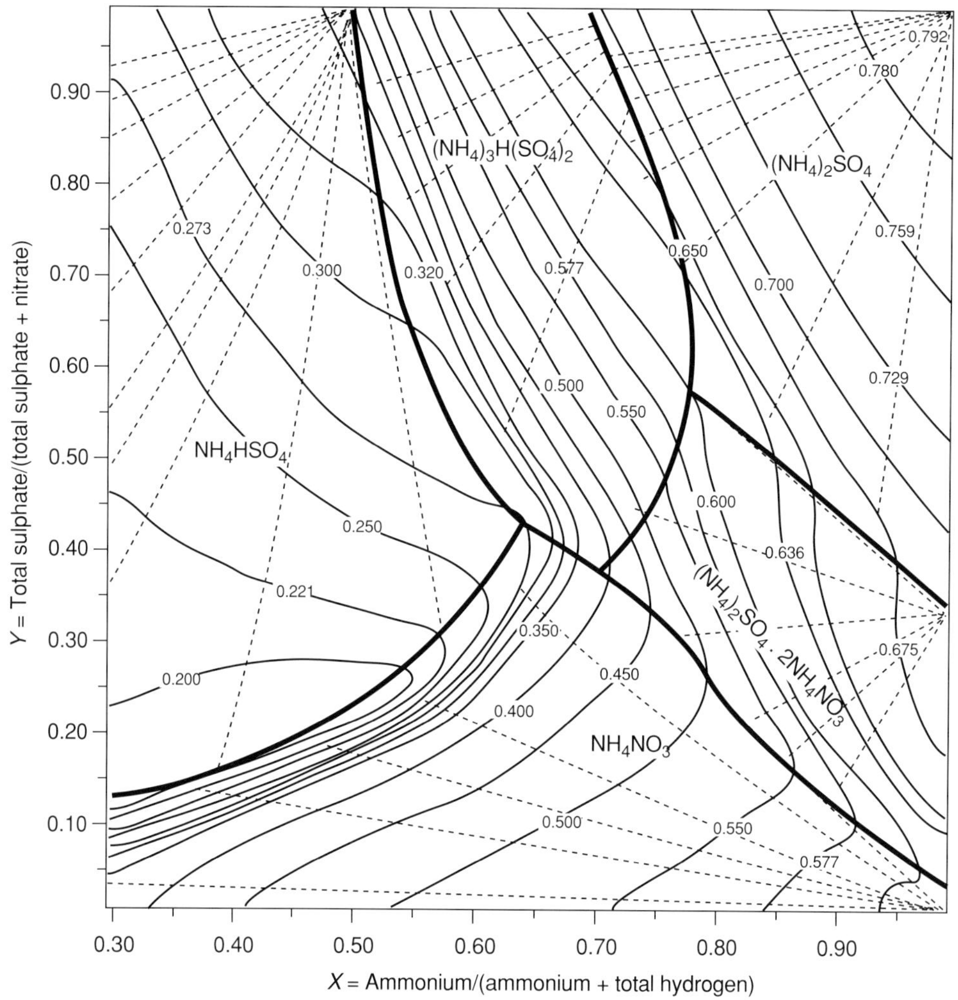

Figure 6.4 Deliquescence relative humidity contours (thin solid lines) for the aqueous solution of H^+–NH_4^+–HSO_4^-–SO_4^{2-}–NO_3^- shown together with the lines (thin dashed lines) describing aqueous-phase composition variation with relative humidity. Phase boundaries are marked with bold lines. All the solid phases are identified and are marked. Labels on the contours represent the DRH values. Total hydrogen is total moles of protons and bisulphate ions. Total sulphate is number of moles of sulphate and bisulphate ions. (Reprinted from [202]. Copyright 1995, with permission from Elsevier.)

with what is known as regular solution theory that interactions were not strongly coupled to mixture composition (see [115] for more details). With this he succeeded in predicting the behaviour of multicomponent systems using only binary data [26]. Following this, models such as the NRTL [116] and UNIQUAC [117] frameworks were developed, which are widely used in the chemical industry today [27]. The problem in the atmosphere is that predictive rather than correlative methods are required. As pointed out by Saxena and Hildemann [25], treating each compound as a molecule would not only be cumbersome but also require binary aqueous data for each and every compound. Indeed, ultimately a technique is required where experimental data are bound to be lacking given the wide range of compounds often identified in aerosols. Fortunately, the problem of characterising activity coefficients in multiple organic solutions has been previously faced in many engineering processes and we now have at our disposal group contribution methods (GCMs). UNIFAC [118] is perhaps the most widely used and successful technique for calculating activity coefficients in mixed organic systems [119] and has been used in various atmospheric applications involving organics in mixed aqueous solutions. The basic logic behind GCMs is that the number of functional groups that constitute organic compounds is much smaller than the actual number of such species. Therefore, if one assumes that a physical property of a fluid is the sum of contributions made by the molecules functional groups, we obtain a possible technique for correlating properties of a very large number of fluids in terms of a much smaller number of parameters that characterise the contributions of individual groups [118]. UNIFAC, like the CSB inorganic technique (Section 6.6.1), is based on a model for the excess Gibbs free energy. It follows the analytical solution of groups method [120] by utilising the 'solution-of-groups' concept, where activity coefficients are related to interactions between structural groups. The UNIQUAC model contains a combinatorial part, essentially due to differences in size and shape of the molecules in the mixture, and a residual part, essentially due to energy interactions [118]. In a multicomponent mixture, the UNIQUAC equation for the activity coefficient of the molecular component 'i' is

$$\ln \gamma_i = \ln \gamma_i^{\text{combinatorial}} + \ln \gamma_i^{\text{residual}} \tag{6.29}$$

Originally designed for chemical separation processes where the chemical components of interest are short-chain monofunctional compounds, UNIFAC's estimates for properties of multifunctional compounds that are common in the atmosphere have in the past been noted to be uncertain [121]. In addition, UNIFAC is a first-order technique whereby positions of functional groups within a molecule are not considered, thus precluding its use for distinction between isomers. Indeed, recently Marcolli and Peter [122] developed a modified UNIFAC parameterisation to discriminate between three types of alkyl groups for polyol/water systems depending on their position on the molecule and found a marked improvement for the predicted behaviour compared to the original model. There are, however, ways of improving UNIFAC. There have been numerous attempts to modify the original equations. These include the modified UNIFAC (Dortmund) [123–125] and modified UNIFAC (Lyngby) [126] models. There are other derivatives of the original model and the user is referred to the literature for more information (see, e.g., [127] and references therein). However, the interaction parameter matrices of the modified versions are smaller than the original version and the temperature at which the new parameters have been derived may make them unsuitable for typical atmospheric conditions. Bypassing the need for model equation adjustments, one can also readjust the interaction parameters and improve predictions considerably. Ninni *et al.* [128] improved predictions for polyol components by readjusting some parameters. More recently, Peng *et al.* [129] used an EDB to modify the functional group interaction parameters of the COOH/H_2O, OH/H_2O and OH/COOH pairs. These groups were chosen as they are most affected by hydrogen bonding [129]. Similarly, Chan *et al.* [130] revised the interaction parameters between the OH, COOH, CH_2NH_2, H_2O and the amide groups (apart from the OH/amide group pair since the compounds studied did not possess these functional groups together), complementing the work of Peng *et al.* [129]. A comparison between the original UNIFAC model with both old and new interaction parameters is made between the Dortmund model and experimental data

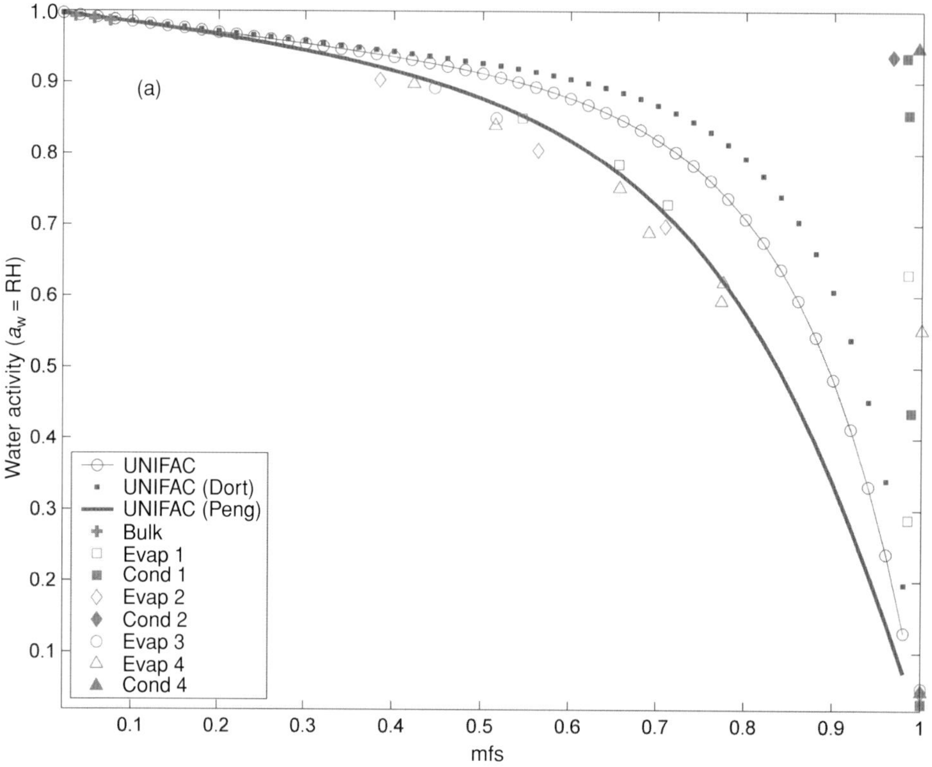

Figure 6.5 Two examples of water activity of a binary aqueous solution versus mass fraction of solute (mfs = mass solute/ mass solute + mass water) for (a) succinic acid and (b) glutaric acid. The lines designated 'UNIFAC' represent predictions from variation of the UNIFAC model (Dort – Dortmund model, Peng – original UNIFAC with modified parameters of Peng *et al.* [129], see text), whereas the single points are the experimentally determined results provided by Peng *et al.* [129], who used an EDB. The solid single data points correspond to condensational water uptake where the particle is taken from a dry state and allowed to deliquesce. The open symbols however represent evaporation measurements.

of Peng *et al.* [129] in Figure 6.5. The improvement is clear for all systems studied. It is also evident that the Dortmund model does not reproduce measured results at all well.

Recently, Choi and Chan [68] found good agreement with measured water activity data for mixed maleic/malic acid and malonic/glutaric acid systems, using these improved interaction parameters, even at low RH (ca. 20%). Data are sparse; however, Topping *et al.* [21] compared these data with the water activity data of Marcolli *et al.* [31], who measured the water uptake for five- and six-component organic mixtures. They found that the modified parameters improved predictions significantly for all systems studied, even at low RH. Similarly, however, Topping *et al.* [21] found that an analysis with typical biomass burning compounds and a proxee for humic-like material, that the use of the 'old' interaction parameters, as published by Hansen *et al.* [131], reproduces the experimental data better than the use of UNIFAC with the modified interaction parameters of Peng *et al.* [129]. Whilst this may be due to the choice of chemical

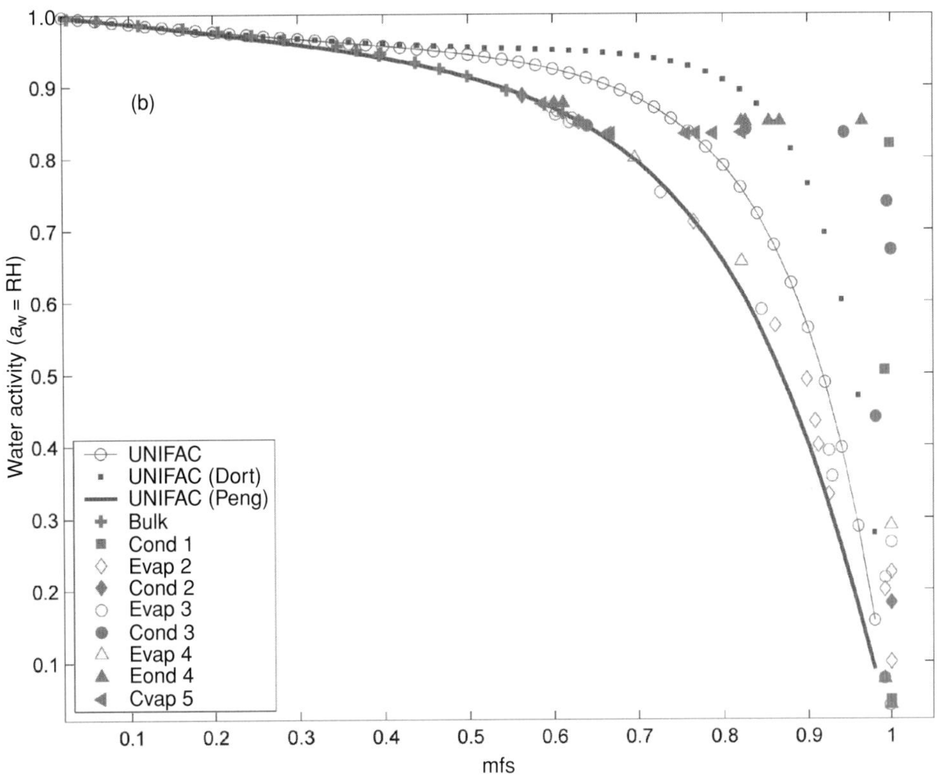

Figure 6.5 *(cont.)*

structure employed, this is an interesting result. Indeed, this sensitivity to revised databases is discussed more in Section 6.8, concerning mixed inorganic/organic aqueous systems.

6.7.2 Bulk organic contributions to particulate water content in the regime subsaturated with respect to water vapour

Previous laboratory studies in the subsaturated humid regime have mostly focused on pure and mixed low-molecular-weight organic acids, such as carboxylic acids and dicarboxylic acids, and multifunctional organic acids or their salts [31, 64, 68, 132–134]. Recent studies have attributed a large fraction of water-soluble organic carbon to macromolecular compounds including proteins and humic-like substances [135–138]. However, only a couple of experiments have investigated the effect of humic-like substances [139, 140] and proteinaceous material [141] on aerosol hygroscopicity. Saxena and Hildemann [25] analysed the uptake behaviour of a wide range of oxygenated multifunctional organic compounds using published laboratory data. They found that a number of identified species absorbed less water per unit mass of solute than NH_4HSO_4 and NH_4NO_3. Those compounds with finite solubility exhibited general deliquescence behaviour, whereas completely miscible organics absorbed water at all RHs

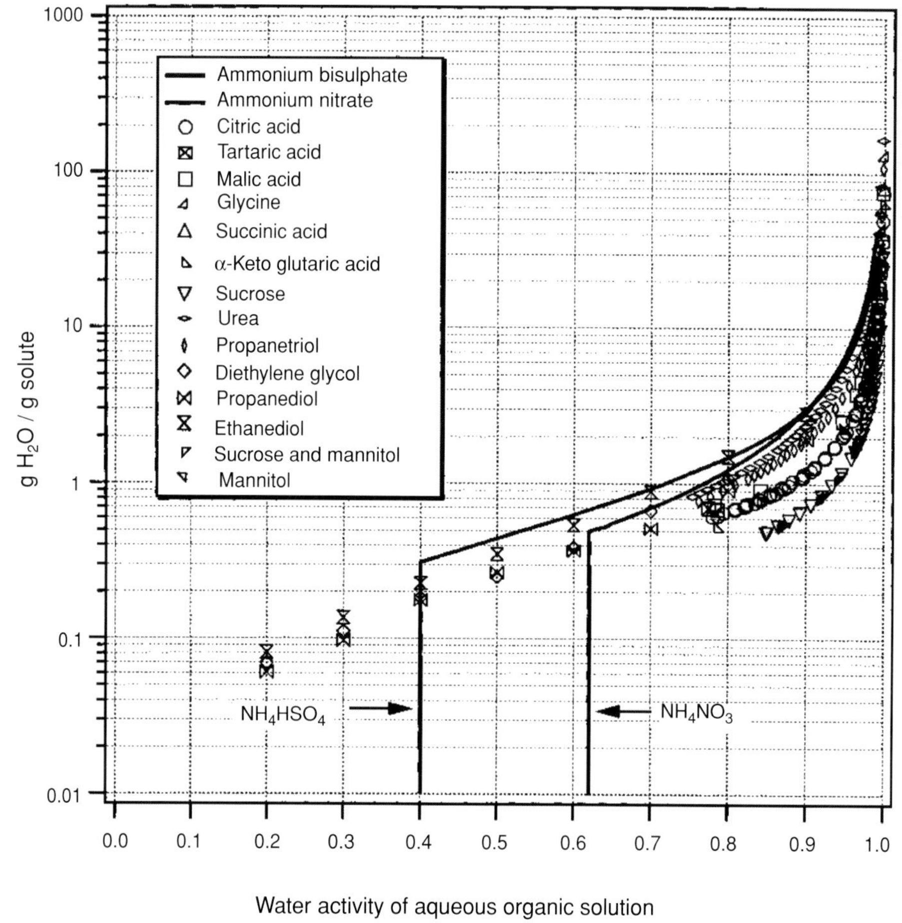

Figure 6.6 Grams of water per gram of solute for various organic solutions and two common inorganic salts versus water activity of aqueous solution. (Reprinted with permission from [25]. Copyright 1997 American Chemical Society.)

(Figure 6.6). Such results can explain the observation that ambient aerosols are often found to retain water at RH as low as 20%. However, despite the abundance of condensed-phase organic material in the atmosphere, little is known about the possible phase transitions of organic particles [142].

Table 6.3 shows reported growth factors for laboratory studies on carbonaceous aerosols for both pure species and smog chamber studies on the oxidation products of various volatile organic species of atmospheric interest. It is interesting to note that, apart from biomass fuel and a limited number of individual compounds, reported growth factors are quite low. Also included in the list are substances used as representative compounds for humic-like material often identified in aerosols. Again, measured growth factors are low for these substances. Recently, Chan *et al.* [130] measured the water uptake behaviour for compounds attributed to biomass burning

Table 6.3 Measured hygroscopic growth factors for various organic aerosol particles conducted in the laboratory using a HTDMA (see text).

Aerosol type	Conditions	GF	RH%	Reference
Nordic reference fulvic acid	Commercial acids and salts	1.13	90%	Gysel *et al.* [172]
Nordic reference humic acid		1.06	90%	
Humic acid sodium salt		1.06	90%	
Fluka chemical humic acid		1.66	90%	Brooks *et al.* [169]
Pahokee peat reference humic acid		1.12	90%	
Leonardite standard humic acid		1.09	90%	
Suwannee River fulvic acid		1.08	90%	
Polyacrylic acid (2000 MW)		1.16	90%	
Polyacrylic acid (3 000 000 MW)		1.12	90%	
Pthalic acid		1.19	90%	
Glycerol		1.36	85%	Choi and Chan [157]
Succinic acid		1.0	85%	
Malonic acid		1.4	85%	
Citric acid		1.28	85%	
Glutaric acid		1.09	85%	
Isolated organic matter	Solid-phase-extracted organic matter from water-soluble fraction in ambient samples	1.08–1.17	90%	Gysel *et al.* [143]
Diesel soot	1-kW load, no fuel additives	<1 .01	90%	Weingartner *et al.* [61]
	2-kW load, sulphur enhanced	1.015–1.025	90%	
Biomass fuel (moist forest residue)	Moving grate boiler	1.65	90%	Pagels *et al.* [200]
Secondary organic aerosol				
α-Pinene/O_3		1.04	50%	Cocker *et al.* [57]
α-Pinene/O_3		1.09	85%	
α-Pinene/O_3		1.07	85%	Virkkula *et al.* [201]
α-Pinene/NO_x		1.09	85%	
β-Pinene/NO_x		1.1	85%	
Limonene/O_3		1.09	85%	
Oxidation products of toluene	Dry NO_x free	1.22	90%	Mareli *et al.*[a]
	Dry high NO_x	1.15	90%	
	RH = 50% high NO_x	1.17	90%	
Cycloalkanes/O_3	Dry	1.105–1.159	85%	Varutbangkul *et al.* [173]
Oxidation products of sesquiterpenes	RH = 47–54%	1.01–1.04	85%	
Substituted cycloalkanes and related compounds / O_3	Dry	1.087–1.108	85%	

[a] Not published.

Reproduced from Kanakidou *et al.* [56].

(levoglucosan, mannosan and galactosan). These stereoisomers were also found to exhibit relatively low growth factors, whilst they also absorbed water at all RHs. Indeed, the authors found that these species had similar growth factors of around 1.2–1.25 at 90% RH. More extensive results are summarised by Kanakidou *et al.* [56].

6.8 Mixed inorganic/organic systems

6.8.1 Activity coefficients – coupled thermodynamics

The crux of the problem regarding calculating mixed inorganic/organic activity coefficients in aqueous systems stems from not being able to treat the complex interactions taking place in solution between the inorganic and organic fraction. Clegg *et al.* [12] highlights the two main limitations. The first is a lack of experimental data from which important interaction parameters can be derived. The second is the lack of suitable theoretical approaches. Attempts have been made to formulate a coupled thermodynamic approach for atmospheric purposes, namely the models of Ming and Russell [18] and Clegg *et al.* [12]. Most of the few organic-electrolyte models are designed for industrial purposes, and the compounds in these models are usually different from the major species found in atmospheric aerosols [114]. With regards to the theoretical frameworks employed, the CSB model, already discussed in Section 6.6.1, can include neutral solutes. However, it differs from alternative thermodynamic models such as those of Chen *et al.* [147], Chen and Evans [148] and Ananth and Ramachandran [149], as it is based on a Margules expansion for the excess Gibbs free energy. The latter models use the same Debye–Huckel term as used by Pitzer and Simonson [107] but account for all the short-range interactions by expressions based upon the 'local composition' concept [24]. Similar frameworks are discussed shortly. Firstly however, as discussed by Clegg *et al.* [12], the CSB mole fraction model has been developed to include neutral solutes as solution components, yet it is difficult to apply as the presence of neutral solutes leads to the mixture being treated as a mixed solvent system. This in turn necessitates the correction of calculated activity coefficients that requires data for pure organic components that are unlikely to be available [12].

The Pitzer-molality-based equations do not require any reference state corrections, yet the high concentrations found at low RH can lead to erroneous predictions, as discussed in Section 6.6.1. However, in the formulation of Pitzer's equations, the contributions from purely ionic components to the activity coefficient are expressed independently of each other and therefore need not be calculated using the same model [12]. Indeed, Clegg *et al.* [12] present the theory behind this model and, using appropriate data, apply it to sodium chloride/sucrose water, letovicite/2-butenedioic acid/water and sodium chloride/butanoic acid/water systems at 298.15 K. Parsons *et al.* [142] have used this model for organic compounds saturated with respect to ammonium sulphate. They found that calculated DRH points for mixed organic/$(NH_4)_2SO_4$ particles agreed with experimental values up to an organic mole fraction of 0.4.

A scheme based on the ZSR relationship, developed by Clegg *et al.* [76], is an alternative to previous approaches. As discussed by Clegg and Seinfeld [29], for this model, whilst the CSB model could be used for the inorganic components and UNIFAC the non-electrolytes, the calculations would be carried out for each component at the water activity of the mixture and not the actual solution concentration. A recent study by Clegg and Seinfeld [29] tested both of the above models against extensive water activity and DRH for multicomponent solutions of dicarboxylic acids, treating the acids as non-dissociating compounds. The authors found that the extended ZSR model produced more accurate predictions and that measured DRH and water activities of supersaturated aqueous droplets were generally quite well represented by the model but often required the use of additional parameters for the interactions between pairs of solutes.

As noted by Liu *et al.* [150, 151], perhaps the most popular practice with a framework dissimilar to the models of Clegg *et al.* [24, 97] is to combine the electrostatic theory of Debye and Huckel with modifications of well-known methods for non-electrolyte solutions. Here, Liu *et al.* [150, 151] focused on local composition models for the non-electrolyte contribution (e.g. NRTL, Wilson models). However, most of these schemes have been designed for industrial purposes. The first electrolyte model combined with the notion of using functionality, rather than individual species, to describe the organic species was designed by Kikic *et al.* [152] and used to describe vapour–liquid equilibrium for various solvent–water–salt mixtures. Here the authors modified the model presented by Sander *et al.* [153] by using a modified Debye–Huckel term to account for the long-range interactions and the UNIFAC model to account for the short-range interactions. The model presented by Achard *et al.* [154], also known as modified UNIFAC [114] for water–alcohol–salt mixtures, combined a modified form of the Debye–Huckel model presented by Pitzer [96], the modified UNIFAC model [126], and was based on the solvation concept. The model of Li *et al.* [155] and Yan *et al.* [144] differed from those mentioned above, as it accounted for charge-dipole and charge-induced dipole interactions between solvent groups and ions. Indeed, LIFAC [144] has three terms for the excess Gibbs energy: (1) a Debye–Huckel term that represents the long-range interactions, (2) a virial term that accounts for the midrange (MR) interactions caused by ion-dipole effects and (3) the UNIFAC term that accounts for the short-range interactions [144]. It is similar to LIQUAC [155] but all species are defined using functional groups in UNIFAC instead of UNIQUAC.

In the model of Ming and Russell [18] ions were treated as new groups within UNIFAC, whereas Clegg *et al.* [12] formulated a molality-based model for coupling the two separate fractions. Both utilised the same activity coefficient models for the separate fractions (organic – UNIFAC and inorganic – mole-fraction-based activity model of Clegg *et al.* [24, 97]). When ion–organic interactions are neglected, both models should give equal results, as both utilise the same activity coefficient models for treating the inorganic and organic fractions, respectively. In the model of Ming and Russell [18], interaction parameters between different groups in UNIFAC were calculated using binary inorganic/organic systems at limited concentrations for sugars, carboxylic acids, $(NH_4)_2SO_4$ and NaCl. Whilst the model reproduced measured growth factor data well for simple inorganic/organic systems where data were available, the interactions between organics and NH_4^+ and SO_4^{2-} were assumed to be equal to those of Na^+ and Cl^-.

Raatikainen and Laaksonen [114] compared some existing activity coefficient models that could be potentially suitable for modelling the hygroscopic properties of organic-electrolyte particles. The authors used predictive models described above to analyse the behaviour of dicarboxylic acids and acids including the hydroxyl groups, focusing on water activity and solubility predictions. The theoretical basis of such models are different as are the matrix of interaction parameters available for both organic/electrolyte and organic/electrolyte solutions. Indeed, out of the above frameworks only that of Ming and Russell [18] had existing parameters available for atmospherically relevant systems [114]. Even then, the parameter matrix is quite small and considers only a few organic–ion interactions. Where appropriate inorganic-organic data were available to calculate required interaction parameters that could then be used to calculate the hygroscopic properties of the mixtures, the authors found the selected predictive models to work sufficiently well. This is perhaps to be expected given the simplified systems studied. However, laboratory and theoretical studies on such simplified systems do not necessarily yield parameters that can be used over the entire composition space of organic–inorganic atmospheric constituents. This is discussed in more detail in the following section that focuses on predictions of aerosol water content for mixed inorganic/organic systems.

6.8.2 Activity coefficients – uncoupled thermodynamics

It is expected that more simplified approaches to modelling the inorganic/organic aqueous system is required. Indeed, Clegg *et al.* [12] concluded that only simple approaches to modelling ion-organic effects on thermodynamic properties of mixed inorganic/organic soluble aerosols seem justified at present, given the uncertainties that could exist even in simple binary mixed inorganic/organic systems. In the model of Clegg *et al.* [12], when there are no interactions between the inorganic and organic species, the water activity of the mixture is equal to

$$a_w = a_w^{inorg} a_w^{org} \tag{6.30}$$

where a_w^{inorg} and a_w^{org} represent the water activity of the inorganic and organic fractions, respectively, and solute concentrations are calculated using the total water content of the mixture. In the additive approach of Topping *et al.* [19, 21], the water activity of both the inorganic and organic fraction is equal to the ambient saturation ratio of water vapour:

$$a_w = a_w^{inorg} = a_w^{org} = RH/100\% \tag{6.31}$$

In this manner, the water contents from both fractions are calculated using an additive approach and the solute concentrations, and thus activity coefficients, are calculated using the water contents from the respective fractions. This was erroneously referred to by Raatikainen and Laaksonen [114] as a method that relies on data from binary systems analogous to the ZSR method. However, in the above method the thermodynamics of the two separate fractions are dealt with explicitly using two separate activity coefficient models, following the technique of Clegg *et al.* [12]. For systems with only one inorganic and one organic solute, in this instance the method is analogous to the ZSR technique.

6.8.3 Bulk mixed inorganic/organic contributions to particulate water content in the regime subsaturated with respect to water vapour

Deciphering the influence of organic species on the hygroscopic growth of aerosols is difficult. Ambient measurements are useful for purposes of closure, provided the appropriate modelling tools are available, and can be used to infer the influence of the organic fraction as a whole. Laboratory measurements on the other hand are more refined and can provide quantitative results in controlled environments, though these often include simplified mixtures using only one representative organic species. In either case the need for accurate modelling tools is clear, both from a validation and also from a predictive perspective.

With regards to the effect of organics on the inorganic fraction, results are often contradictory in the literature, which points to different overall effects depending on the organic components, particle morphology and chemistry [156]. Some studies have suggested that organics have a negative effect on the growth factor of inorganic compounds. These studies have focused primarily on organics that are slightly soluble or behave as surfactants [157]. For example, Lightstone *et al.* [158] studied the deliquescence curve of NH_4NO_3/succinic acid particles and found a pronounced reduction in the particle growth factor at deliquescence, indicating that succinic acid does not take up a significant amount of water. This may be expected and suggests that slightly soluble organic compounds that have not completely dissolved exhibit the general behaviour of insoluble solid components [158]. However, both negative and positive contributions have been observed. As stated by Kanakidou *et al.* [56], organics have been observed to decrease the water uptake by NaCl but increase that of $(NH_4)_2SO_4$ [156, 157, 159]. Specifically, Cruz and Pandis [156] performed hygroscopic growth measurements for four internally mixed aerosol mixtures of NaCl, $(NH_4)_2SO_4$, glutaric acid and pinonic acid. The authors found that measured growth factors suggest a complex interaction between the organic and inorganic ions, which can be positive or negative depending on the type of salt and organic volume fraction. In most cases it was observed to enhance water absorption by $(NH_4)_2SO_4$ by as much as a factor of 2–3 for particles with 80% mass comprising organics. The NaCl mixtures, however, presented evidence of positive and negative interactions depending on the mass fraction, ranging from a 40% decrease to a 20% increase in water uptake [156]. Similarly, Choi and Chan [157], using an EDB, found that glycerol, succinic acid, malonic acid, citric acid and glutaric acid reduced the water absorption of NaCl but enhanced that of $(NH_4)_2SO_4$ relative to the pure inorganic salts. Chan and Chan [139] studied the interaction between two-model humic-like substances, including Suwannee River fulvic acid (FA), and two inorganic salts. They found that the FA–$(NH_4)_2SO_4$ mixtures had a larger uptake than the sum of the individual contributions at 90% RH that also increased as the RH decreased. Similarly, for mixtures with NaCl they found a reduced water content compared with the sum of the individual uptakes. Interestingly, this proved to be a function of RH as found earlier. Extending the limited analysis of macromolecular compounds, Mikhailov *et al.* [141] also studied the behaviour of the protein bovine serum albumin (BSA), chosen as a representative compound for proteins and other macromolecular organics, and the inorganic salts sodium chloride and ammonium nitrate by laboratory experiments and model calculations. They found that pure BSA particles showed deliquescence and efflorescence transitions at around 35% RH and a diameter growth factor of around 1.1 at 90% RH.

In the atmosphere, it is generally accepted now that organic compounds in ambient aerosols contribute to water uptake most significantly at low RH, while at high RH the inorganic compounds seem to dominate [56]. For example, Kotchenruther and Hobbs [160] measured light-scattering data and hygroscopic growth of aerosols from biomass burning in Brazil. They found that aerosol dominated by smoke in Brazil grew less than aerosols sampled on the east coast of the US (ca. 1.12 at 90% – which is lower than the growth factors measured for three biomass-burning-derived organic compounds studied by [130]). Rissler *et al.* [161] measured the hygroscopic properties of aerosols over the Amazon rainforest for a selected 'clean' period, period when 'young' biomass burning plumes had influenced the measurements site and a period subject to aged biomass burning. They found that the hygroscopic growth factors often showed a bimodal structure, though one clearly dominated with a relatively stable growth factor of around 1.29 for 100-nm particles. They also report the presence of a 'hydrophobic'

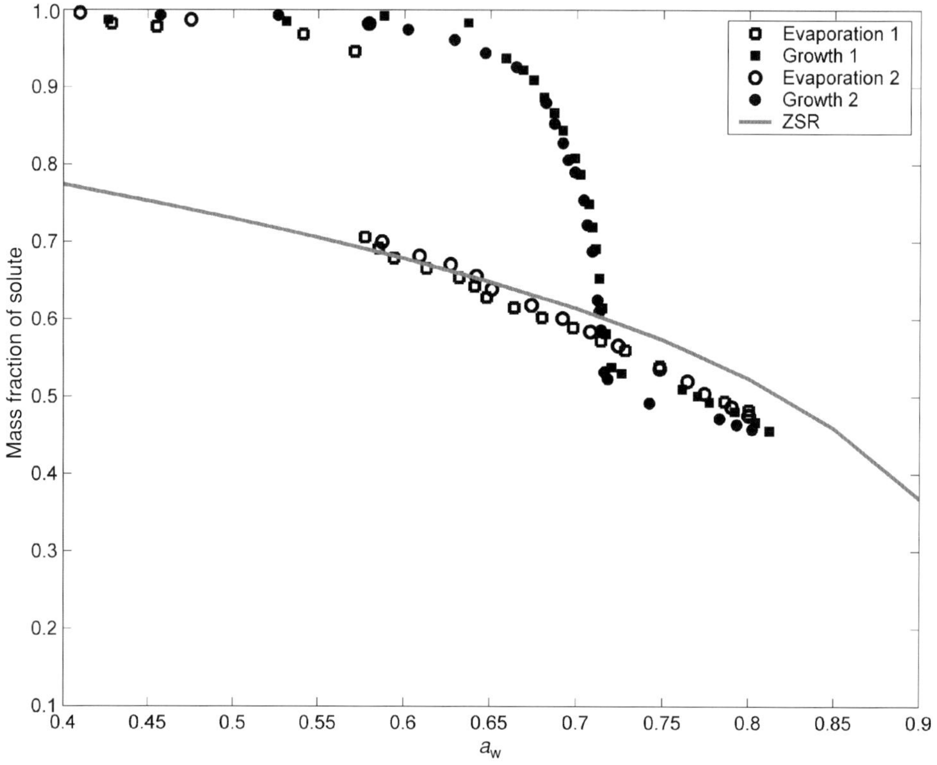

Figure 6.7 Mass fraction of solute (mfs = mass solute/mass solute + mass water) versus water activity for a mixed NaCl/glutaric acid particle at a mass ratio of 1:1. The solid line represents predictions using the additive scheme employed in ADDEM (see text). The single data points represent the experimental data of Choi and Chan [157].

mode. However, this mode had a growth factor of around 1.12 for all periods at sizes greater than 100 nm, which is similar to both the previous study and the humic-like material studied in the laboratory.

Various studies in the literature, all focused on simple binary mixed inorganic/organic aerosols, have used the ZSR assumption to analyse measured data with varying success. Experimental results derived from H-TDMAs often indicate reasonable agreement with ZSR predictions (e.g. [156, 159, 162]), even for complex components such as humic-like substances [140]. It is better to compare with water activity data from instruments such as the EDB, as one largely removes uncertainties from kinetic effects and density information of both the solution and the dry state. Such uncertainties to growth calculations are discussed by Topping *et al.* [21]. An example of such a comparison is shown in Figure 6.7 for a mixture of glutaric acid with NaCl (mass ratio of 1:1), using the combination of the two separate thermodynamic models in ADDEM [19, 21]. There are other simplified test cases where coupled thermodynamic reproduces the water uptake behaviour of simplified systems in the subsaturated humid regime with greater accuracy simply because data were available to derive the appropriate interaction parameters [12, 18, 29, 163]. Thus, where possible, explicit treatment of the ion–organic interactions offers an advantage as one might expect, though a wider range of systems need to be analysed to assess whether this advantage remains within the boundaries of experimental uncertainty. Thus, whilst studies such as that of Raatikainen and Laaksonen [114], for example, highlight the possible benefit of coupled thermodynamic systems, there is no grounding for trusting applicability to more atmospherically representative mixtures using current parameter matrices derived from a small subset of systems. Also, assumed types of interaction, between components for which there is no direct experimental data, may lead to erroneous results where a simpler additive scheme [19], or the methods of Clegg *et al.* [12] and Ming and Russell [18], with ion–organic interaction parameters set to zero, could potentially provide more accurate results. Further testing is required for a wide range of species as are comparisons with ambient data. Indeed, recently the additive approach employed in ADDEM was used to model hygroscopic growth factors for comparison with ambient data with success [164].

6.8.4 Deliquescence

Whilst the behaviour of simple mixed inorganic/organic aerosols is often described using the properties of the individual components (e.g. [165, 166]), the role of organics on the crystallisation RH and the DRH of inorganic aerosols is not well understood [157]. Also, it is likely that aerosols harbour a wider range of organics than the simple systems often studied in the laboratory (e.g. [167, 168]). Several groups have studied the deliquescence properties of mixed inorganic/organic particles, and results often suggest an altered deliquescence point compared to the binary inorganic system when the organic dry mole fraction decreased beyond a certain value [142]. However, again the range of organics likely to exist in the atmosphere heavily outweighs the components studied in the laboratory. Brooks *et al.* [169] found that in the case of soluble dicarboxylic acids, adding molar fractions up to and including eutonic proportions lowers the DRH below that found for pure $(NH_4)_2SO_4$. Similarly, Marcolli *et al.* [31] found that as the number of miscible dicarboxylic acids increased, the DRH of the mixed system, including $(NH_4)_2SO_4$, dropped considerably, as shown in Figure 6.8. The DRH for a multicomponent mixture can be calculated by using the following:

$$\mathrm{RH} = 100\% \cdot a_\mathrm{w} = 100\% \exp\left(-M_\mathrm{w}\phi \sum_j m_j\right) \tag{6.32}$$

where a_w is the water activity, ϕ the osmotic coefficient of the solution and m_j the molality of solute 'j', all calculated for the saturated solution. One option, used by Parsons *et al.* [142] is to assume ideality and use the solubility data (thus molality) from the respective binary systems. Parsons *et al.* [142] found that this approach substantially underestimated the DRH for all systems even when the organic mole fraction is zero. Similarly, Marcolli *et al.* [31] used single-component aqueous solubility to calculate the mixed DRH of multicomponent dicarboxylic acid solutions of increasing complexity. They found that the measured DRH could be modelled relatively well. However, on comparison with a five-component organic mixture and three inorganic salts, a significantly higher deviation was found. Parsons *et al.* [142] used the model of Clegg *et al.* [12] to calculate the inorganic contribution to ϕ and UNIFAC for the organic contribution using the parameter matrix of Hansen *et al.* [131]. They generally found good agreement up to an organic mole fraction of 0.4. Clegg *et al.* [12], using a coupled

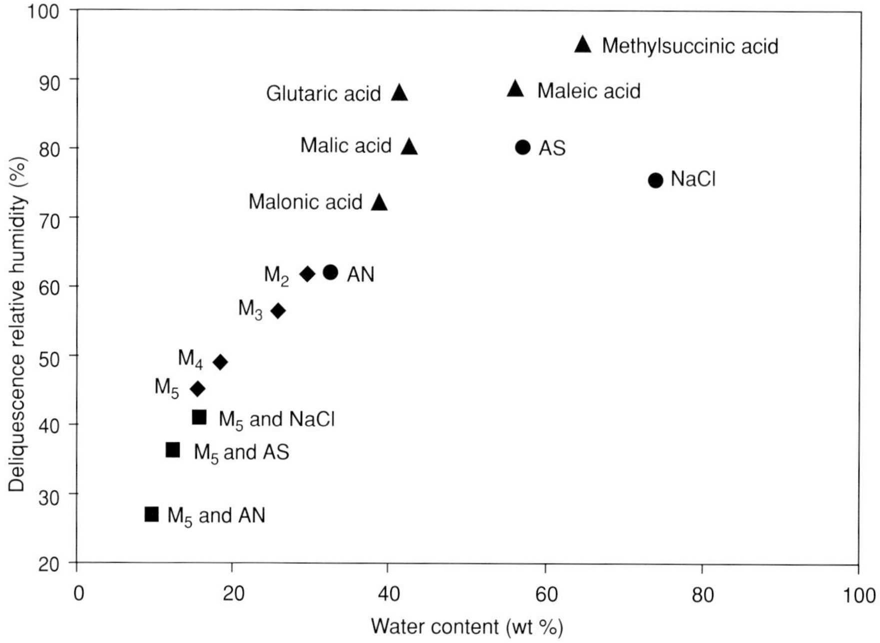

Figure 6.8 Deliquescence relative humidity and water content at the DRH for the pure organic (solid triangles) and the pure inorganic (solid circles) substances, as well as for the organic mixtures M_2–M_5 (solid diamonds) and the organic–inorganic mixtures (solid squares). M_2: malic+malonic acids, M_3: malic+malonic+maleic acids, M_4: malic+malonic+maleic+glutaric acids, M_5: malic+malonic+maleic+glutaric+methylsuccinic acids. All mixtures are eutonic, that is saturated with respect to all components. (Reprinted with permission from [31]. Copyright 2004 American Chemical Society.)

thermodynamic model and derived inorganic–organic interaction parameters, found that for the sodium chloride/sucrose/water system, a simplified approach to treating inorganic–organic interactions resulted in a 'satisfactory' reproduction of the deliquescence curve (see [12]). However, data were readily available for deriving mixed inorganic–organic interaction parameters in this case. In calculations of the deliquescence properties for the letovicite/2-butanedioic acid/water system utilising different organic activity coefficient methods, differences in the predicted RH over the eutectic mixture of 13% were found. It is difficult to extrapolate uncertainties to a vast range of systems, as results appear to be specie specific.

However, results from field measurements suggest that aerosols may remain aqueous at low RH [143, 146, 170, 171]. Figure 6.9 shows results from the study of Weingartner *et al.* [143] highlighting the response of aged ambient aerosols to varying RH at low temperatures, indicating growth even at low RH. Similarly, the laboratory study of Marcolli *et al.* [31] showed that the liquid phase may be the most stable phase for mixtures involving inorganic and organic components at low RH. Fulvic acids, often used as a proxee for humic-like material, seem to continually absorb water at all RHs [139, 140, 172]. Similarly, smog chamber studies on oxidation products of gas-phase organics often show hygroscopic response at low RH (e.g. [173] and references therein).

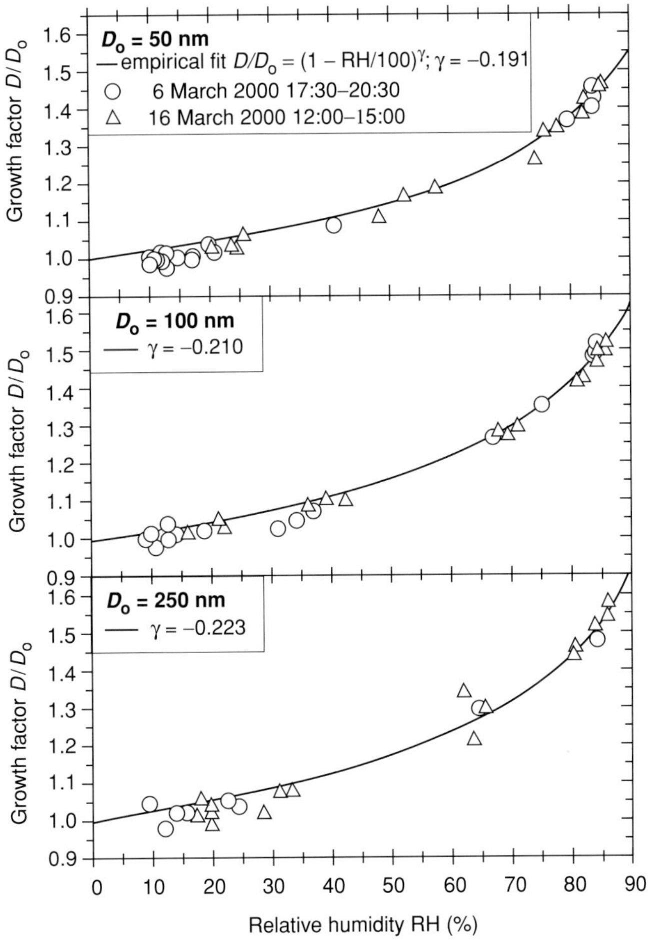

Figure 6.9 Humidograms of 50-, 100- and 250-nm particles measured at −10°C during increasing RH scans at the high alpine site Jungfraujoch located in the free troposphere. (Reprinted with permission from [170]. Copyright 2002 American Chemical Society.)

6.9 Temperature-dependent equilibrium

Equilibrium compositions respond to changes in concentrations of reactants and products, pressure and temperature. Capturing any temperature dependence is advantageous since the physical state and even existence of a particle can be determined entirely by temperature even for a system of very simple composition [17, 88].

6.9.1 Deliquescence

The deliquescence point of an aerosol varies with temperature. Similarly, the stable solid existing prior to deliquescence may also vary, depending on the composition being studied. Using

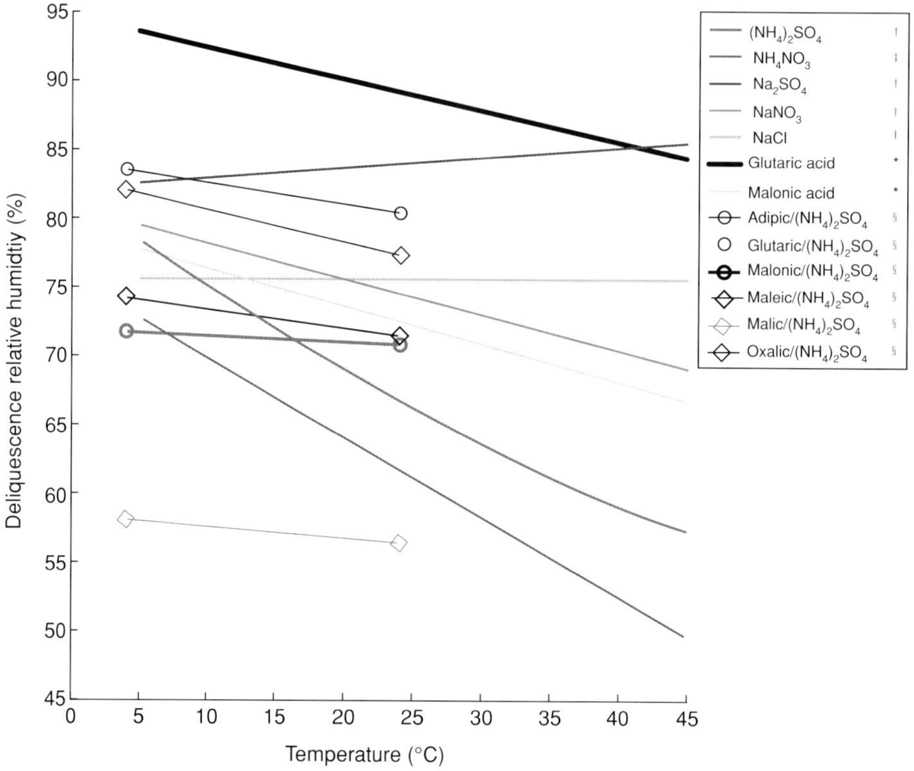

Figure 6.10 Measured and calculated DRH as a function of temperature for common inorganic salts, two organic acids and various binary inorganic–organic mixtures. * Parsons *et al.* [177]; † Onasch *et al.* [176]; ‡ calculated using AIM online (Clegg *et al.* [17, 88]); § Brooks *et al.* [169].

thermodynamic relations, Tang and Munkelwitz [174] showed that the variation of the DRH with temperature can be calculated from

$$\mathrm{LnDRH}(T) = \mathrm{LnDRH}(T^{*}) + \frac{1}{R}\int_{T^{*}}^{T} \frac{n(T)\Delta H_{\mathrm{s}}(T)}{T^{2}}\,\mathrm{d}T \tag{6.33}$$

where DRH(T^{*}) is the measured DRH at a temperature T^{*}, R the ideal gas constant and ΔH_{s} the integral heat of solution. Previous studies which used (6.33) to fit parameterisations for DRH(T) commonly assumed a constant value for ΔH_{s}. Tabazadeh and Toon [175], Onasch *et al.* [176] and more recently Parsons *et al.* [142, 177] bypassed this problem by using the AIM model to calculate the temperature dependence of inorganic DRH points (the model automatically accounting for the variation in ΔH_{S}), and by fitting laboratory data at atmospherically relevant temperatures for organics to the following functional form:

$$\mathrm{Ln}(\mathrm{DRH}(T)) = A + \frac{B}{T} + \frac{C}{T^{2}} + \frac{D}{T^{3}} \tag{6.34}$$

Figure 6.10 shows the predicted and measured variation for various inorganic salts, various organic systems and mixed inorganic/organic systems. The variation in DRH with temperature for various inorganic systems can be analysed interactively by utilising AIM online [176]. Qualitatively, it is best to illustrate firstly how the DRH varies using a simple inorganic system. Take, for example, NH_4NO_3. At the DRH point, the solid is in equilibrium with the saturated solution. Thus, the concentrations at this point for the various temperatures represent the solubility at these temperatures, which for NH_4NO_3 varies greatly. For example, it is 15 mol kg^{-1} at 273.15 K to 42.7 mol kg^{-1} at 323.15 K [17, 88]. For $(NH_4)SO_4$, this varies from 5.2 to 6.2 mol kg^{-1} [17, 88]. Since the equilibrium RH above a salt solution decreases with concentration, the DRH drops accordingly. Therefore, since the solubility of NH_4NO_3 is much higher at 323.15 K and the concentration of the saturated aqueous solution almost three times greater, the DRH is much lower. For $(NH_4)SO_4$ the change between the selected temperatures is minimal, reflecting the small change in solubility. Clegg *et al.* [17, 88], using the CSB model, show that for a single strong electrolyte like NH_4NO_3 or $(NH_4)SO_4$, the primary influence controlling the DRH point is the variation of the solubility of the solid salt with temperature since the water activity coefficient of both NH_4NO_3 and $(NH_4)SO_4$ binary systems has only a small dependence upon temperature. More temperature-dependent studies on ambient aerosols are required, yet as discussed earlier and highlighted in Figure 6.9, heavily aged aerosols can remain aqueous at low RH even at low temperatures.

6.9.2 Gas-to-particle partitioning

Gas-to-particle partitioning is dependent on various factors, including composition, phase state and ambient conditions. Analysing the effect of temperature for a wide range of compositions is difficult for restrictive reasons already mentioned, that is lack of appropriate fundamental data. Qualitatively, it is quite simple to get a feel for this varying response by looking at a simple system. As the volatility of the participating gaseous species increases with temperature, the possibility of formation of a solid phase, or indeed a liquid phase, depending on the RH region of interest, decreases. For example, take the formation of NH_4NO_3. There are various possible phase transitions between the solid, aqueous and gas phase. Figure 6.11 shows the variation in composition and hygroscopic growth factor (radius aqueous aerosol/radius original dry NH_4NO_3 aerosol) for a system with a relative molar composition of 1:1:2 for H^+:NH_4^+:NO_3^-. The equilibrium composition was calculated using AIM online and the growth factor calculated using density information provided by Tang and Munkelwitz [72]. Firstly, solid NH_4NO_3 exists up to a temperature of around 284 K. At this temperature, the DRH drops to around 70%, that at which the model was employed, initiating the presence of an aqueous solution. This is highlighted by the sharp increase in NH_4^+ and NO_3^-. With increasing temperature, the volatility of ammonia and nitric acid increase, and it is only in the lower temperatures of the range that we find a particle phase. Indeed, above 300 K the amount of water present in the particle declines steeply because the dissolved $NH4^+(aq)$, $H^+(aq)$ and $NO3^-(aq)$ ions partition increasingly into the gas phase (as $HNO_3(g)$ and $NH_3(g)$) leaving smaller quantities in the particle. The amount of liquid water therefore also decreases, to maintain equilibrium with the ambient RH (which requires that the ion *concentrations* in the particle remain about the same) [17, 88]. At 315 K, all of the ions have transferred to the gas phase and no particulate-phase NH_4NO_3 exists. Equilibrium constants as a function of temperature for various species can be found in the literature (e.g. [14, 69]).

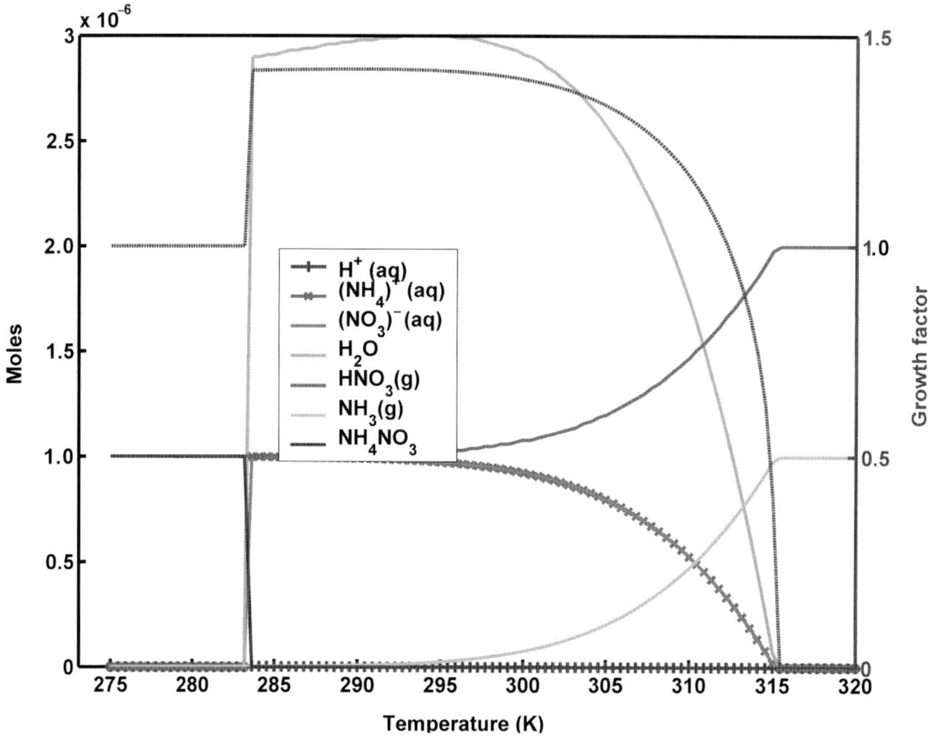

Figure 6.11 Concentrations of gas/solid/liquid phase components in the $H^+/NH_4^+/NO_3^-$ (1:1:2) system at 298.15 K (left *y*-axis) calculated using AIM online [17, 88]. Growth factor predictions (dashed line) calculated assuming the Kelvin effect is negligible using the density information of Tang and Munkelwitz [72] (right *y*-axis).

6.9.3 Activity

There are contrasting arguments as to the importance of capturing this variation found in the literature. The empirical parameterisations employed within EQSAM [93] do not include any temperature dependence. The authors claim that the temperature affects the solute activity coefficients most noticeably due to changes in aerosol composition, that is when evaporation or crystallisation occurs. However, as stated by Ansari and Pandis [16], Clegg *et al.* [178] showed that the solute activity of aqueous $(NH_4)_2SO_4$ varied by as much as 5% over the temperature range of 0–50°C. Capturing this variation can be hindered by a lack of experimental data. In the CSB model, interaction parameters have been derived over a range of tropospheric temperatures [17]. Indeed, the authors state that the model is valid from 328 to <200 K dependent on the liquid-phase composition. The ability of UNIFAC to model systems of atmospheric importance has only been done at room temperature (ca. 298.15 K) since data are severely lacking in the literature. The Dortmund UNIFAC model, though including temperature dependence, was found to be particularly bad at modelling simple atmospheric systems (see Figure 6.5). Thus, existing models can essentially be considered reliable at a range of lower tropospheric conditions, yet further proof is inhibited by a lack of data in the literature.

6.10 The influence of curvature

The influence of curvature is an aspect of aerosol science that can distinguish it from more general equilibrium problems. In all solution systems, the curvature will affect the solvent activity in both sub- and supersaturated conditions; thus, in the special case when water is the solvent, the curvature affects its hygroscopicity. Although this effect, also known as the Kelvin effect, may be neglected on larger aerosol (greater than ca.100-nm dry radius) not treating the Kelvin effect restricts the range of applicability of any equilibrium model with respect to both ambient conditions and the aerosol population being studied.

6.10.1 *The Kelvin effect*

The Kelvin relation predicts the increase in vapour pressure that a small drop of a given substance would experience over that of a bulk liquid, thus directly influencing the partitioning of volatile compounds between the liquid and gaseous phase across a curved interface:

$$p = p^o \exp\left(\frac{2\sigma \upsilon}{\mathrm{RT}r}\right) \tag{6.35}$$

where p^o is the vapour pressure of the pure species above a flat surface, p the vapour pressure above the new curved surface, σ the surface tension and υ the partial molar volume. Equation (6.35) is known as the Kelvin equation. It tells us that the vapour pressure over a curved interface always exceeds that of the same substance over a flat surface. Thus, for any given substance equilibrating between the liquid and gaseous phase, as the size of the droplet decreases then the vapour pressure required to maintain equilibrium between the droplet and its environment increases. In the case of a pure substance, the partial molar quantities ($\upsilon_{\mathrm{water}}$) are identical with the molar quantities (V_{water}). For mixed solutions, Chylek and Wong [179] and Shulman *et al.* [166] state that the above relationship can be simplified to include the calculation for the partial molar volume of water in solution [166, 179]:

$$p = p^o \exp\left(\frac{2\sigma M_{\mathrm{w}}}{\mathrm{RT}r\rho_{\mathrm{solution}}}\right) \tag{6.36}$$

where M_{w} is the molecular weight of water and ρ_{solution} the density of the solution. However, a recent study by Kreidenweis *et al.* [180] pointed out the errors associated with the above form that leads to low estimates for $\upsilon_{\mathrm{water}}$. Rather, the partial molar volume of water can be estimated from

$$\nu = \frac{M_{\mathrm{w}}}{\rho_{\mathrm{solution}}}\left[1 + \frac{x}{\rho_{\mathrm{solution}}}\frac{\mathrm{d}\rho_{\mathrm{solution}}}{\mathrm{d}x}\right] \tag{6.37}$$

where x is the mass percentage of the solute. In ideal solutions the partial molar volume is equal to the molar volume of pure water, also known as the volume additivity assumption, and (6.36) can be written as [14]

$$p = p^o \exp\left(\frac{2\sigma M_{\mathrm{w}}}{\mathrm{RT}r\rho_{\mathrm{water}}}\right) \tag{6.38}$$

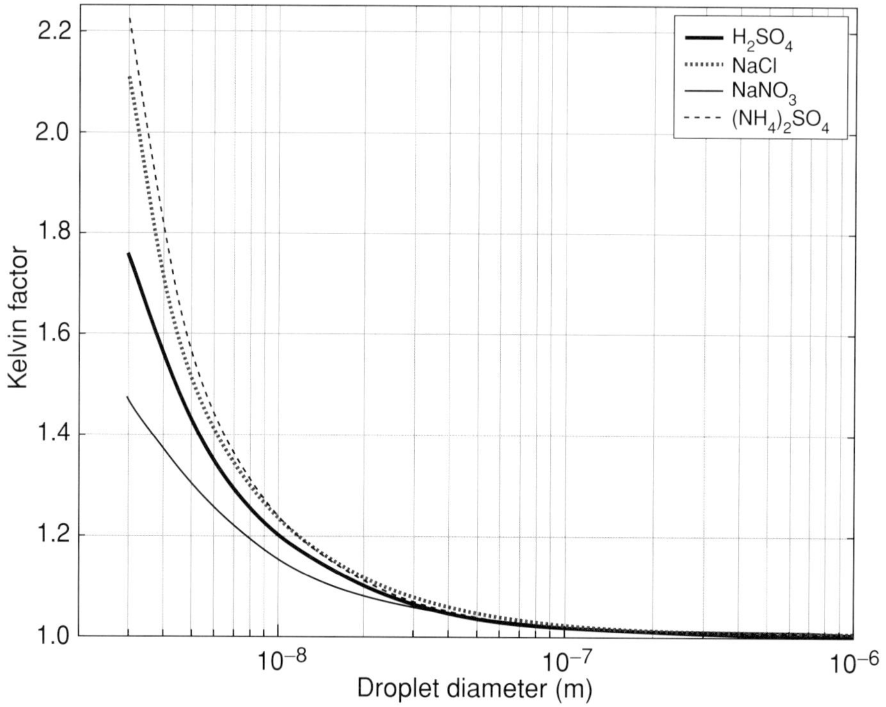

Figure 6.12 Kelvin factor versus aqueous droplet size for common inorganic components. Surface tension calculations were made using the model of Li and Lu [181].

When the volume additivity assumption is employed for calculating solution densities where experimental does not exist, for example with many organic solutions, then it is of course consistent to use (6.38). Figure 6.12 shows the Kelvin factor as a function of aerosol wet size for various inorganic compositions. It serves to illustrate that as the aerosol size decreases the Kelvin factor increases exponentially. Similarly, it is sensitive to the chemical composition of the droplet, thus requiring models for the surface tension that can capture this variability. Again, inorganic systems are relatively well understood. For single-salt systems, empirical fits are commonly used, though there can be inconsistencies between different fits that inevitably introduce sensitivities to growth factor calculations (see [182]). Indeed, Hameri *et al.* [182] found that below 10 nm, the variation in surface tension for $(NH_4)_2SO_4$ derived from different parameterisations produced a noticeable difference in predicted water uptake behaviour. Above 10 nm, the difference was negligible. For multicomponent systems, three groups of calculation have been presented in the literature [183]. These include simple additive methods (e.g. [184–187]) and conventional methods based on the Gibbs adsorption equation [181]. It is well known that some organic compounds are surface active and their presence in solution can significantly affect the surface tension of cloud droplets [166, 188, 189]. Whereas for non-aqueous solutions the mixture surface tension in some cases can be approximated by a linear dependence, aqueous systems show pronounced non-linear characteristics. This is typical of organic/aqueous systems, in which small concentrations of the organic material may significantly affect the mixture surface tension [190]. Whilst predictive methods are available, capturing this variability is complex and many uncertainties exist. Interestingly, however, Topping [191] found that deviations between predicted and measured surface tensions are not likely to introduce large errors for predicted aerosol growth in the subsaturated humid regime around 90% RH. Since the sensitivity seems to increase as the aerosol dry size decreases and the RH increases, the difference is likely to be more important for cloud activation studies or indeed extension of the model into or approaching the supersaturated humid regime in general.

For a *single*-component drop, the Kelvin relation corresponds to an unstable equilibrium. For homogenous nucleation of water vapour to occur using this above theory alone in the atmosphere would require unattainable supersaturations [42]. The problem of unstable equilibrium was solved to some extent and experimentally verified by Lamer and Gruen [192], who added an involatile solute (dioctyl phthalate) to a volatile solvent (toluene). However, the fact that the unstable equilibrium could be converted to a stable one by the influence of an involatile solute seems to have been first realised by Köhler [45], who used this idea to explain the difference between stabilities of droplets in mists on the one hand and in clouds on the other hand [193]. Here, the two competing curvature (Kelvin) and solute (Raoult) effects are considered in parallel. This leads to the derivation of the well-known Köhler equation that describes the behaviour of curved particles in both the sub- and supersaturated humid regime.

6.10.2 The Köhler equation – describing the equilibrium between a curved particle and its humid environment: basic equilibrium considerations

Combining the two competing Raoult (solute) and Kelvin (curvature) effects leads to the Köhler equation [14, 42, 45]. Used extensively in cloud physics to predict the activation capabilities of aerosols (e.g. [134, 186, 187, 194]), it is presented in the literature under various guises. These include revisions due to the creation of new assumptions or corrections to the original form in order to make it more applicable for subsaturated hygroscopic growth predictions. A thorough review was recently published by McFiggans *et al.* [53] with regards to warm cloud activation and the reader is referred to this source for comparing the different modelling techniques and theoretical considerations. The effect of neglecting the solid/gas phase boundary used in the general Köhler formulation is discussed shortly.

$$\frac{P'}{P_{\mathrm{sat,w}}} = \mathrm{RH}/100\% = a_{\mathrm{w}} \exp\left(\frac{2v\sigma}{\mathrm{RT}r}\right) \tag{6.39}$$

Equation (6.39) is the general form of the Köhler equation that provides us with the new definition of equilibrium above a curved aqueous surface. For a full derivation using two different approaches the reader is referred to sources in the literature [42, 191, 195]. One can see that for a water activity of 1 (pure water) the Kelvin equation derived in Section 6.10.1 is recovered. For a flat surface (Kelvin factor of 1), our original definition of equilibrium used solely in the subsaturated humid regime discussed in Section 6.4 is recovered. Thus, the competing curvature and solute influences are combined in a relatively simple algebraic formula. Rearranging (6.39) one can obtain a difference relationship to clarify the influence of curvature on the hygroscopic growth (or indeed the equilibrium of any volatile species between the gas and aerosol phase) [19, 21]:

$$(\mathrm{RH}/100\%)/a_{\mathrm{w}} - \exp\left(\frac{2\nu\sigma}{\mathrm{RT}r}\right) = 0 \tag{6.40}$$

Thus, for a smaller particle (larger exponential term), the water activity has to decrease for a given RH for (6.40) to be satisfied. Thus, the water content would decrease, leading to a smaller aqueous particle. Few models designed for analysing equilibrium in the subsaturated humid regime treat the influence of curvature. Sole use of the Köhler equation to calculate the complete phase equilibrium composition for multiple solutes, in systems where deliquescence is expected to occur, in the subsaturated regime cannot be justified due to the invalid assumption of infinite solubility used in its derivation. Indeed, modifications are continually being made in order to make this simple formulation more applicable to hygroscopic growth regions (e.g. [165, 166, 196]). These include the introduction of solubility limitations, thus allowing deliquescence to be considered. There may also be complicating issues regarding an accurate representation of the water activity as a function of solute concentration. For example, a common representation is often given as follows [14]:

$$\mathrm{Ln}(\mathrm{RH}/100\%) = \frac{A}{D_{\mathrm{p}}^{3}} - \frac{B}{D_{\mathrm{p}}^{3}} \tag{6.41}$$

$$A = \frac{4M_{\mathrm{w}}\sigma_{\mathrm{w}}}{RT\rho_{\mathrm{w}}} \tag{6.42}$$

$$B = \frac{6n_{\mathrm{s}}M_{\mathrm{w}}}{\pi\rho_{\mathrm{w}}} \tag{6.43}$$

where D_{p} is the diameter of the droplet, n_{s} the number of moles of solute in the dry particle (involatile salt) and σ_{w} the surface tension of pure water. However, this assumes solution ideality and thus cannot be trusted for mixtures that are not extremely dilute. (for more information, see [53] and references therein.) Importantly, the Köhler equation now gives us the ability to analyse the behaviour of particulate matter in both the sub- and supersaturated humid regime.

Figure 6.13 shows growth factor predictions in the subsaturated humid regime calculated by finding a solution to (6.40) using ADDEM [19, 21] for a mixed $NH_4NO_3/(NH_4)_2SO_4$ aerosol. Interestingly, the deliquescence point of an aerosol is predicted to increase with decreasing particle size. By using Equation (6.39), one explicitly takes into account the influence of the liquid–vapour interface through the solution surface tension. Indeed, consideration of this interface is explicit in the derivation of (6.39) by including the work required to create this surface in the Gibbs energy summation. Since the size of the aerosol particles can range from just a few molecules to particles with radii larger than 100 μm, it is important to understand the role the surface effects have on any possible deliquescence process by which small drops are formed [197]. There have been few attempts to predict the influence of curvature on the hygroscopic properties, specifically in relation to the change in deliquescence patterns. There are various complicating features surrounding the point of deliquescence. It is the norm to consider the transition from an uncoated dry aerosol to a dissolved salt droplet. However, several studies have shown that it may be more suitable to consider the transition from a wetted particle to a dissolved droplet [58, 198, 199]. Mirabel *et al.* [197] developed a model to predict the prompt deliquescence, and its dependence on particle size, for a generic crystal with an

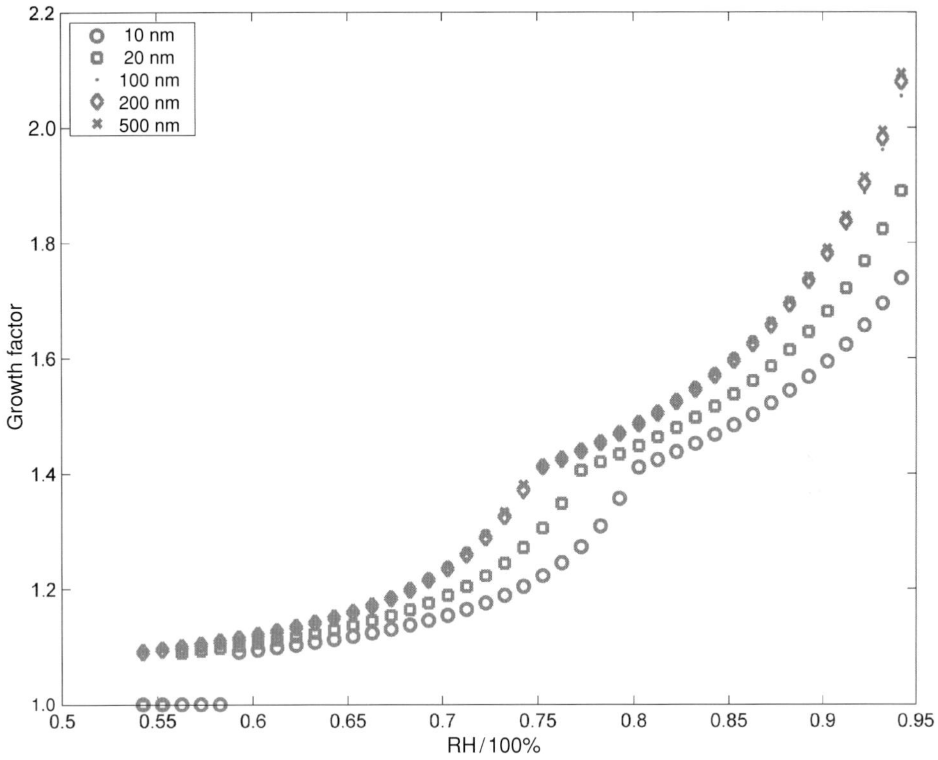

Figure 6.13 Diameter-dependent growth factor predictions for an equimolar $(NH_4)_2SO_4$:NH_4NO_3 aerosol at 298.15 K using ADDEM [19, 21].

ideal solution. More recently, Russell and Ming [59] developed a similar model for treating wetted particles and non-ideal effects. Similarly, the same authors incorporated the curvature effect into a larger thermodynamic equilibrium model used to treat mixed inorganic/organic systems [18]. Topping [191] found that whilst the lack of any change in DRH for the sub 100-nm $(NH_4)_2SO_4$ aerosol studied by [182] is an interesting result, both ADDEM [19, 21] and the model of Ming and Russell [18] predict an increase should occur. As stated by Russell and Ming [59], additional experiments might be able to not only confirm the increase in DRH with decreasing particle size, but also allow size-dependent surface tension measurements to be made. Figure 6.14 shows a schematic of the interfaces that are taken into account by the latter model and the possible influences they can exert on deliquescence behaviour. Russell and Ming [59] probed the sensitivity to the calculated DRH point for various values of the surface tensions of the solid–vapour and solid–liquid interface. As shown in Figure 6.14, it is possible that a wetted particle may be more stable than the aqueous solution, thus increasing the DRH point. Unfortunately, knowledge of surface tensions of crystalline structures is limited and it is unlikely that data are available for mixed inorganic/organic particles.

Also shown in Figure 6.15 are predictions of the equilibrium curve for mixed inorganic aerosols in the supersaturated humid regime using ADDEM [19, 21]. The maximum critical point determines the likelihood of an aerosol 'activating' into a cloud drop. It is important

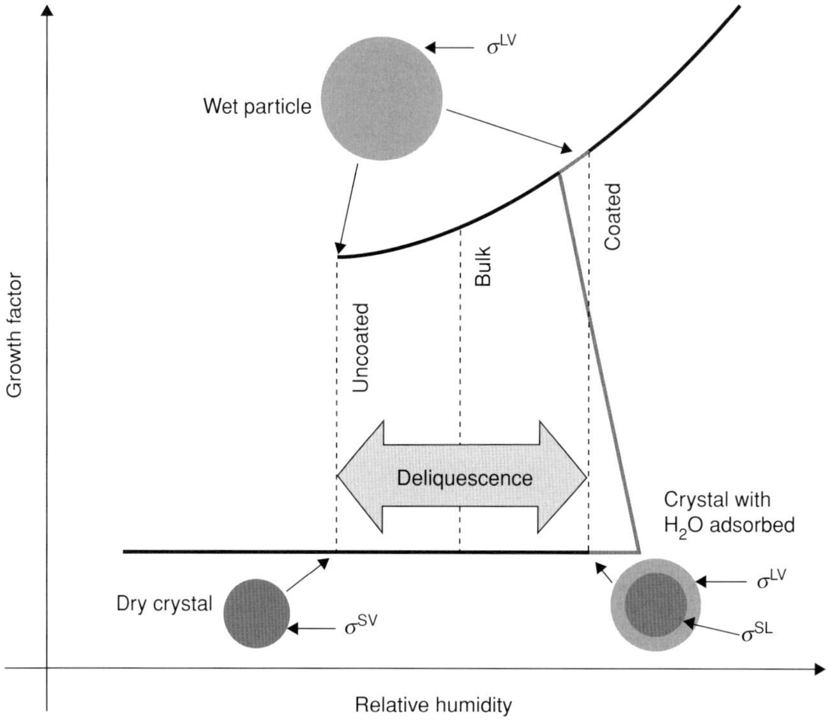

Figure 6.14 Schematic diagram of deliquescence of a salt crystal with two models: the coated model assumes that the prior to deliquescence multiple water layers will be adsorbed on the crystal surface, whereas the uncoated model assumes that the vapour interfaces with the solid crystal. The difference for dry particles smaller than 100-nm dry diameter is that, in this instance, when compared with the deliquescence of a bulk phase, the uncoated model predicts lower DRHs and the coated model predicts higher DRH. The solid lines represent dry (thin) and wet (thick) states and the dotted lines show deliquescence. The grey line shows unstable equilibria for partially wet states resulting from the role of surface tension in the coated model. However there are sensitivities to these perturbations, as discussed by Russell and Ming [59]. (Reprinted with permission from [59]. Copyright 2002, American Institute of Physics.)

to note that there are regions of varying stability on the full Köhler curve. To the left of the critical point, the droplets are in a stable equilibrium with the environment. That is to say, if a small amount of water molecules were added to the drop, its equilibrium pressure would be larger than the fixed ambient value. Thus, the drop would evaporate, eventually returning to its equilibrium state. Similarly, if a small amount was evaporated from the droplet, its equilibrium pressure would be lower than the ambient and water would condense onto the drop until returning to its equilibrium state. To the right of the critical point, the droplet is in an unstable equilibrium. If it grew slightly, the equilibrium pressure would be lower than the ambient. Thus, water would condense onto the droplet. However, the more the droplet grows, the larger this difference becomes and the droplet would grow hindered only by the supply of water vapour. Similarly, if the droplet shrank a small amount, its equilibrium vapour pressure would be higher than the ambient and water would start to evaporate from the droplet. However, as more water evaporated, this difference would increase and water would continue to evaporate

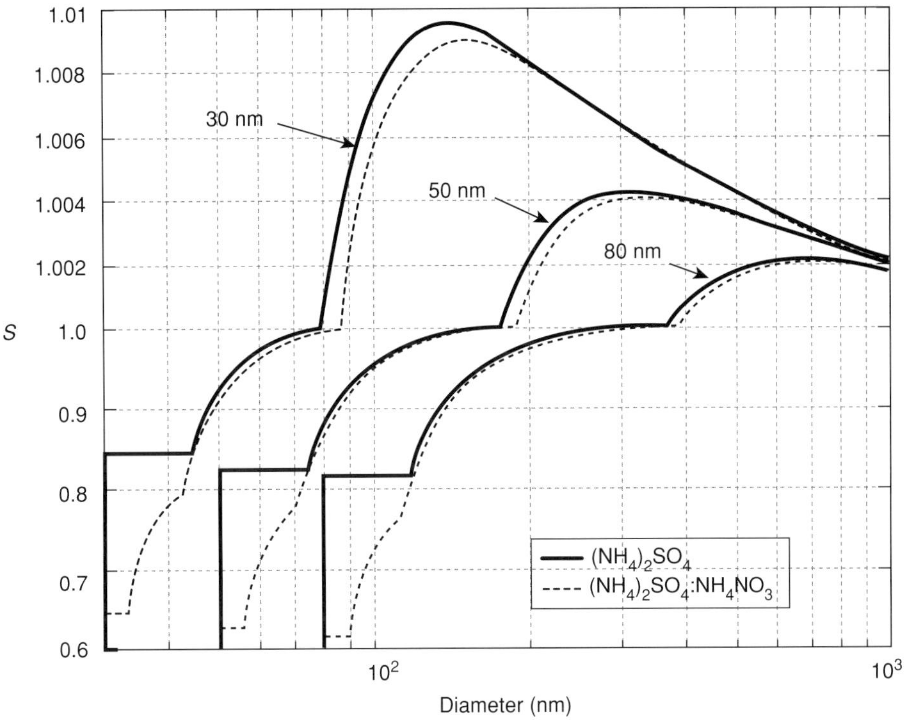

Figure 6.15 Full equilibrium curves for a $(NH_4)_2SO_4$ and equimolar $(NH_4)_2SO_4$:NH_4NO_3 aerosol particle at three different diameters and 298.15 K using ADDEM [19, 21].

until it reaches its stable equilibrium point on the Köhler curve. If the ambient saturation ratio exceeds the critical point, then there is no feasible equilibrium size and the particle will try to grow indefinitely [14].

References

1 Denbigh, K.G. *The Principles of Chemical Equilibrium: With Applications in Chemistry and Chemical Engineering*. Cambridge University Press, Cambridge: 1981.

2 Prausnitz, J.M. *Molecular Thermodynamics of Fluid-Phase Equilibria*, 2nd edn. Prentice-Hall Publishers, Englewood Cliffs, NJ: 1986.

3 Chang, D.P.Y. and Hill, R.C. Retardation of aqueous droplet evaporation by air-pollutants. *Atmospheric Environment* 1980; **14**(7): 803–807.

4 Barnes, G.T. The effects of monolayers on the evaporation of liquids. *Advances in Colloid and Interface Science* 1986; **25**(2): 89–200.

5 Xiong, J.Q., Zhong, M.H., Fang, C.P., Chen, L.C. and Lippmann, M. Influence of organic films on the hygroscopicity of ultrafine sulfuric acid aerosol. *Environmental Science & Technology* 1998; **32**(22): 3536–3541.

6 Hansson, H.-C., Rood, M.J., Koloutsou-Vakakkis, S., Hameri, K., Orsini, D. and Wiedensohler, A. NaCl aerosol particle hygroscopicity dependence on mixing with organic compounds. *Journal of Atmospheric Chemistry* 1998; **31**: 321–346.

7 Chen, Y.Y. and Lee, W.M.G. Hygroscopic properties of inorganic-salt aerosol with surface-active organic compounds. *Chemosphere* 1999; **38**(10): 2431–2448.

8 Cubison, M.J. *Design, Construction and Use of a Hygroscopic Tandem Differential Mobility Analyser*. PhD Thesis. UMIST, Manchester: 2004.

9 Chuang, P.Y. Measurement of the timescale of hygroscopic growth for atmospheric aerosols. *Journal of Geophysical Research – Atmospheres* 2003; **108**(D9): 4282.

10 Giddings, W.P. and Baker, M.B. Sources and effects of monolayers on atmospheric water droplets. *Journal of the Atmospheric Sciences* 1977; **34**: 1957–1964.

11 Wexler, A.S. and Clegg, S.L. Atmospheric aerosol models for systems including the ions H+, NH_4+, Na+, SO_4^{2-}, NO_3^-, Cl^-, Br^-, and H_2O. *Journal of Geophysical Research – Atmospheres* 2002; **107**(D14): 4207.

12 Clegg, S.L., Seinfeld, J.H. and Brimblecombe, P. Thermodynamic modelling of aqueous aerosols containing electrolytes and dissolved organic compounds. *Journal of Aerosol Science* 2001; **32**(6): 713–738.

13 Sorjamaa, R., Svenningsson, B., Raatikainen, T., Henning, S., Bilde, M. and Laaksonen, A. The role of surfactants in Kohler theory reconsidered. *Atmospheric Chemistry and Physics* 2004; **4**: 2107–2117.

14 Seinfeld, J.H. and Pandis, S.N. *Atmospheric Chemistry and Physics*. Wiley-Interscience Publication, New York: 1998.

15 Jacobson, M.Z. Studying the effects of calcium and magnesium on size-distributed nitrate and ammonium with EQUISOLV II. *Atmospheric Environment* 1999; **33**(22): 3635–3649.

16 Ansari, A.S. and Pandis, S.N. Prediction of multicomponent inorganic atmospheric aerosol behavior. *Atmospheric Environment* 1999; **33**(5): 745–757.

17 Clegg, S.L., Brimblecombe, P. and Wexler A.S. Thermodynamic model of the system H^+-NH_4-SO_4^{2-}-NO_3^--H_2O at tropospheric temperatures. *Journal of Physical Chemistry A* 1998; **102**(12): 2137–2154.

18 Ming, Y. and Russell, L.M. Thermodynamic equilibrium of organic-electrolyte mixtures in aerosol particles. *Aiche Journal* 2002; **48**(6): 1331–1348.

19 Topping, D.O., McFiggans, G.B. and Coe, H. A curved multi-component aerosol hygroscopicity model framework: Part 1 – Inorganic compounds. *Atmospheric Chemistry and Physics* 2005; **5**: 1205–1222.

20 Amundson, N.R., Caboussat, A., He, J.W., Martynenko, A.V., Savarin, V.B., Seinfeld JH and Yoo, K.Y. A new inorganic atmospheric aerosol phase equilibrium model (UHAERO). *Atmospheric Chemistry and Physics* 2006; **6**: 975–992.

21 Topping, D.O., McFiggans, G.B. and Coe, H. A curved multi-component aerosol hygroscopicity model framework: Part 2 – Including organic compounds. *Atmospheric Chemistry and Physics* 2005; **5**: 1223–1242.

22 Marrero, J. and Gani, R. Group-contribution based estimation of pure component properties. *Fluid Phase Equilibria* 2001; **183**: 183–208.

23 Atkins, P.W. and DePaula J. *Physical Chemistry*. W.H. Freeman and Company, Oxford: 2001.

24 Clegg, S.L. and Pitzer, K.S. Thermodynamics of multicomponent, miscible, ionic-solutions – generalized equations for symmetrical electrolytes. *Journal of Physical Chemistry* 1992; **96**(8): 3513–3520.

25 Saxena, P. and Hildemann, L.M. Water absorption by organics: survey of laboratory evidence and evaluation of UNIFAC for estimating water activity. *Environmental Science & Technology* 1997; **31**(11): 3318–3324.

26 Gmehling, J. Present status of group-contribution methods for the synthesis and design of chemical processes. *Fluid Phase Equilibria* 1998; **144**(1–2): 37–47.

27 Chen, C.C. and Mathias, P.M. Applied thermodynamics for process modeling. *Aiche Journal* 2002; **48**(2): 194–200.

28 Vaslow, F. *Water and Aqueous Solutions: Structure, Thermodynamics and Transport Processes.* John Wiley, New York: 1972.
29 Clegg, S.L. and Seinfeld, J.H. Thermodynamic models of aqueous solutions containing inorganic electrolytes and dicarboxylic acids at 298.15K. I. The acids as non-dissociating compounds. *Journal of Physical Chemistry A* 2006; **110**: 5692–5717.
30 Marcolli, C. and Krieger, U.K. Phase changes during hygroscopic cycles of mixed organic/inorganic model systems of tropospheric aerosols. *Journal of Physical Chemistry A* 2006; **110**(5): 1881–1893.
31 Marcolli, C., Luo, B.P. and Peter, T. Mixing of the organic aerosol fractions: liquids as the thermodynamically stable phases. *Journal of Physical Chemistry A* 2004; **108**(12): 2216–2224.
32 Erdakos, G.B. and Pankow, J.F. Gas/particle partitioning of neutral and ionizing compounds to single- and multi-phase aerosol particles. 2. Phase separation in liquid particulate matter containing both polar and low-polarity organic compounds. *Atmospheric Environment* 2004; **38**(7): 1005–1013.
33 Pankow, J.F. Gas/particle partitioning of neutral and ionizing compounds to single and multi-phase aerosol particles. 1. Unified modeling framework. *Atmospheric Environment* 2003; **37**(24): 3323–3333.
34 Pankow, J.F., Seinfeld, J.H., Asher, W.E. and Erdakos, G.B. Modeling the formation of secondary organic aerosol. 1. Application of theoretical principles to measurements obtained in the alpha-pinene/, beta-pinene/, sabinene/, delta(3)-carene/, and cyclohexene/ozone systems. *Environmental Science & Technology* 2001; **35**(6): 1164–1172.
35 Seinfeld, J.H. and Pankow, J.F. Organic atmospheric particulate material. *Annual Review of Physical Chemistry* 2003; **54**: 121–140.
36 Seinfeld, J.H., Erdakos, G.B., Asher, W.E. and Pankow, J.F. Modeling the formation of secondary organic aerosol (SOA). 2. The predicted effects of relative humidity on aerosol formation in the alpha-pinene-, beta-pinene-, sabinene-, Delta(3)-Carene-, and cyclohexene-ozone systems. *Environmental Science & Technology* 2001; **35**(9): 1806–1817.
37 Pankow, J.F. An absorption-model of the gas aerosol partitioning involved in the formation of secondary organic aerosol. *Atmospheric Environment* 1994; **28**(2): 189–193.
38 Johnson, D., Utembe, S.R., Jenkin, M.E., Derwent, R.G., Hayman, G.D., Alfarra, M.R., Coe, H. and McFiggans, G. Simulating regional scale secondary organic aerosol formation during the TORCH 2003 campaign in the southern UK. *Atmospheric Chemistry and Physics* 2006; **6**: 403–418.
39 Johnson, D., Utembe, S.R. and Jenkin, M.E. Simulating the detailed chemical composition of secondary organic aerosol formed on a regional scale during the TORCH 2003 campaign in the southern UK. *Atmospheric Chemistry and Physics* 2006; **6**: 419–431.
40 Griffin, R.J., Nguyen, K., Dabdub, D. and Seinfeld, J.H. A coupled hydrophobic-hydrophilic model for predicting secondary organic aerosol formation. *Journal of Atmospheric Chemistry* 2003; **44**(2): 171–190.
41 Wexler, A.S. and Seinfeld, J.H. 2nd-generation inorganic aerosol model. *Atmospheric Environment Part A –General Topics* 1991; **25**(12): 2731–2748.
42 Pruppacher, H.R. and Klett, J.D. *Microphysics of Clouds and Precipitation.* Kluwer, Dordrecht: 1997.
43 Rudich, Y. Laboratory perspectives on the chemical transformations of organic matter in atmospheric particles. *Chemical Reviews* 2003; **103**: 5097–5124.
44 Chughtai, A.R., Brooks, M.E. and Smith, D.M. Hydration of black carbon. *Journal of Geophysical Research –Atmospheres* 1996; **101**(D14): 19505–19514.
45 Köhler, H. The nucleus in the growth of hygroscopic droplets. *Transactions of the Faraday Society* 1936; **32**: 1152–1161.
46 McMurry, P.H. and Stolzenburg, M.R. On the sensitivity of particle-size to relative-humidity for Los-Angeles aerosols. *Atmospheric Environment* 1989; **23**(2): 497–507.
47 Zhang, X.Q., McMurry, P.H., Hering, S.V. and Casuccio, G.S. Mixing characteristics and water-content of submicron aerosols measured in Los-Angeles and at the Grand-Canyon. *Atmospheric Environment Part A – General Topics* 1993; **27**(10): 1593–1607.

48 Pitchford, M.L. and McMurry, P.H. Relationship between measured water-vapor growth and chemistry of atmospheric aerosol for Grand-Canyon, Arizona, in winter 1990. *Atmospheric Environment* 1994; **28**(5): 827–839.
49 Swietlicki, E., Zhou, J.C., Berg, O.H., Martinsson, B.G., Frank, G., Cederfelt, S.I., Dusek, U., Berner, A., Birmili, W., Wiedensohler, A., Yuskiewicz, B. and Bower, K.N. A closure study of sub-micrometer aerosol particle hygroscopic behaviour. *Atmospheric Research* 1999; **50**(3–4): 205–240.
50 Svenningsson, B., Hansson, H.C., Wiedensohler, A., Ogren, J., Noone, K.J. and Hallberg, A. Hygroscopic growth of aerosol particles in the Po Valley. *Tellus* 1992; **44B**: 556–569.
51 Svenningsson, B., Hansson, H.C., Wiedensohler, A., Noone, K.J., Ogren, J., Hallberg, A. and Colvile, R. Hygroscopic growth of aerosol particles and its influence on nucleation scavenging in cloud: experimental results from Kleiner Feldberg. *Journal of Atmospheric Chemistry* 1994; **19**: 129–152.
52 Berg, O.H., Swietlicki, E. and Krejci, R. Hygroscopic growth of aerosol particles in the marine boundary layer over the Pacific and Southern Oceans during the First Aerosol Characterization Experiment (ACE 1). *Journal of Geophysical Research – Atmospheres* 1998; **103**(D13): 16535–16545.
53 McFiggans, G., Artaxo, P., Baltensperger, U., Coe, H., Facchini, C., Feingold, G., Fuzzi, S., Gysel, M., Laaksonen, A., Lohmann, U., Mentel, T.F., Murphy, D.M., O'Dowd, C.D., Snider, J.R. and Weingartner, E. The effect of physical and chemical aerosol properties on warm cloud droplet activation. *Atmospheric Chemistry and Physics Discussions* 2006; **5**: 8507–8647.
54 Swietlicki, E., Zhou, J.C., Covert, D.S., Hameri, K., Busch, B., Vakeva, M., Dusek, U., Berg, O.H., Wiedensohler, A., Aalto, P., Makela, J., Martinsson, B.G., Papaspiropoulos, G., Mentes, B., Frank, G. and Stratmann, F. Hygroscopic properties of aerosol particles in the northeastern Atlantic during ACE-2. *Tellus Series B – Chemical and Physical Meteorology* 2000; **52**(2): 201–227.
55 Vakeva, M., Kulmala, M., Stratmann, F. and Hameri, K. Field measurements of hygroscopic properties and state of mixing of nucleation mode particles. *Atmospheric Chemistry and Physics* 2002; **2**: 55–66.
56 Kanakidou, M., Seinfeld, J.H., Pandis, S.N., Barnes, I., Dentener, F.J., Facchini, M.C., van Dingenen, R., Ervens, B., Nenes, A., Nielsen, C.J., Swietlicki, E., Putaud, J.P., Balkanski, Y., Fuzzi, S., Horth, J., Moortgat, G.K., Winterhalter, R., Myhre, C.E.L., Tsigaridis, K., Vignati, E., Stephanou, E.G. and Wilson, J. Organic aerosol and global climate modelling: a review. *Atmospheric Chemistry and Physics Discussions* 2004; **4**: 5855–6024.
57 Cocker, D.R., Whitlock, N.E., Flagan, R.C. and Seinfeld J.H. Hygroscopic properties of Pasadena, California aerosol. *Aerosol Science and Technology* 2001; **35**(2): 637–647.
58 Weis, D.D. and Ewing, G.E. Water content and morphology of sodium chloride aerosol particles. *Journal of Geophysical Research – Atmospheres* 1999; **104**(D17): 21275–21285.
59 Russell, L.M. and Ming, Y. Deliquescence of small particles. *Journal of Chemical Physics* 2002; **116**(1): 311–321.
60 Weingartner, E., Baltensperger, U. and Burtscher, H. Growth and structural-change of combustion aerosols at high relative-humidity. *Environmental Science & Technology* 1995; **29**(12): 2982–2986.
61 Weingartner, E., Burtscher, H. and Baltensperger, U. Hygroscopic properties of carbon and diesel soot particles. *Atmospheric Environment* 1997; **31**(15): 2311–2327.
62 Liu, B.Y.H., Pui, D.Y.H., Whitby, K.T., Kittelson, D.B., Kousaka, Y. and Mckenzie, R.L. Aerosol mobility chromatograph – new detector for sulfuric-acid aerosols. *Atmospheric Environment* 1978; **12**(1–3): 99–104.
63 McMurry, P.H., Litchy, M., Huang, P.F., Cai, X.P., Turpin, B.J., Dick, W.D. and Hanson, A. Elemental composition and morphology of individual particles separated by size and hygroscopicity with the TDMA. *Atmospheric Environment* 1996; **30**(1): 101–108.
64 Peng, C.G., Chow, A.H.L. and Chan, C.K. Hygroscopic study of glucose, citric acid, and sorbitol using an electrodynamic balance: comparison with UNIFAC predictions. *Aerosol Science and Technology* 2001; **35**(3): 753–758.

65 Stokes, R.H. and Robinson, R.A. Interactions in aqueous nonelectrolyte solutions. I. Solute-solvent equilibria. *Journal of Physical Chemistry* 1966; **70**(7): 2126–2130.

66 Ha, Z.Y., Choy, L. and Chan, C.K. Study of water activities of supersaturated aerosols of sodium and ammonium salts. *Journal of Geophysical Research – Atmospheres* 2000; **105**(D9): 11699–11709.

67 Chan, C.K., Ha, Z.Y. and Choi, M.Y. Study of water activities of aerosols of mixtures of sodium and magnesium salts. *Atmospheric Environment* 2000; **34**(28): 4795–4803.

68 Choi, M.Y. and Chan, C.K. Continuous measurements of the water activities of aqueous droplets of water-soluble organic compounds. *Journal of Physical Chemistry A* 2002; **106**(18): 4566–4572.

69 Jacobson, M.Z. *Fundamentals of Atmospheric Modelling*. Cambridge University Press, Cambridge: 1999.

70 Cohen, M.D., Flagan, R.C. and Seinfeld, J.H. Studies of concentrated electrolyte-solutions using the electrodynamic balance. 1. Water activities for single-electrolyte solutions. *Journal of Physical Chemistry* 1987; **91**(17): 4563–4574.

71 Cohen, M.D., Flagan, R.C. and Seinfeld, J.H. Studies of concentrated electrolyte-solutions using the electrodynamic balance. 2. Water activities for mixed-electrolyte solutions. *Journal of Physical Chemistry* 1987; **91**(17): 4575–4582.

72 Tang, I.N. and Munkelwitz, H.R. Water activities, densities, and refractive-indexes of aqueous sulfates and sodium-nitrate droplets of atmospheric importance. *Journal of Geophysical Research – Atmospheres* 1994; **99**(D9): 18801–18808.

73 Tang, I.N. Thermodynamic and optical properties of mixed-salt aerosols of atmospheric importance. *Journal of Geophysical Research – Atmospheres* 1997; **102**(D2): 1883–1893.

74 Chan, C.K., Flagan, R.C. and Seinfeld, J.H. Water activities of $NH_4NO_3/(NH_4)_2SO_4$ solutions. *Atmospheric Environment Part A – General Topics* 1992; **26**(9): 1661–1673.

75 Decesari, S., Facchini, M.C., Fuzzi, S., McFiggans, G.B., Coe, H. and Bower, K.N. The water-soluble organic component of size-segregated aerosol, cloud water and wet depositions from Jeju Island during ACE-Asia. *Atmospheric Environment* 2005; **39**(2): 211–222.

76 Clegg, S.L., Seinfeld, J.H. and Edney, E.O. Thermodynamic modelling of aqueous aerosols containing electrolytes and dissolved organic compounds. II. An extended Zdanovskii-Stokes-Robinson approach. *Journal of Aerosol Science* 2003; **34**(6): 667–690.

77 Bassett, M. and Seinfeld, J.H. Atmospheric equilibrium-model of sulfate and nitrate aerosols. *Atmospheric Environment* 1983; **17**(11): 2237–2252.

78 Bassett, M.E. and Seinfeld, J.H. Atmospheric equilibrium-model of sulfate and nitrate aerosols. 2. Particle-size analysis. *Atmospheric Environment* 1984; **18**(6): 1163–1170.

79 Saxena, P., Hudischewskyj, A.B., Seigneur, C. and Seinfeld, J.H. A comparative-study of equilibrium approaches to the chemical characterization of secondary aerosols. *Atmospheric Environment* 1986; **20**(7): 1471–1483.

80 Pilinis, C. and Seinfeld, J.H. Continued development of a general equilibrium-model for inorganic multicomponent atmospheric aerosols. *Atmospheric Environment* 1987; **21**(11): 2453–2466.

81 Kim, Y.P. and Seinfeld, J.H. Atmospheric gas-aerosol equilibrium. 3. Thermodynamics of crustal elements Ca^2+, K+, and Mg^2+. *Aerosol Science and Technology* 1995; **22**(1): 93–110.

82 Kim, Y.P., Seinfeld, J.H. and Saxena, P. Atmospheric gas aerosol equilibrium. 1. Thermodynamic model. *Aerosol Science and Technology* 1993; **19**(2): 157–181.

83 Kim, Y.P., Seinfeld, J.H. and Saxena, P. Atmospheric gas-aerosol equilibrium. 2. Analysis of common approximations and activity-coefficient calculation methods. *Aerosol Science and Technology* 1993; **19**(2): 182–198.

84 Meng, Z.Y., Seinfeld, J.H., Saxena, P. and Kim, Y.P. Atmospheric gas-aerosol equilibrium. 4. Thermodynamics of carbonates. *Aerosol Science and Technology* 1995; **23**(2): 131–154.

85 Binkowski, F.S. and Shankar, U. The regional particulate matter model. 1. Model description and preliminary results. *Journal of Geophysical Research – Atmospheres* 1995; **100**(D12): 26191–26209.

86 Jacobson, M.Z., Tabazadeh, A. and Turco, R.P. Simulating equilibrium within aerosols and nonequilibrium between gases and aerosols. *Journal of Geophysical Research – Atmospheres* 1996; **101**(D4): 9079–9091.
87 Wexler, A.S. and Seinfeld, J.H. The distribution of ammonium-salts among a size and composition dispersed aerosol. *Atmospheric Environment Part A – General Topics* 1990; **24**(5): 1231–1246.
88 Clegg, S.L., Brimblecombe, P. and Wexler, A.S. Thermodynamic model of the system H+-NH_4+-Na+-SO_4^{2-}-NB^{3-}-Cl^--H_2O at 298.15 K. *Journal of Physical Chemistry A* 1998; **102**(12): 2155–2171.
89 Nenes, A., Pandis, S.N. and Pilinis, C. ISORROPIA: a new thermodynamic equilibrium model for multiphase multicomponent inorganic aerosols. *Aquatic Geochemistry* 1998; **4**(1): 123–152.
90 Nenes, A., Pandis, S.N. and Pilinis, C. Continued development and testing of a new thermodynamic aerosol module for urban and regional air quality models. *Atmospheric Environment* 1999; **33**(10): 1553–1560.
91 Metzger, S. *Gas/Aerosol Partitioning: A Simplified Method for Global Modelling*. PhD Thesis. Universiteit Utrecht, Utrecht: 2000.
92 Metzger, S., Dentener, F., Krol, M., Jeuken, A. and Lelieveld, J. Gas/aerosol partitioning – 2. Global modeling results. *Journal of Geophysical Research – Atmospheres* 2002; **107**(D16): 4313.
93 Metzger, S., Dentener, F., Pandis, S. and Lelieveld, J. Gas/aerosol partitioning: 1. A computationally efficient model. *Journal of Geophysical Research – Atmospheres* 2002; **107**(D16): 4312.
94 Zhang, Y., Seigneur, C., Seinfeld, J.H., Jacobson, M., Clegg, S.L. and Binkowski, F.S. A comparative review of inorganic aerosol thermodynamic equilibrium modules: similarities, differences, and their likely causes. *Atmospheric Environment* 2000; **34**(1): 117–137.
95 Pilinis, C. Modelling atmospheric aerosols using thermodynamic arguments: a review. *Global Nest: The International Journal* 1999; **1**: 5–13.
96 Pitzer, K.S. Thermodynamics of electrolytes. 1. Theoretical basis and general equations. *Journal of Physical Chemistry* 1973; **77**(2): 268–277.
97 Clegg, S.L., Pitzer, K.S. and Brimblecombe, P. Thermodynamics of multicomponent, miscible, ionic-solutions. 2. Mixtures including unsymmetrical electrolytes. *Journal of Physical Chemistry* 1992; **96**(23): 9470–9479.
98 Carslaw, K.S., Luo, B.P. and Peter, T. An analytic-expression for the composition of aqueous HNO_3-H_2SO_4 stratospheric aerosols including gas-phase removal of HNO_3. *Geophysical Research Letters* 1995; **22**(14): 1877–1880.
99 Zaveri, R.A., Easter, R.C. and Wexler, A.S. A new method for multicomponent activity coefficients of electrolytes in aqueous atmospheric aerosols. *Journal of Geophysical Research – Atmospheres* 2005; **110**(D2): D02201.
100 Carslaw, K.S., Peter, T. and Clegg, S.L. Modeling the composition of liquid stratospheric aerosols. *Reviews of Geophysics* 1997; **35**(2): 125–154.
101 Clegg, S.L. and Brimblecombe, P. Equilibrium partial pressures of strong acids over concentrated saline solutions. 2. HCL. *Atmospheric Environment* 1988; **22**(1): 117–129.
102 Pitzer, K.S. and Mayorga, G. Thermodynamics of electrolytes. 2. Activity and osmotic coefficients for strong electrolytes with one or both ions univalent. *Journal of Physical Chemistry* 1973; **77**(19): 2300–2308.
103 Pitzer, K.S. Thermodynamics of electrolytes. 5. Effects of higher-order electrostatic terms. *Journal of Solution Chemistry* 1975; **4**(3): 249–265.
104 Pitzer, K.S. and Silvester, L.F. Thermodynamics of electrolytes. 6. Weak electrolytes including H_3PO_4. *Journal of Solution Chemistry* 1976; **5**(4): 269–278.
105 Pilinis, C., Pandis, S.N. and Seinfeld, J.H. Sensitivity of direct climate forcing by atmospheric aerosols to aerosol-size and composition. *Journal of Geophysical Research – Atmospheres* 1995; **100**(D9): 18739–18754.
106 Millero, F.J. and Pierrot, D. A chemical equilibrium model for natural waters. *Aquatic Chemistry* 1998; **4**: 153–199.

107 Pitzer, K.S. and Simonson, J.M. Thermodynamics of multicomponent, miscible, ionic systems – theory and equations. *Journal of Physical Chemistry* 1986; **90**(13): 3005–3009.
108 Clegg, S.L. and Brimblecombe, P. Application of a multicomponent thermodynamic model to activities and thermal-properties of 0-40-Mol Kg(-1) aqueous sulfuric-acid from less-than-200-K to 328-K. *Journal of Chemical and Engineering Data* 1995; **40**(1): 43–64.
109 Baltensperger, U., Streit, N., Weingartner, E., Nyeki, S., Prevot, A.S.H., Van Dingenen, R., Virkkula, A., Putaud, J.P., Even, A., Ten Brink, H., Blatter, A., Neftel, A. and Gäggeler, H.W. Urban and rural aerosol characterisation of summer smog events during the PIPAPO field campaign in Milan, Italy. *Journal of Geophysical Research – Atmospheres* 2002; **107**: 8193.
110 Stelson, A.W. and Seinfeld, J.H. Relative-humidity and temperature-dependence of the ammonium-nitrate dissociation-constant. *Atmospheric Environment* 1982; **16**(5): 983–992.
111 Harrison, R.M. and Sturges, W.T. Physicochemical speciation and transformation reactions of particulate atmospheric nitrogen and sulfur-compounds. *Atmospheric Environment* 1984; **18**(9): 1829–1833.
112 Pilinis, C., Seinfeld, J.H. and Grosjean, D. Water-content of atmospheric aerosols. *Atmospheric Environment* 1989; **23**(7): 1601–1606.
113 Crocker, L.S., Varsolona, R.J. and McCauley, J.A. Two methods for the measurement of the dissociation pressure of a crystalline hydrate. *Journal of Pharmaceutical and Biomedical Analysis* 1997; **15**(11): 1661–1665.
114 Raatikainen, T. and Laaksonen, A. Application of several activity coefficient models to water-organic-electrolyte aerosols of atmospheric interest. *Atmospheric Chemistry and Physics* 2005; **5**: 2475–2495.
115 Elliot, J.R. and Lira, C.T. *Introductory Chemical Engineering Thermodynamics*. Prentice Hall, Englewood Cliffs, NJ: 1998.
116 Renon, H. and Prausnitz, J.M. Local compositions in thermodynamic excess functions for liquid mixtures. *Aiche Journal* 1968; **14**(1): 135–144.
117 Abrams, D.S. and Prausnitz, J.M. Statistical thermodynamics of liquid-mixtures – new expression for excess Gibbs energy of partly or completely miscible systems. *Aiche Journal* 1975; **21**(1): 116–128.
118 Fredenslund, A., Jones, R.L. and Prausnitz, J.M. Group-contribution estimation of activity-coefficients in nonideal liquid-mixtures. *Aiche Journal* 1975; **21**(6): 1086–1099.
119 Liu, Q.L. and Cheng, Z.F. A modified UNIFAC model for the prediction of phase equilibrium for polymer solutions. *Journal of Polymer Science Part B – Polymer Physics* 2005; **43**(18): 2541–2547.
120 Derr, E.L. and Deal, C.H. Analytical solution of groups: correlation of activity coefficients through structural group parameters. *Institution of Chemical Engineers Symposium Series* 1969; **32**: 44–51.
121 Koo, B., Ansari, A.S. and Pandis, S.N. Integrated approaches to modelling the organic and inorganic atmospheric aerosol components. *Atmospheric Environment* 2003; **37**(34): 4757–4768.
122 Marcolli, C. and Peter, T. Water activity in polyol/water systems: new UNIFAC parameterization. *Atmospheric Chemistry and Physics* 2005; **5**: 1501–1527.
123 Gmehling, J., Li, J.D. and Schiller, M. A modified unifac model. 2. Present parameter matrix and results for different thermodynamic properties. *Industrial & Engineering Chemistry Research* 1993; **32**(1): 178–193.
124 Gmehling, J., Lohmann, J., Jakob, A., Li, J.D. and Joh, R. A modified UNIFAC (Dortmund) model. 3. Revision and extension. *Industrial & Engineering Chemistry Research* 1998; **37**(12): 4876–4882.
125 Weidlich, U. and Gmehling, J. A modified unifac model. 1. Prediction of Vle, He, and gamma-infinity. *Industrial & Engineering Chemistry Research* 1987; **26**(7): 1372–1381.
126 Larsen, B.L., Rasmussen, P. and Fredenslund, A. A modified unifac group-contribution model for prediction of phase-equilibria and heats of mixing. *Industrial & Engineering Chemistry Research* 1987; **26**(11): 2274–2286.
127 Zhang, S.J., Hiaki, T., Hongo, M. and Kojima, K. Prediction of infinite dilution activity coefficients in aqueous solutions by group contribution models: a critical evaluation. *Fluid Phase Equilibria* 1998; **144**(1–2): 97–112.

128 Ninni, L., Camargo, M.S. and Meirelles, A.J.A. Water activity in polyol systems. *Journal of Chemical and Engineering Data* 2000; **45**(4): 654–660.
129 Peng, C., Chan, M.N. and Chan, C.K. The hygroscopic properties of dicarboxylic and multifunctional acids: measurements and UNIFAC predictions. *Environmental Science & Technology* 2001; **35**(22): 4495–4501.
130 Chan, M.N., Choi, M.Y., Ng, N.L. and Chan, C.K. Hygroscopicity of water-soluble organic compounds in atmospheric aerosols: amino acids and biomass burning derived organic species. *Environmental Science & Technology* 2005; **39**(6): 1555–1562.
131 Hansen, H.K., Rasmussen, P., Fredenslund, A., Schiller, M. and Gmehling, J. Vapor-liquid-equilibria by unifac group contribution. 5. Revision and extension. *Industrial & Engineering Chemistry Research* 1991; **30**(10): 2352–2355.
132 Na, H.S., Arnold, S. and Myerson, A.S. Water activity in supersaturated aqueous-solutions of organic solutes. *Journal of Crystal Growth* 1995; **149**(3–4): 229–235.
133 Peng, C.G. and Chan, C.K. The water cycles of water-soluble organic salts of atmospheric importance. *Atmospheric Environment* 2001; **35**(7): 1183–1192.
134 Prenni, A.J., DeMott, P.J., Kreidenweis, S.M., Sherman, D.E., Russell, L.M. and Ming, Y. The effects of low molecular weight dicarboxylic acids on cloud formation. *Journal of Physical Chemistry A* 2001; **105**(50): 11240–11248.
135 Mukai, H. and Ambe, Y. Characterization of a humic acid-like brown substance in airborne particulate matter and tentative identification of its origin. *Atmospheric Environment* 1986; **20**(5): 813–819.
136 Havers, N., Burba, P., Lambert, J. and Klockow, D. Spectroscopic characterization of humic-like substances in airborne particulate matter. *Journal of Atmospheric Chemistry* 1998; **29**(1): 45–54.
137 Zappoli, S., Andracchio, A., Fuzzi, S., Facchini, M.C., Gelencser, A., Kiss, G., Krivacsy, Z., Molnar, A., Meszaros, E., Hansson, H.-C., Rosman, K. and Zebuhr Y. Inorganic, organic and macromolecular components of fine aerosol in different areas of Europe in relation to their water solubility. *Atmospheric Environment* 1999; **33**(17): 2733–2743.
138 Decesari, S., Facchini, M.C., Matta, E., Lettini, F., Mircea, M., Fuzzi, S., Tagliavini, E. and Putaud J.-P. Chemical features and seasonal variation of fine aerosol water-soluble organic compounds in the Po Valley, Italy. *Atmospheric Environment* 2001; **35**(21): 3691–3699.
139 Chan, M.N. and Chan, C.K. Hygroscopic properties of two model humic-like substances and their mixtures with inorganics of atmospheric importance. *Environmental Science & Technology* 2003; **37**(22): 5109–5115.
140 Brooks, S.D., DeMott, P.J. and Kreidenweis, S.M. Water uptake by particles containing humic materials and mixtures of humic materials with ammonium sulfate. *Atmospheric Environment* 2004; **38**(13): 1859–1868.
141 Mikhailov, E., Vlasenko, S., Niessner, R. and Poschl, U. Interaction of aerosol particles composed of protein and salts with water vapor: hygroscopic growth and microstructural rearrangement. *Atmospheric Chemistry and Physics* 2004; **4**: 323–350.
142 Parsons, M.T., Knopf, D.A. and Bertram, A.K. Deliquescence and crystallization of ammonium sulfate particles internally mixed with water-soluble organic compounds. *Journal of Physical Chemistry A* 2004; **108**(52): 11600–11608.
143 Weingartner, E., Gysel, M. and Baltensperger, U. Hygroscopicity of aerosol particles at low temperatures. 1. New low-temperature H-TDMA instrument: setup and first applications. *Environmental Science & Technology* 2002; **36**(1): 55–62.
144 Yan, W.D., Topphoff, M., Rose, C. and Gemhling, J. Prediction of vapor-liquid equilibria in mixed-solvent electrolyte systems using the group contribution concept. *Fluid Phase Equilibria* 1999; **162**(1–2): 97–113.
145 Ferron, G.A., Karg, E., Busch, B. and Heyder, J. Hygroscopicity of ambient particles. *Journal of Aerosol Science* 1999; **30**: S19–S20.

146 Massling, A., Wiedensohler, A., Busch, B., Neususs, C., Quinn, P., Bates, T. and Covert, D. Hygroscopic properties of different aerosol types over the Atlantic and Indian Oceans. *Atmospheric Chemistry and Physics* 2003; **3**: 1377–1397.
147 Chen, C.C., Britt, H.I., Boston, J.F. and Evans, L.B. Local composition model for excess Gibbs energy of electrolyte systems. 1. Single solvent, single completely dissociated electrolyte systems. *Aiche Journal* 1982; **28**(4): 588–596.
148 Chen, C.C. and Evans, L.B. A local composition model for the excess Gibbs energy of aqueous-electrolyte systems. *Aiche Journal* 1986; **32**(3): 444–454.
149 Ananth, M.S. and Ramachandran, S. Self-consistent local composition model of electrolyte-solutions. *Aiche Journal* 1990; **36**(3): 370–386.
150 Liu, Y., Harvey, A.H. and Prausnitz, J.M. Thermodynamics of concentrated electrolyte-solutions. *Chemical Engineering Communications* 1989; **77**: 43–66.
151 Liu, Y., Wimby, M. and Gren, U. An activity-coefficient model for electrolyte systems. *Computers & Chemical Engineering* 1989; **13**(4–5): 405–410.
152 Kikic, I., Fermeglia, M. and Rasmussen, P. Unifac prediction of vapor-liquid-equilibria in mixed-solvent salt systems. *Chemical Engineering Science* 1991; **46**(11): 2775–2780.
153 Sander, B., Fredenslund, A. and Rasmussen, P. Calculation of vapor-liquid-equilibria in mixed-solvent salt systems using an extended uniquac equation. *Chemical Engineering Science* 1986; **41**(5): 1171–1183.
154 Achard, C., Dussap, C.G. and Gros, J.B. Representation of vapor-liquid-equilibria in water-alcohol electrolyte mixtures with a modified unifac group-contribution method. *Fluid Phase Equilibria* 1994; **98**: 71–89.
155 Li, J.D., Polka, H.M. and Gmehling, J. A G(E) model for single and mixed-solvent electrolyte systems. 1. Model and results for strong electrolytes. *Fluid Phase Equilibria* 1994; **94**: 89–114.
156 Cruz, C.N. and Pandis, S.N. Deliquescence and hygroscopic growth of mixed inorganic-organic atmospheric aerosol. *Environmental Science & Technology* 2000; **34**(20): 4313–4319.
157 Choi, M.Y. and Chan, C.K. The effects of organic species on the hygroscopic behaviors of inorganic aerosols. *Environmental Science & Technology* 2002; **36**(11): 2422–2428.
158 Lightstone, J.M., Onasch, T.B., Imre, D. and Oatis, S. Deliquescence, efflorescence, and water activity in ammonium nitrate and mixed ammonium nitrate/succinic acid microparticles. *Journal of Physical Chemistry A* 2000; **104**(41): 9337–9346.
159 Prenni, A.J., De Mott, P.J. and Kreidenweis, S.M. Water uptake of internally mixed particles containing ammonium sulfate and dicarboxylic acids. *Atmospheric Environment* 2003; **37**(30): 4243–4251.
160 Kotchenruther, R. and Hobbs, P.V. Humidification factors of aerosols from biomass burning in Brazil. *Journal of Geophysical Research – Atmospheres* 1998; **103**: 32081–32090.
161 Rissler, J., Swietlicki, E., Zhou, J., Roberts, G., Andreae, M.O., Gatti, L.V. and Artaxo, P. Physical properties of the sub-micrometer aerosol over the Amazon rain forest during the wet-to-dry season transition - comparison of modeled and measured CCN concentrations. *Atmospheric Chemistry and Physics* 2004; **4**: 2119–2143.
162 Hameri, K., Charlson, R. and Hansson, H.C. Hygroscopic properties of mixed ammonium sulfate and carboxylic acids particles. *Aiche Journal* 2002; **48**(6): 1309–1316.
163 Clegg, S.L. and Seinfeld, J.H. Thermodynamic models of aqueous solutions containing inorganic electrolytes and dicarboxylic acids at 298.15K. II. Systems including dissociation equilibria. *Journal of Physical Chemistry* 2006; **110**: 5718–5734.
164 McFiggans, G., Alfarra, M.R., Allan, J., Bower, K., Coe, H., Cubison, M., Topping, D, Williams, P., Decesari, S., Facchini, C. and Fuzzi S Simplification of the representation of the organic component of atmospheric particulates. *Faraday Discussions* 2005; **130**: 341–362.
165 Bilde, M. and Svenningsson, B. CCN activation of slightly soluble organics: the importance of small amounts of inorganic salt and particle phase. *Tellus Series B – Chemical and Physical Meteorology* 2004; **56**(2): 128–134.

166 Shulman, M.L., Jacobson, M.C., Charlson, R.J., Synovec, R.E. and Young, T.E. Dissolution behavior and surface tension effects of organic compounds in nucleating cloud droplets. *Geophysical Research Letters* 1996; **23**: 277–280. [*Geophysical Research Letters* 1996; **23**(5): 603–603.]
167 Rogge, W.F., Mazurek, M.A., Hildemann, L.M., Cass, G.R. and Simoneit, B.R.T. Quantification of urban organic aerosols at a molecular-level – identification, abundance and seasonal-variation. *Atmospheric Environment Part A – General Topics* 1993; **27**(8): 1309–1330.
168 Decesari, S., Facchini, M.C., Fuzzi, S. and Tagliavini, E. Characterization of water-soluble organic compounds in atmospheric aerosol: a new approach. *Journal of Geophysical Research – Atmospheres* 2000; **105**(D1): 1481–1489.
169 Brooks, S.D., Wise, M.E., Cushing, M. and Tolbert, M.A. Deliquescence behavior of organic/ammonium sulfate aerosol. *Geophysical Research Letters* 2002; **29**(19): doi: 10.1029/2002GL014733.
170 Saxena, P. and Hildemann, L.M. Water-soluble organics in atmospheric particles: a critical review of the literature and application of thermodynamics to identify candidate compounds. *Journal of Atmospheric Chemistry* 1996; **24**(1): 57–109.
171 Dick, W.D., Saxena, P. and McMurry, P.H. Estimation of water uptake by organic compounds in submicron aerosols measured during the Southeastern Aerosol and Visibility Study. *Journal of Geophysical Research – Atmospheres* 2000; **105**(D1): 1471–1479.
172 Gysel, M., Weingartner, E., Nyeki, S., Paulsen, D., Baltensperger, U., Galambos, I. and Kiss, G. Hygroscopic properties of water-soluble matter and humic-like organics in atmospheric fine aerosol. *Atmospheric Chemistry and Physics* 2004; **4**: 35–50.
173 Varutbangkul, V., Brechtel, F.J., Bahreini, R., Ng, N.L., Keywood, M.D., Kroll, J.H., Flagan, R.C., Seinfeld, J.H., Lee, A. and Goldstein, A.H. Hygroscopicity of secondary organic aerosols formed by oxidation of cycloalkanes, monoterpenes, sesquiterpenes, and related compounds. *Atmospheric Chemistry and Physics Discussions* 2006; **6**: 1121–1177.
174 Tang, I.N. and Munkelwitz, H.R. Composition and temperature-dependence of the deliquescence properties of hygroscopic aerosols. *Atmospheric Environment Part A – General Topics* 1993; **27**(4): 467–473.
175 Tabazadeh, A. and Toon, O.B. The role of ammoniated aerosols in cirrus cloud nucleation. *Geophysical Research Letters* 1998; **25**(9): 1379–1382.
176 Onasch, T.B., Siefert, R.L., Brooks, S.D., Prenni, A.J., Murray, B., Wilson, M.A. and Tolbert M.A. Infrared spectroscopic study of the deliquescence and efflorescence of ammonium sulfate aerosol as a function of temperature. *Journal of Geophysical Research – Atmospheres* 1999; **104**(D17): 21317–21326.
177 Parsons, M.T., Mak, J., Lipetz, S.R. and Bertram, A.K. Deliquescence of malonic, succinic, glutaric, and adipic acid particles. *Journal of Geophysical Research – Atmospheres* 2004; **109**(D6): D06212, doi:10.1029/2003JD004075.
178 Clegg, S.L., Milioto, S. and Palmer, D.A. Osmotic and activity coefficients of aqueous $(NH_4)_{(2)}SO_4$ as a function of temperature, and aqueous $(NH_4)_2SO_{(4)}$-H_2SO_4 mixtures at 298.15 K and 323.15 K. *Journal of Chemical and Engineering Data* 1996; **41**(3): 455–467.
179 Chylek, P. and Wong, J.G.D. Erroneous use of the modified Kohler equation in cloud and aerosol physics applications. *Journal of the Atmospheric Sciences* 1998; **55**(8): 1473–1477.
180 Kreidenweis, S.M., Koehler, K., DeMott, P., Prenni, A.J., Carrico, C. and Ervens, B. Water activity and activation diameters from hygroscopicity data. Part 1: theory and application to inorganic salts. *Atmospheric Chemistry and Physics Discussions* 2005; **5**: 287–323.
181 Li, Z.B. and Lu, B.C.Y. Surface tension of aqueous electrolyte solutions at high concentrations – representation and prediction. *Chemical Engineering Science* 2001; **56**(8): 2879–2888.
182 Hameri, K., Vakeva, M., Hansson, H.C. and Laaksonen, A. Hygroscopic growth of ultrafine ammonium sulphate aerosol measured using an ultrafine tandem differential mobility analyzer. *Journal of Geophysical Research – Atmospheres* 2000; **105**(D17): 22231–22242.

183 Hu, Y.F. and Lee, H. Prediction of the surface tension of mixed electrolyte solutions based on the equation of Patwardhan and Kumar and the fundamental Butler equations. *Journal of Colloid and Interface Science* 2004; **269**(2): 442–448.
184 Hanel, G. The properties of atmospheric aerosol particles as functions of the relative humidity at thermodynamic equilibrium with the surrounding moist air. *Advances in Geophysics* 1976; **19**: 73–188.
185 Chen, J.P. Theory of deliquescence and modified Kohler curves. *Journal of the Atmospheric Sciences* 1994; **51**(23): 3505–3516.
186 Brechtel, F.J. and Kreidenweis, S.M. Predicting particle critical supersaturation from hygroscopic growth measurements in the humidified TDMA. Part I: theory and sensitivity studies. *Journal of the Atmospheric Sciences* 2000; **57**(12): 1854–1871.
187 Brechtel, F.J. and Kreidenweis, S.M. Predicting particle critical supersaturation from hygroscopic growth measurements in the humidified TDMA. Part II: laboratory and ambient studies. *Journal of the Atmospheric Sciences* 2000; **57**(12): 1872–1887.
188 Tuckermann, R. and Cammenga, H.K. The surface tension of aqueous solutions of some atmospheric water-soluble organic compounds. *Atmospheric Environment* 2004; **38**(36): 6135–6138.
189 Facchini, M.C., Decesari, S., Mircea, M., Fuzzi, S. and Loglio, G. Surface tension of atmospheric wet aerosol and cloud/fog droplets in relation to their organic carbon content and chemical composition. *Atmospheric Environment* 2000; **34**(28): 4853–4857.
190 Poling, B.E., Prausnitz, J.M. and O'Connell, J.P. *The Properties of Gases and Liquids*. McGraw-Hill Professional, New York: 2000.
191 Topping, D. *Modelling the Hygroscopic Properties of Atmospheric Aerosols*. University of Manchester, Manchester: 2005.
192 Lamer, V.K. and Gruen, R. A direct test of Kelvin equation connecting vapour pressure and radius of curvature. *Transactions of the Faraday Society* 1952; **48**(5): 410–415.
193 Reiss, H. and Koper, G.J.M. The kelvin relation – stability, fluctuation, and factors involved in measurement. *Journal of Physical Chemistry* 1995; **99**(19): 7837–7844.
194 Raymond, T.M. and Pandis, S.N. Cloud activation of single-component organic aerosol particles. *Journal of Geophysical Research – Atmospheres* 2002; **107**(D24): 4787.
195 Brechtel, F.J. *Predicting Particle Critical Supersaturation From Hygroscopic Growth Measurements in the Humidified Tandem Differential Mobility Analyzer*. Colorado State University, Colorado: 1998.
196 Laaksonen, A., Korhonen, P., Kulmala, M. and Charlson, R.J. Modification of the Kohler equation to include soluble trace gases and slightly soluble substances. *Journal of the Atmospheric Sciences* 1998; **55**(5): 853–862.
197 Mirabel, P., Reiss, H. and Bowles, R.K. A theory for the deliquescence of small particles. *Journal of Chemical Physics* 2000; **113**(18): 8200–8205.
198 Ghosal, S. and Hemminger, J.C. Effect of water on the HNO_3 pressure dependence of the reaction between gas-phase HNO_3 and NaCl surfaces. *Journal of Physical Chemistry A* 1999; **103**(25): 4777–4781.
199 Finlayson-Pitts, B.J. and Hemminger, J.C. Physical chemistry of airborne sea salt particles and their components. *Journal of Physical Chemistry A* 2000; **104**(49): 11463–11477.
200 Pagels, J., Strand, M., Rissler, J., Szpila, A., Gudmundsson, A., Bohgard, M., Lillieblad, L., Sanati, M. and Swietlicki, E. Characteristics of aerosol particles formed during grate combustion of moist forest residue. *Journal of Aerosol Science* 2003; **34**(8): 1043–1059.
201 Virkkula, A., Van Dingenen, R., Raes, F. and Hjorth, J. Hygroscopic properties of aerosol formed by oxidation of limonene, α-pinene, and β-pinene. *Journal of Geophysical Research* 1999; **104**(D3): 3569–3580.
202 Potukuchi, S. and Wexler, A.S. Identifying solid-aqueous-phase transitions in atmospheric aerosols .2. Acidic solutions. *Atmospheric Environment* 1995; **29**(22): 3357–3364.

Chapter 7
Stratospheric Chemistry: Aerosols and the Ozone Layer

Rob MacKenzie

7.1 Introduction

Aerosol particles make the stratosphere evident. At times, particularly after explosive volcanic eruptions, this part of the atmosphere – remote from our everyday experience and from the vigorous atmospheric cleaning of the tropospheric weather systems – becomes visible or evident even to the naked eye [1, 2]. It is surprising, therefore, that the presence of aerosol in the stratosphere seems to have played little or no part in the discovery of the stratosphere [3]. In fact, some accounts of the upper atmosphere just prior to discovery of the stratosphere gave a quite misleading impression: W.M. Davies commented that

> *the upper air, pure and dry,* free from clouds and dust, *far from the surface of the Earth and out of reach of ordinary convectional action, must possess a low temperature and must change its temperature slowly and by small amounts*' ([4], cited from, [5]).

In the century since this statement was made it has become abundantly clear how wrong Davies was, and that atmospheric chemistry and climate are inextricably bound to the chemistry and physics of aerosol and cloud in the stratosphere. The idea that temperature decreased continuously throughout the whole depth of the atmosphere was falsified by the careful measurements of Léon Tesserenc de Bort and Richard Assmann at the beginning of the twentieth century [3, 4]. The careful measurements required to prove the existence of a continuous, ever-present, aerosol layer in the stratosphere had to wait for Christian Junge, six decades later. The stratospheric aerosol layer is also known as the *Junge layer*, in recognition of his pioneering observations [6].

Figure 7.1 shows example temperature, aerosol and ozone profiles – in this case from the HALOE instrument on board the UARS satellite (see [7] for an overview). The stratosphere is a layer of increasing temperature and, hence, increasing atmospheric stability: convection is suppressed because warm, less dense air sits atop colder, denser air. Vertical transport in the stratosphere occurs not by convection, as often in the troposphere, but by a pumping action due to the interaction between planetary scale waves and the mean stratospheric flow (e.g. [8, 9]). This slow, global-scale overturning of the stratosphere is called the *Brewer–Dobson circulation*, having been deduced by Alan Brewer and G.M.B. Dobson from measurements of ozone, helium and water in the stratosphere. The Brewer–Dobson circulation transports air upwards in the

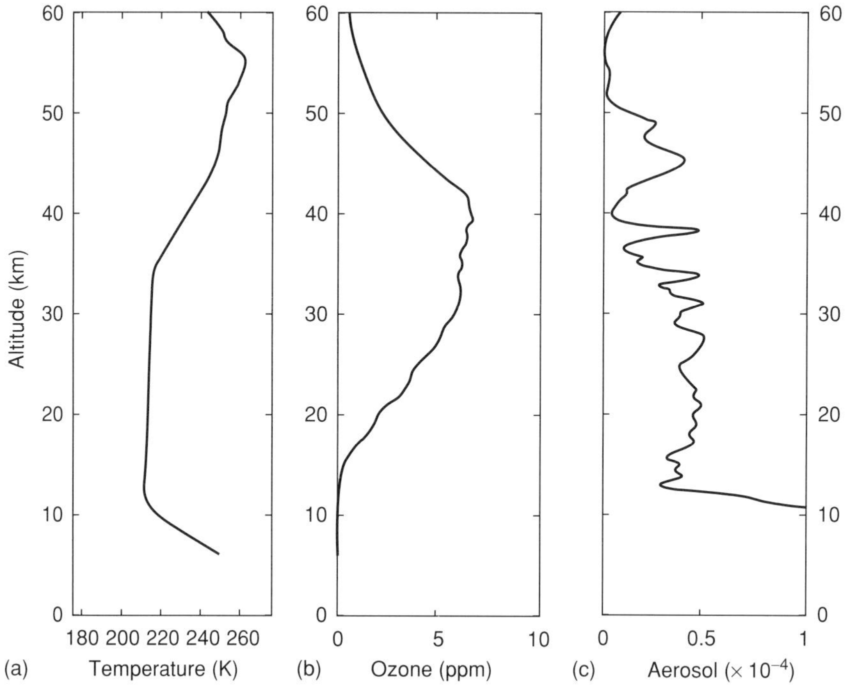

Figure 7.1 (a) Temperature (K), (b) ozone mixing ratio (ppm) and (c) aerosol amount ($m^2\ kg^{-1}$) from a single profile of the HALOE instrument on board the UARS satellite. The time of the profile is 15:45:29 UTC on 21 November 2005, and the location is 45.88°N, 5.33°E. The aerosol metric shown is the aerosol extinction at 5.26 μm, divided by the atmospheric density, a metric which helps to emphasise the stratospheric aerosol layer. Individual 'spikes' and layers in the aerosol plot are sensitive to the retrieval algorithm and to the exact meteorology on this day, but the general shape of the aerosol layer is more robust. The slightly 'stepped' ozone profile below its maximum is presumably also an artefact of the retrieval. The troposphere is at about 11 km; below that altitude the aerosol amount increases rapidly due to tropospheric cloud. (The data used in this figure were acquired using the GES-DISC Interactive Online Visualization and Analysis Infrastructure (Giovanni) as part of the NASA's Goddard Earth Sciences (GES) Data and Information Services Center (DISC) – http://disc.sci.gsfc.nasa.gov/techlab/giovanni/.)

tropical stratosphere, polewards in the middle latitudes and downwards in the polar regions. The circulation is particularly strong in the winter hemisphere, when planetary waves are generated by the large meridional temperature gradient, and is stronger in northern hemisphere winter than in southern hemisphere winter, because of the propensity for continental topography, concentrated in the northern hemisphere, to generate waves. The orderly procession of air from tropics to pole is muddled by mixing, particularly in the middle latitude *surf zone*, where planetary waves break in a two-dimensional (quasi-horizontal) analogue of waves breaking on a beach, but the general picture of tropical ascent and polar descent is obvious in zonal-mean constituent maps [7–9]. In the winter hemisphere, the strong meridional temperature gradient establishes strong westerly winds and a low-pressure vortex at high latitudes. The air inside the polar vortex is isolated from that outside and descends throughout the winter period (e.g. [8–12]).

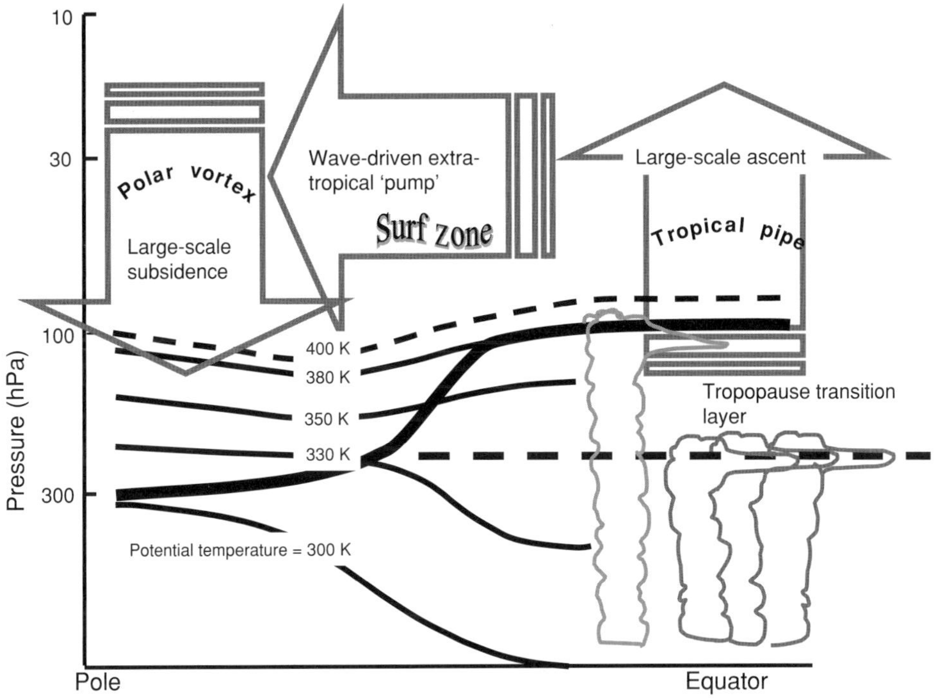

Figure 7.2 A schematic zonal cross-section of a wintertime hemisphere, showing the tropopause (thick line) and the Brewer–Dobson circulation (broad arrows). Thin lines are potential temperature surfaces, or isentropes, labelled in Kelvin. Note that there is the possibility of isentropic (i.e. constant potential temperature) exchange of air between the upper troposphere and the lowermost parts of the extra-tropical stratosphere, whilst transport by the Brewer–Dobson circulation causes cross-isentrope transport in the rest of the stratosphere. Below the tropical tropopause is a region into which the Brewer–Dobson circulation extends, but which is also occasionally perturbed by convection. This region is called the tropopause transition layer (TTL). Trace gases (OCS and SO_2) and ultrafine aerosol particles, which are the source of the stratospheric aerosol layer, enter the stratosphere through the TTL. (The figure is a redrafting and adaptation of Figure 7.3 of Holton *et al.* [13].)

Figure 7.2 tries to capture schematically the major elements of the stratospheric transport described above. One important consequence of the Brewer–Dobson circulation is that air entering the stratosphere in the tropics is more likely to be carried around the stratosphere than air entering in the middle or high latitudes. Aerosol particles formed at, or transported to, the tropical tropopause will have a high chance of entering the stratosphere and staying there for some time (Section 7.2, below). Similarly, injections of gas and aerosol particles from explosive volcanic eruptions in tropical regions are likely to stay longer, and be transported further, than injections at higher latitudes (Section 7.6, below).

7.2 General properties of the stratospheric aerosol layer

The stratospheric aerosol is composed of submicrometer sulphuric acid solution droplets with a total number density of 5–10 cm^{-3} and a geometric mean radius of about 70 nm. Many trace elements other than sulphuric acid have been detected in the aerosol by mass spectrometry [14], but it is a very good approximation in most circumstances to regard the aerosol droplets as pure sulphuric acid solution. Near the equator, mixing with tropospheric aerosol that has been rapidly transported from near the earth surface can account for much of the variability in composition. The trace element composition also indicates an influence of extra-terrestrial material (see Section 7.7, below). Changes in temperature can cause semi-volatile acid gases to condense onto the aerosol, changing its composition (see Section 7.5, below). Setting aside these modifications to the basic properties of the stratospheric aerosol for a moment, two substantive issues remain: the source of the sulphuric acid that makes up the bulk of the aerosol mass, and the region(s) of nucleation of the particles that make up the aerosol layer.

During times of volcanic perturbation of the aerosol layer, the vast majority of the sulphur in the aerosol comes from the sulphur dioxide released in the volcanic plume (see Section 7.6, below). Now, however, 15 years since the last major volcanic injection, it is clear that there is a small non-volcanic contribution, presumably from the oxidation of sulphur-containing compounds that are transported to the stratosphere. Sulphuric acid has a very short tropospheric lifetime, due to rapid uptake into clouds and washout. Less soluble sulphur-containing gases are more likely to contribute to the background aerosol sulphur: carbonyl sulphide (OCS) and sulphur dioxide (SO_2) being the most important of these [15]. The other major component of the aerosol particles, water, is taken up to produce concentrated sulphuric acid solutions of a specific composition at a given temperature, corresponding to a temperature-dependent minimum in the free energy of the system [16, 17].

The presence of an aerosol phase requires either a continuous source of particles from outside the stratosphere or regions of the stratosphere where nucleation of the aerosol phase takes place. Seed particles may enter the stratosphere from below through explosive injections by volcanoes or through troposphere-to-stratosphere transport of air parcels; particles may enter from above through the sedimentation of extra-terrestrial material. Volatility studies – using a pair of condensation nuclei counters, one of which has a heated channel – suggest that approximately half the stratospheric aerosol particles contain refractory cores [18]. These cores could be heterogeneous nuclei upon which sulphuric acid condenses, or could be the result of coagulation between sulphuric acid particles and solid, perhaps meteoritic, particles. In any case, it is clear that not all the stratospheric aerosol is nucleated on solid seed particles and some homogeneous nucleation must also occur. The tropical tropopause region has been identified as a region where such nucleation can occur and can transport newly formed aerosol into the stratosphere [19]. Under volcanically quiescent conditions, therefore, the troposphere may supply the majority of the aerosol *number* to the stratosphere, through nucleation in, and subsequent transport through, the tropical tropopause, while the major part of the aerosol *mass* is supplied by in situ oxidation of insoluble, sulphur-containing gases, such as OCS, in the stratosphere. The major steps in the oxidation of OCS are (e.g. [20])[1]

$$OCS + h\nu(\lambda < 300\,\text{nm}) \rightarrow CO + S \tag{7.1}$$

$$S + O_2 \rightarrow SO + O \tag{7.2}$$

[1] IUPAC summaries for most of the chemical reactions discussed in this chapter are available online at http://www.iupac-kinetic.ch.cam.ac.uk/. The IUPAC database and a similar database organised by NASA's Jet Propulsion Laboratory (http://jpldataeval.jpl.nasa.gov/) form the consensus set of chemical reactions from which numerical models of stratospheric chemistry are constructed.

$$SO + O_2 \rightarrow SO_2 + O \quad (7.3)$$

$$SO_2 + OH + M \rightarrow HSO_3 + M \quad (7.4)$$

$$HSO_3 + O_2 \rightarrow HO_2 + SO_3 \quad (7.5)$$

$$SO_3 + H_2O \rightarrow H_2SO_4 \quad (7.6)$$

Once formed, the stratospheric aerosol is persistent. The only significant loss processes in the lower stratosphere are coagulation and bulk transport into the troposphere as a result of atmospheric dynamics. The half-life for coagulation of stratospheric particles at 50 hPa is greater than 1 year. Similarly, transport to the troposphere is slow away from the extra-tropical tropopause. For aerosol particles in the (initially tropical) air masses that are carried to higher altitudes by the Brewer–Dobson circulation, sedimentation becomes significant, even for 10- to 100-nm particles, because of the decrease of atmospheric density with height [21]. Also, in the upper stratosphere temperatures rise, so that the sulphuric acid (and water) in the aerosol must volatilise in order to maintain equilibrium between aerosol and gas phase. At heights above 35 km ($\sim$3 hPa, 255 K), aerosol particles will evaporate completely from an air parcel within 1 day [21]. When this air subsequently descends into the polar stratosphere, nucleation of new sulphuric acid particles can occur.

7.3 General stratospheric chemistry

The presence of ozone in the stratosphere can be deduced knowing how oxygen molecules interact with short-wavelength, high-energy, solar photons:

$$O_2 + h\nu(\lambda < 243\text{ nm}) \rightarrow O + O \quad (7.7)$$

$$O + O_2 + M \rightarrow O_3 + M \quad (7.8)$$

$$O_3 + h\nu(\lambda < 1180\text{ nm}) \rightarrow O + O_2 \quad (7.9)$$

$$O + O_3 \rightarrow O_2 + O_2 \quad (7.10)$$

(A fifth reaction, the self-reaction of oxygen atoms, is too slow to be important in the stratosphere.) However, when reaction and photolysis rates from the laboratory are used with this set of reactions and known solar photon fluxes, the calculated steady-state ozone concentration is much higher than observations. Reaction (7.4) turns out to be too slow to account for all the removal of ozone and atomic oxygen. (Because of their tight coupling we can group together ozone and atomic oxygen and call them *odd oxygen*.)

7.3.1 Catalysed ozone loss

If reaction (7.10) is too slow to be the sole loss mechanism for odd oxygen, there must be other reaction partners for ozone and atomic oxygen. There are no important stoichiometric titration reactants for odd oxygen in the atmosphere (i.e. reactants for the one-way loss process A + B $\rightarrow$ products). Instead, catalytic destruction of odd oxygen occurs:

$$X + O_3 \rightarrow XO + O_2 \quad (7.11)$$

$$XO + O \rightarrow X + O_2 \quad (7.12)$$

$$\text{Net } O + O_3 \rightarrow 2O_2$$

Reactions (7.11) and (7.12) accomplish reaction (7.10) in two steps, with reagent X acting as a catalyst. Since X is not destroyed, but is rather regenerated to take part in another two-reaction odd-oxygen-destroying cycle, chemical species present at much lower concentrations than odd oxygen can effectively reduce the odd-oxygen concentration. A great deal of laboratory and atmospheric observation has gone into finding species that can act like X in reactions (7.11) and (7.12) above: important examples are H, OH, NO, Cl and Br. Note that these examples are generally highly reactive free radicals and that the key elements are hydrogen, nitrogen and the halogens. Work in understanding the role of these trace gas catalysts in the distribution of ozone won Paul Crutzen, Mario Molina and Sherry Rowland the Nobel Prize in 1995.

It is often convenient to group the X/XO species together in *chemical families*, since they are interconverted rapidly. Below, we shall discuss $NO_x (= N + NO + NO_2 + NO_3$, where the abundance of compounds is counted by number, i.e. moles, not mass), ClO_x $(= Cl + ClO + 2Cl_2O_2)$, HO_x $(= H + HO + HO_2)$ and so on. It is also useful to consider the total amount of potentially reactive nitrogen, which is collectively known as $NO_y (= NO_x + HONO + 2N_2O_5 + HNO_3 + HNO_4 + ClONO_2)$. Similarly, the total amount of potentially reactive chlorine, the *inorganic chlorine*, is collectively known as $Cl_y (= ClO_x + HOCl + HCl + ClONO_2 + 2Cl_2 + BrCl)$. Quantities like $\{NO_y - NO_x\}$ give the amount of potentially reactive compound not (yet) in reactive form; this is the *reservoir* of potentially reactive compound, and the non-X/XO compounds (e.g. N_2O_5, HNO_3, HNO_4, HCl and $ClONO_2$) are called *reservoir compounds*. Notice that some compounds contribute to more than one reservoir.

The decrease in atmospheric density with height makes the ozone production reaction (7.8) less efficient as altitude increases. The increase in the intensity of short-wave solar radiation with height makes the O-atom-production reactions (7.7) and (7.9) more efficient as altitude increases. Hence, odd oxygen is mostly O atoms at high altitude (above ~85 km) and mostly O_3 molecules at lower altitudes. In the middle of the ozone layer, daytime O-atom mixing ratios are about one-millionth of the total odd oxygen mixing ratio and, at night, the O-atom mixing ratio rapidly falls to zero. Therefore, reaction cycles that destroy odd oxygen, without requiring the presence of O atoms, are very important in the main stratospheric ozone layer. These ozone-specific cycles can take a variety of forms: (i) formation of an XO compound which itself reacts with ozone, or (ii) formation of a compound from two XO species and subsequent elimination of O_2 by the breaking of a bond other than the O—O bond that formed when the two XO species combined (see [22, 23] for more details). An important example of an ozone-specific cycle is the chlorine dimer cycle that is responsible for most of the ozone destruction in the 'ozone hole':

$$2(Cl + O_3 \rightarrow ClO + O_2) \tag{7.13}$$

$$ClO + ClO + M \rightarrow Cl_2O_2 + M \tag{7.14}$$

$$Cl_2O_2 + h\nu \rightarrow Cl + ClOO \tag{7.15}$$

$$ClOO + M \rightarrow Cl + O_2 + M \tag{7.16}$$

$$\text{Net } 2O_3 + h\nu \rightarrow 3O_2$$

7.3.2 Initiation, null cycles and termination of stratospheric chain reactions

All stratospheric chain reactions are initiated by the action of sunlight. Usually the process is direct photolysis of a *source gas*; for example

$$CF_2Cl_2 + h\nu(\lambda < 227\text{nm}) \rightarrow CF_2Cl + Cl \quad (7.17)$$

where CF_2Cl_2 is CFC-12.[2] Alternatively, initial attack on a source gas may be by an electronically excited[3] O atom; for example

$$N_2O + O(^1D) \rightarrow NO + NO \quad (7.18)$$

$$H_2O + O(^1D) \rightarrow OH + OH \quad (7.19)$$

Since these reactions require a significant mixing ratio of O atoms, their efficiency increases with altitude. The result of this high-altitude loss and the upward atmospheric transport (Section 7.1) is that for CFC-12, for instance, mixing ratios fall from 0.5 to 0.05 pptv in the first 20 km of the tropical stratosphere. Reactions (7.17)–(7.19) show that X/XO species are formed directly from the reaction of source gases, and so the odd-oxygen destruction cycles described above can proceed immediately.

Along with odd-oxygen-destroying chain reactions, the reactive compounds released from source gases can take part in other cycles. Reaction cycles which regenerate the reactive compounds – but have no direct effect on odd oxygen – are called null cycles, and serve only to slow down the overall rate of odd-oxygen destruction by diverting reactive species from the odd-oxygen-destroying cycles. Other reactions are not cyclic, but instead form stable compounds that do not take part directly in odd-oxygen-destroying cycles. These are termination reactions, two important examples of which are

$$OH + NO_2 + M \rightarrow HNO_3 + M \quad (7.20)$$

which terminates HO_x and NO_x cycles, and

$$Cl + CH_4 \rightarrow CH_3 + HCl \quad (7.21)$$

which terminates ClO_x cycles. Termination reactions generally form reservoir compounds, from which reactive compounds can be regenerated by a variety of routes, including gas-phase (homogeneous) reactions and heterogeneous reactions involving aerosol.[4] Reservoir compounds may also be removed from the stratosphere by atmospheric transport to the troposphere or, for soluble compounds, by the settling of aerosol particles.

---

[2] See the USEPA site http://www.epa.gov/ozone/geninfo/numbers.html for details on the naming convention for ozone-depleting substances, such as CFCs and their bromine-containing cousins, the halons.

[3] See Refs. [21] (Chapter 3) and [24] (Chapter 2), and for a discussion of energy levels in atoms and molecules and their importance in atmospheric chemistry.

[4] In the atmospheric science community, the term 'heterogeneous chemistry' is generally taken to include any reactions occurring on or *in* particles in the atmosphere. Reactions occurring in liquid particles are, of course, homogeneous liquid-phase reactions, *sensu stricto*, but atmospheric chemistry models usually fold the gas–particle partitioning into the rate expression (see below), so there is some justification for the terminology.

7.4 Heterogeneous stratospheric chemistry

Aerosol particles provide an alternative medium for chemical reactions in the stratosphere. Since reactant molecules dissolved in, or sorbed onto, aerosol particles will interact with (i.e. form bonds with) the condensed phase, the energetics of reaction are modified, and so reactions can occur in or on aerosol particles that do not occur in the gas phase. Given the aqueous acid character of the aerosol phase in the stratosphere, it is not surprising that many of the important heterogeneous reactions are acid-catalysed hydrolysis reactions and acid displacement reactions. As far as ozone chemistry is concerned, the overall effect of heterogeneous reactions is to convert the chlorine and hydrogen in reservoir compounds into more active forms, whilst converting NO_x into nitric acid.[5] Some of the most important heterogeneous reactions are

$$N_2O_5 + H_2O \rightarrow 2HNO_3 \tag{7.22}$$

$$ClONO_2 + H_2O \rightarrow HOCl + HNO_3 \tag{7.23}$$

$$ClONO_2 + HCl \rightarrow Cl_2 + HNO_3 \tag{7.24}$$

$$HOCl + HCl \rightarrow Cl_2 + H_2O \tag{7.25}$$

$$BrONO_2 + H_2O \rightarrow HOBr + HNO_3 \tag{7.26}$$

$$HOBr + HCl \rightarrow BrCl + H_2O \tag{7.27}$$

The HNO_3 and H_2O formed in these reactions will often remain in the aerosol, depending on temperature, but the halogen-containing products will partition into the gas phase and quickly breakdown by photolysis to form halogen radicals. In the Arctic, and at the edge of the Antarctic, atmospheric dynamics induces enough meridional transport to ensure that there is always enough light to photolyse the halogen products of heterogeneous reactions [24]; deep in the Antarctic polar vortex the halogen photolysis takes place at the end of polar night, leading to the sudden depletion of ozone in austral spring [25].

In principle, the rate of heterogeneous reactions, like (7.23)–(7.27) above, is a function of transport processes in the gas phase, accommodation of gaseous molecules at the aerosol particle surface, dissolution of species in liquid aerosol particles, transport in the liquid phase and the liquid-phase reaction rate coefficient [20, 23]. For slow reactions – that is, where the liquid-phase reaction rate coefficient is below about $10^5\ M^{-1}\ s^{-1}$ – the reactants become well mixed in the droplets, so that the transport terms can be neglected and the rate of reaction calculated straightforwardly using a standard rate expression:

$$-\frac{d[A]}{dt} = k[A][B]$$

where k is the liquid-phase reaction rate coefficient and the liquid-phase concentrations $[A]$ and $[B]$ can be related to the gas-phase mixing ratios by Henry's law. Reaction (7.25) on cold sulphuric acid aerosol exhibits this type of behaviour [26]. For faster heterogeneous reactions,

[5] Strictly speaking, reaction (7.22) converts one nitrogen reservoir into another, but N_2O_5 converts much more readily to NO_x than HNO_3.

the rate of loss of a compound is governed by the aerosol surface area density, S, the rate of molecular collisions with the aerosol surface and the reaction probability, γ:

$$-\frac{\mathrm{d}A}{\mathrm{d}t} = \frac{\gamma \bar{c}}{4} S$$

The reaction probability is itself the combination of two processes acting in series: rate limitation due to accommodation effects, and rate limitation due to aerosol-phase transport and reaction. Hence, the reaction probability can be cast in the form

$$\frac{1}{\gamma} = \frac{1}{\alpha} + \frac{1}{\Gamma_{\text{bulk}} + \Gamma_{\text{surf}}}$$

where α is the accommodation coefficient, and Γ_i is the effective rate coefficient in the bulk or surface of the aerosol droplet. The result of this competition between reaction and transport is that many of the heterogeneous reactions of stratospheric importance are highly temperature dependent (through the temperature dependence of reactant solubilities and diffusivities) [27]. Reaction (7.23), for example, requires this kind of sophisticated kinetic treatment to capture its behaviour with temperature. At the other extreme from the liquid-reaction-controlled kinetics of reaction (7.25), reaction (7.22) occurs very efficiently on aqueous surfaces, so that γ is a constant near unity. (The accepted value is 0.1.)

For solid aerosol particles – such as are found in some polar stratospheric clouds – reactions in the bulk of the condensed phase do not occur. The physical chemistry of the reactions on the surfaces of solid particles depends on the details of the particle surface – surface roughness, mobility on the surface, the existence of a quasi-liquid surface layer, etc. – which are not well determined. It has yet to be established, for example, whether the surface of solid polar stratospheric cloud (PSC) particles is indeed solid, or if it consists of a quasi-liquid layer a few molecular radii deep (e.g. [28–30]). Studies of finely divided aqueous systems, which are stabilised nanometre-scale droplet populations, support the quasi-liquid layer hypothesis [31], at least for concentrated surface solutions of HCl. Without a robust theoretical model of solid PSC particle surfaces, calculated chlorine activation rates, and hence ozone depletion rates, remain very uncertain when the chlorine activation is occurring on solid PSCs.

7.5 Polar stratospheric clouds and the 'ozone hole'

When temperatures fall in the polar winter stratosphere, stratospheric aerosol particles take up water, nitric acid and other condensable gases. If the temperatures fall below the frost point (i.e. the temperature at which the saturation vapour pressure over ice becomes less than the ambient partial pressure of water vapour – about 188 K in the lower stratosphere) then visible water ice clouds can form. This aerosol, composed of water ice particles with sulphuric and nitric acid inclusions, is called a type 2 PSC, nacreous cloud or mother-of-pearl cloud.[6] Between about 195 K and the frost point, the aerosol particles swell in size by uptake of water and nitric acid (Figure 7.3); this type of aerosol is called a type 1 PSC. Type 1 PSCs can be composed of

[6] The beauty of mother-of-pearl clouds is indicated by their name. They are occasionally visible in the more highly populated regions of the northern and southern hemisphere, at latitudes above about 50°N. The fascination of these, and other, clouds is entertainingly evoked by Gavin Pretor-Pinney in his *Cloudspotter's Guide* [32].

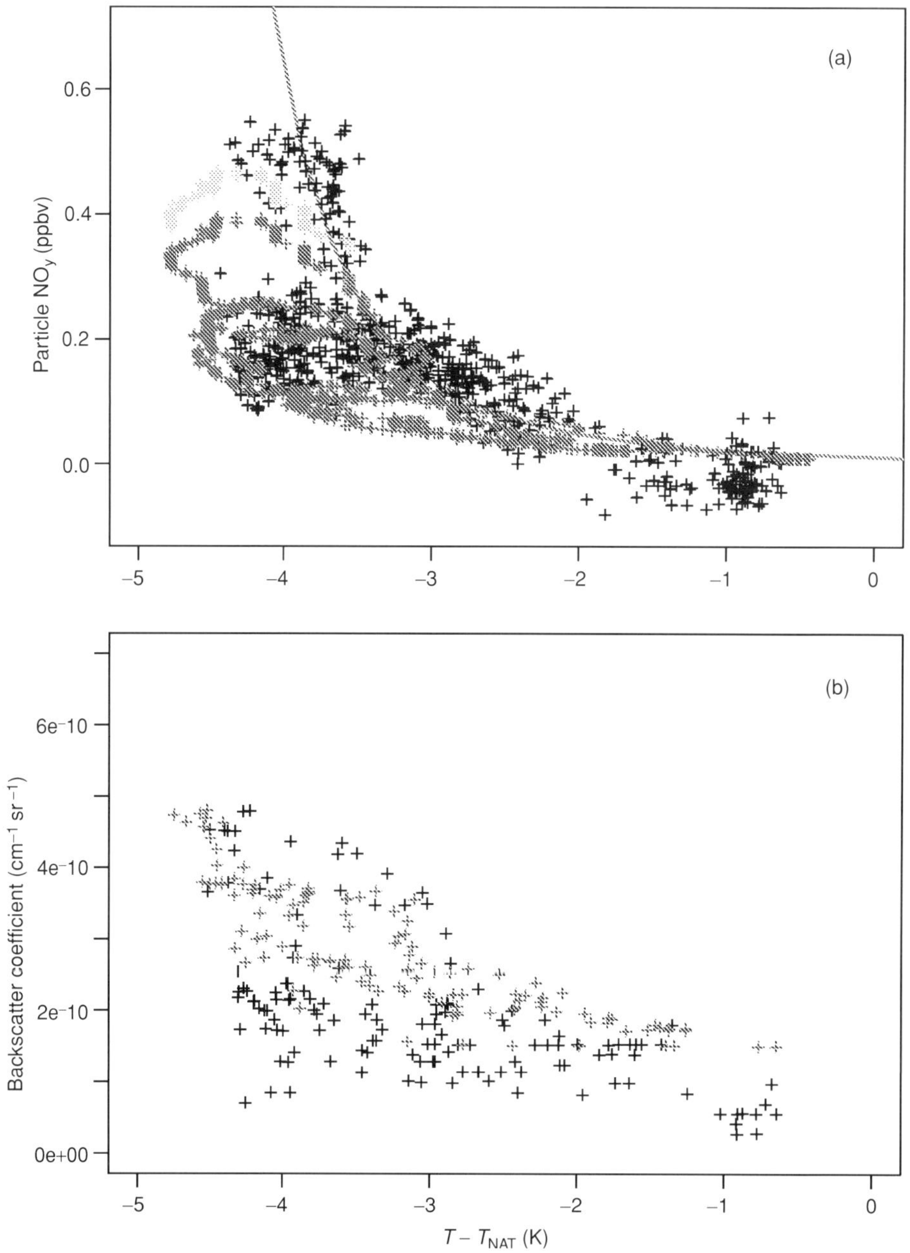

Figure 7.3 Observations and modelling of the variation of aerosol properties as a function of temperature. In (a), the grey line represents the equilibrium composition of aerosol for 4.8 ppmv H_2O, 8.1 ppbv HNO_3, 0.5 ppbv H_2SO_4, at a pressure of 50 hPa, and demonstrates how aerosol composition changes with temperature. The temperature scale used is the amount of cooling below the nitric acid trihydrate stability temperature ($T - T_{NAT}$). (a) also shows measured NO_y content of the aerosol phase (black crosses). The measurements were taken as air parcels were being cooled rapidly by a mountain wave, and may be compared to microphysical model simulations (grey crosses) and the equilibrium curve (grey line). Although the measured behaviour follows the equilibrium line approximately, there is significant deviation due to lag and hysteresis in the absorption of nitric acid. The change in composition is accompanied by an increase in condensed volume, which is exemplified by the change in the visible-light backscattering coefficient in (b) (black crosses). Again the measurements may be compared to microphysical model simulations (grey crosses). (This figure is a black-and-white version of Figure 9 in Lowe *et al.* [33]. For the first observations of this temperature-dependent composition and volume change at polar stratospheric cloud temperatures, see Dye *et al.* [34].)

solid particles (type 1a) or liquid particles (type 1b) [35, 36]. Further subcategorisations have appeared in the literature [37–41], but the three categories above are sufficient to explain most of the important issues. Figure 7.4 shows some of the transformation paths between aerosol and PSC that have been investigated in the laboratory, by field observations and by theory.

Below about 200 K the heterogeneous reactions that release active chlorine take place on all aerosol types. It is the coincidence of PSCs and air parcels containing chlorine reservoir compounds, bound in an isolated polar vortex, which produces the catastrophic ozone depletion that is known as the ozone hole [25]. Figure 7.5 shows satellite measurements illustrating ozone depletion in the Antarctic and the Arctic. The depletion over the Antarctic is much more severe and is more zonal. Both of these effects are due to the increased planetary wave activity in the northern hemisphere winter, resulting in (i) a stronger Brewer–Dobson circulation bringing more ozone-rich air into the Arctic, and (ii) more disturbances to the Arctic polar vortex.

The ozone hole was an environmental change of such dramatic proportions that it prompted urgent international action in the form of the Montreal Protocol, which entered into force in 1989. The protocol and its subsequent amendments required stabilisation of emissions of CFCs and many other ozone-depleting substances and then their phase out. In the political and business spheres the Montreal Protocol and its amendments have been some of the most successful environmental mitigation efforts to date, and in the environment the effect of this is that the total concentration of ozone-depleting substances has declined. The decline was first measured in the troposphere in the early 1990s and then, by 1998, in the stratosphere [59]. It is, however, not yet clear that this reduction in ozone-depleting substances has brought about a recovery in ozone [60]. This lack of an immediate response may seem surprising, but remember that there is a lag built into the ozone system by the Brewer–Dobson circulation, and that there is a non-negligible amount of interannual variability. Moreover, the response of ozone to the emissions of ozone-depleting substances cannot be symmetrical, because while emissions have changed, climate has also changed. Our current best estimate is that substantial Antarctic ozone depletion will continue for decades, returning to 1980s levels by, perhaps, 2050 [60, 61].

A key uncertainty in the evolution of aerosol (and hence chemistry) in the polar stratosphere is the freezing of particles to form type 1a and type 2 PSCs. The thermodynamics of these phase transformations is reasonably well determined: above the frost point nitric acid trihydrate is the stable phase [62–64], although the metastable nitric acid dihydrate may also play a role ([65], and references therein); below the frost point water ice is the stable phase [29, 66, 67]. However, like all phase change (e.g. see [68] for a discussion of the fundamentals of phase change), the kinetics of liquid–solid transition in the stratosphere is rather uncertain [50, 69].

7.6 Volcanic aerosol

The primary source of the Junge layer is oxidation of SO_2 injected by volcanic eruptions. The most dramatic injection of volcanic aerosol in recent years came in June 1991 with the eruption of Mt Pinatubo in the Philippines [70]. The global enhancement of the stratospheric aerosol loading due to the eruption was estimated at 30 Tg (i.e. 30 million tonnes). Atmospheric opacity, or optical depth, peaked at values greater than 0.2 in the tropics shortly after the eruption. Such increases in optical depth produce changes in radiative transfer through the atmosphere that are readily measured and can even saturate some of the more sensitive space-borne measurements (see the SPARC Aerosol Assessment [15] for an extended review of stratospheric aerosol time

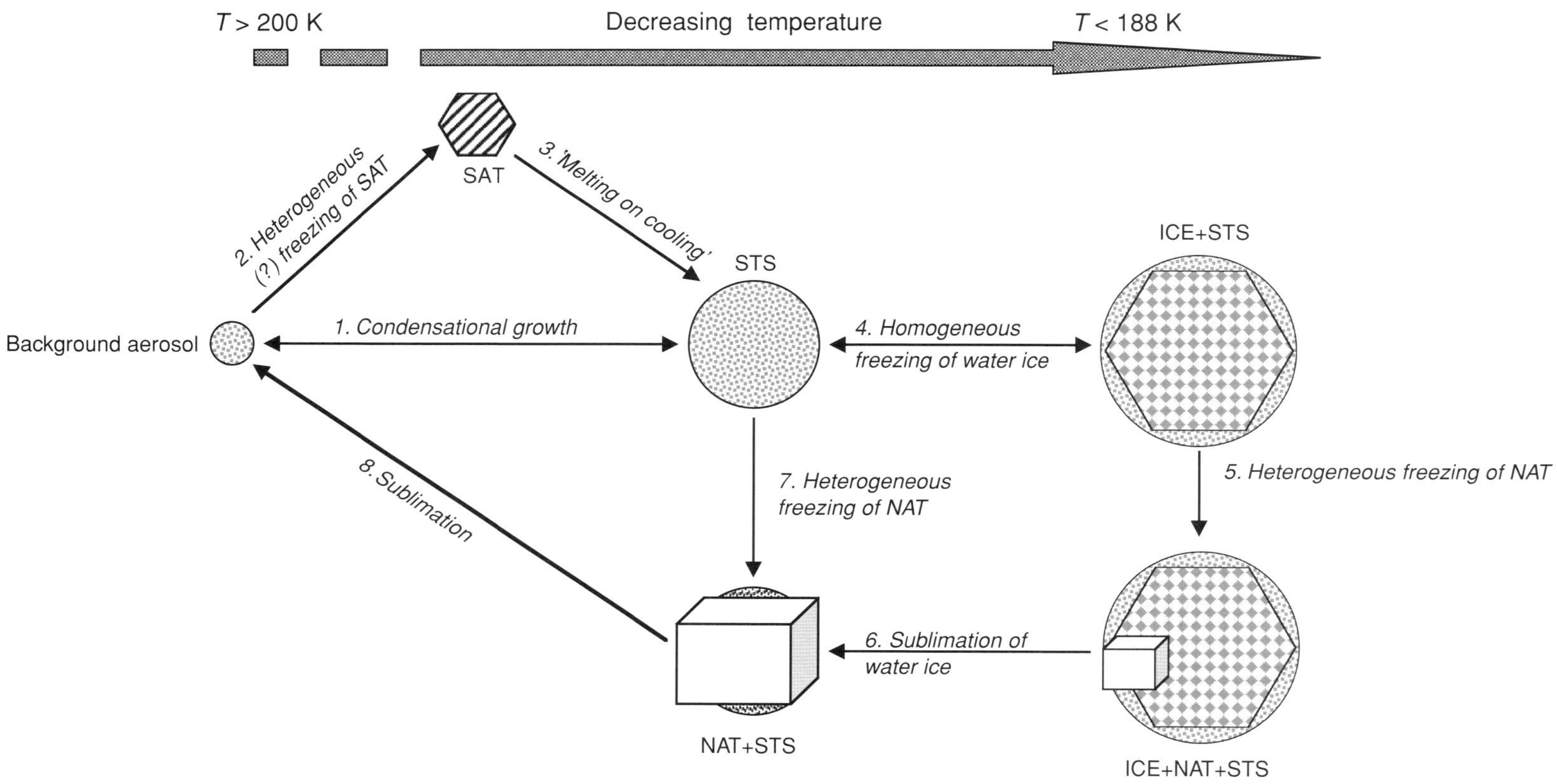

Figure 7.4 Schematic diagram of some of the most likely conversions between aerosol and PSC. STS = supercooled ternary solution (i.e. $H_2SO_4/HNO_3/H_2O$), SAT = sulphuric acid tetrahydrate and NAT = nitric acid trihydrate. (Evidence for the various steps comes from observations, laboratory studies and theoretical studies including: [17, 34, 42–46] for step 1, [47] for step 2, [48] for step 3, [49] for step 4, [50, 51–53] for step 5, [51, 52, 54] for step 6 and [55–57] for step 7. I am not aware of any specialist literature for step 8, which is usually regarded as a barrierless mass transfer problem, following Toon *et al.* [58] and the general ice-cloud literature.)

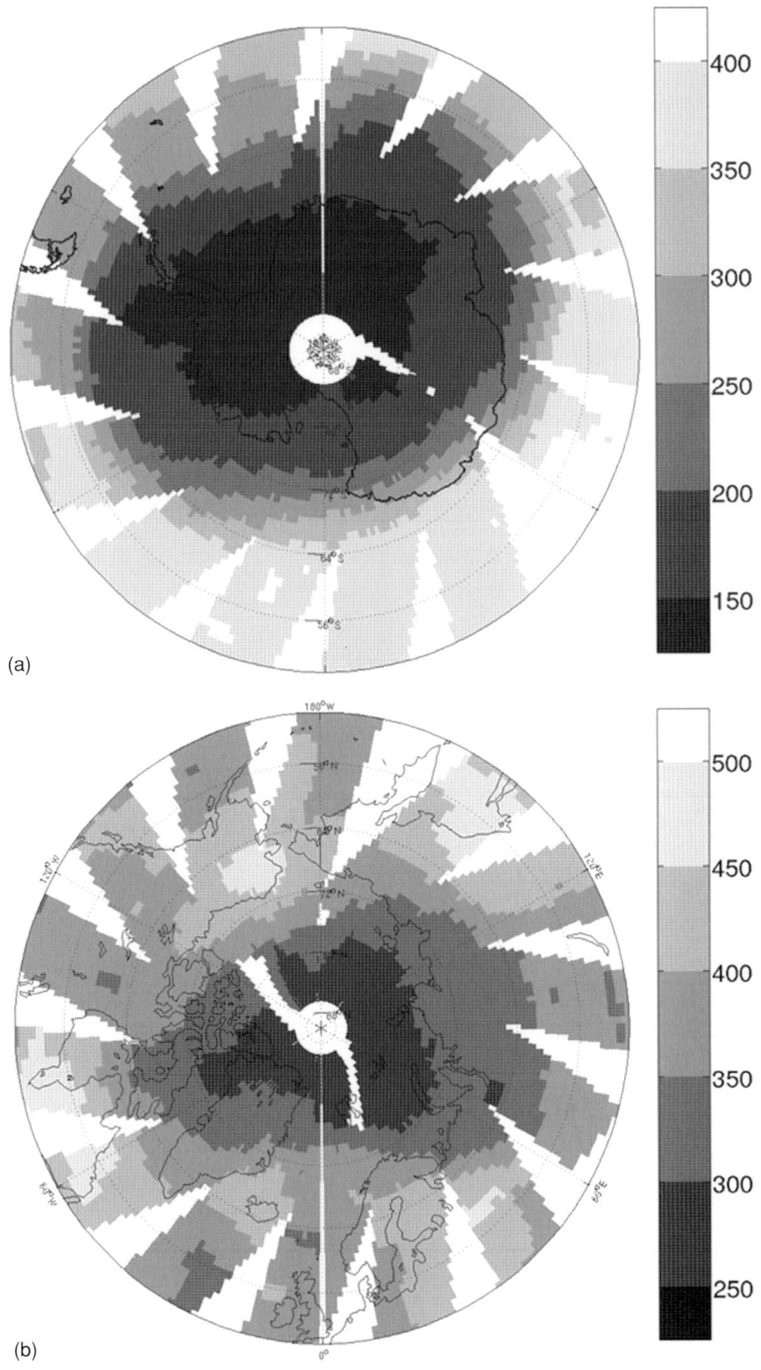

Figure 7.5 Observations of polar ozone from the GOME satellite instrument. (a) Polar stereographic projection of ozone over the Antarctic on 11 October 2001. The Greenwich meridian is directed towards the top of the image. Because GOME is on a polar orbiting satellite, not all longitudes are sampled each day, particularly at lower latitudes. (b) Polar stereographic projection of ozone over the Arctic on 22 March 1997. The Greenwich meridian is directed towards the bottom of the image. Note the change in greyscale between the Antarctic and Arctic images. (GOME data were retrieved from the database at the University of Bremen (http://www-iup.physik.uni-bremen.de/gome/) and the images were drawn by Chuansen Ren, Lancaster University.)

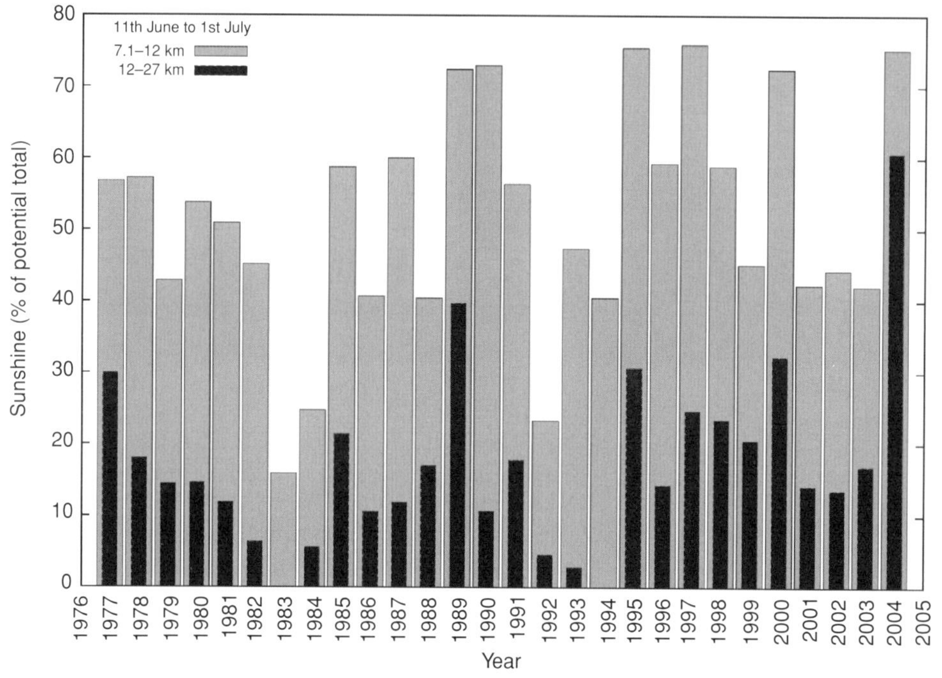

Figure 7.6 Summertime low-elevation sunshine as measured by a Campbell–Stokes sunshine recorder at Hazelrigg near Lancaster, UK. The measure shown, E, is calculated with respect to the overall amount of bright sunshine: $E = (q/((t/d)/w)) \times 100$, where q is the amount of sunshine recorded between given solar zenith angles near sunrise and sunset, t is the total amount of sunshine recorded for a 20-day range around mid-summer, d is the total number of daylight hours between sunrise and sunset for same range of days, that is the maximum potential hours of sunshine, and w is the amount time of the whole template division occupies. For the results shown all these quantities were measured in hours. This measure removes the interannual variability in total sunshine. The solar elevation ranges, plotted in the figure, are labelled by their equivalent path length through a 1-km-deep planetary boundary layer. The reduction in sunshine due to the eruptions of el Chichon (March–April 1982) and Mt Pinatubo (June 1991) is sustained over 2–3 years and is due to an enhancement of the stratospheric aerosol. Such variation in solar irradiance at the surface due to sporadic perturbations of the stratospheric aerosol by volcanic eruptions is likely to significantly affect observed trends in surface measurements, such as those shown here and those used in the 'global dimming' literature [71]. (Figure adapted from Horseman *et al.* [72].)

series). The beautiful sunsets and sunrises months after an explosive eruption are testament to the change in the optical properties of the atmosphere; Figure 7.6 gives an indication of the scale of the reduction in solar radiation transfer through the atmosphere using an analysis of standard sunshine measurements.

The 30 Tg mass injection by Pinatubo in 1991 corresponded to surface area densities of up to 40 $\mu m^2\ cm^{-3}$ [73] and increased the geometric mean radius of the aerosol size distribution from 70 to about 200 nm. The largest injection of volcanic gas and ash into the stratosphere, for which we have evidence, is the eruption of Toba, in Sumatra, 73 500 years before present. It is estimated that the total stratospheric aerosol loading following Toba reached 1000 Tg

[74]. Eruptions of the size of Pinatubo are not at all unprecedented in the geological record, therefore, and it is probably reasonable to consider perturbation by volcanic injection to be the 'normal' state of the stratospheric aerosol. Recently, in contrast, there have been no major injections of volcanic sulphur into the stratosphere since the eruption of Pinatubo in 1991, and stratospheric aerosol surface area densities have slowly decayed to about 0.5 $\mu m^2\ cm^{-3}$ in the middle of the equatorial Junge layer [21].

Sulphur is not, of course, the only component of explosive volcanic plumes; volcanic ash, water vapour, CO_2, CO, H_2, N_2, HCl and HBr are also released in large quantities (e.g. [2]). The ash is emitted as large (supermicrometre) particles that agglomerate efficiently, and so the ash quickly sediments from the volcanic plume [75, 76]. The soluble gases, such as HCl and HBr, partition efficiently into the cloud droplets that form in the expanding volcanic plume, and so are rained out before the plume mixes into the stratosphere. There is therefore no direct chemical depletion of ozone by volcanic chlorine or bromine radicals ([70] and references therein), although there could be an indirect effect on ozone, by the increase of aerosol volume and so the rates of heterogeneous reactions [77, 78]. If the increase of aerosol volume is sufficient, then the effects of polar stratospheric clouds on polar ozone may become less important than direct reaction on liquid aerosol particles [26, 27].

If the chemical effects of volcanic aerosol are subtle, their effects on atmospheric dynamics are more immediate. In September 1991, 6–10 weeks after the eruption of Mt Pinatubo, the monthly mean temperature at 30 hPa (ca. 24 km) was three standard deviations (3.5 K) above the 26-year mean for that month [70]. This warming resulted from the absorption of terrestrial infrared emitted from the surface and troposphere. The warming had a knock-on effect on the circulation in the stratosphere, causing enhanced cross-equatorial transport and enhanced ascent in the tropical branch of the Brewer–Dobson circulation. Given the close coupling between chemistry and circulation in the stratosphere, it is unsurprising to note that the Pinatubo-induced changes to the circulation in turn produced a knock-on effect on ozone; a 6–8% reduction in the tropical ozone column was observed after the eruption of Mt Pinatubo, and much of this can be attributed to the circulation changes [70].

7.7 Meteoritic aerosol

Meteoroids constantly bombard the earth, producing an annual flux of 0.02 Tg into the atmosphere, of which about 1/20 (i.e. 10^{-3} Tg) is composed of rocks larger than 10 cm. That part of the mass flux which does not reach the earth quickly, as meteorites, forms aerosol in the stratosphere and mesosphere. Extra-terrestrial material is rich in metals oxides and silica, materials that are abundant in mass spectra of the stratospheric aerosol [14], albeit at the 1%-by-mass level, thus demonstrating the importance of this source. The fate of meteoroid material entering the atmosphere depends on size [79]. Particles of a few tens of micrometres in size will be slowed by the atmosphere and settle slowly through it; particles larger than several metres in size will hardly be slowed by the atmosphere at all, and will collide explosively with the earth's surface. Between these size limits, it has usually been assumed that meteoroids ablate on heating in the atmosphere [79], releasing their contents as gases that recondense into nanometre-scale meteoritic smoke particles [80]. Recently, direct measurements by lidar of meteoroid breakup suggest that, for larger meteoroids at least, the primary disintegration process may be fragmentation to micrometre-scale dust fragments at altitudes above about

30 km, rather than ablation [81]. If fragmentation rather than ablation is the case for the majority of meteoroids, then the meteoritic aerosol in the stratosphere will presumably have a mineralogical structure similar to the meteoroids from which they have formed, and similar to meteorites found at the earth's surface. If, however, ablation is the dominant process, then the meteoritic aerosol will form by recondensation and coagulation to form chain aggregate particles more like fumed silica and other industrial materials generated by flame synthesis. The structure of meteoritic aerosol in the stratosphere will affect how these aerosol particles interact with gases and other aerosol particles in the stratosphere; for example, fumed silica promotes the freezing of aqueous nitric acid nanometre-scale droplets to NAT above the frost point [82], but meteorite samples retrieved from the earth's surface do not appear to promote NAT formation from bulk aqueous solutions [83].

The importance of the mesosphere as a source of meteoritic material is shown by measurements of the involatile fraction of stratospheric aerosol particles with diameter greater than, say, 10 nm. In this measurement, the aerosol is heated on entry to the instrument, so that water and sulphuric acid are evaporated and only those particles containing refractory components are counted. About two-thirds of aerosol particles in the polar wintertime stratosphere survive the heated inlet, and so must contain solid refractory compounds [18]. This is a much higher fraction than in the stratosphere in general, and comes about because the general circulation in the stratosphere funnels mesospheric air downwards through the wintertime polar vortex (see Section 7.1 and Figure 7.2). These involatile aerosol particles are presumably the nuclei upon which the sulphuric acid recondenses in air that has evaporated its sulphuric acid aerosol on its way around the Brewer–Dobson circulation; a small subset of the involatile aerosol particles may also be the ice nuclei that allow low number densities of large NAT particles ('NAT rocks') to form [55, 82, 84]. However, these remain speculations, because data on the distribution and behaviour of the meteoritic aerosol in the stratosphere are still rather meagre.

7.8 Interactions between stratospheric chemistry and climate

Stratospheric ozone absorbs ultraviolet solar radiation – it is often referred to as the biosphere's external 'UV shield' [85] – but it also absorbs visible light (see reaction (7.9), above). For both these reasons, ozone is a significant component of the earth's climate system. Changes in stratospheric ozone will produce a tendency for climate to change, the so-called radiative forcing of stratospheric ozone on climate. The current consensus estimate for the radiative forcing by stratospheric ozone change is (−0.05 to 0.05) W m^{-2} [86], which implies an overall cooling effect at the earth's surface, due to more efficient emission to space of infrared radiation from the surface and lower atmosphere. This cooling can be compared with a warming from the well-mixed greenhouse gases (GHG: CO_2, CH_4, N_2O and halocarbons) of 2.64 (2.37 to 2.91) W m^{-2}, of which 0.34 (0.31 to 0.37) W m^{-2} is attributable to ozone-depleting halocarbons, but it should be borne in mind that the radiative forcing due to halocarbons and stratospheric ozone depletion has acted over a much shorter time than, say, the warming due to CO_2. The cooling due to stratospheric ozone decreases is connected of course, through atmospheric chemistry, to the warming due to halocarbons. As the Montreal Protocol and its amendments take effect, both the ozone-cooling and the halocarbon-warming effects will diminish [87].

As with most components of the earth system, there are significant feedbacks between climate and stratospheric ozone. Just as changes in ozone produce a forcing on climate, so changes in climate produce a 'forcing' on ozone [87, 88]. It is important first to note the direction of anthropogenic climate change in the stratosphere: global warming at the surface is associated with cooling in the stratosphere. This is because GHG trap terrestrial infrared near the earth's surface and reduce the flux entering the stratosphere. Cooling of the stratosphere has in fact been observed since the 1970s, but is currently mostly due directly to ozone depletion itself, rather than the radiative effect of GHG [89].

Cooling of the stratosphere will directly affect the gas-phase ozone chemistry described in Sections 7.3.1 and 7.3.2, above. The rates of bimolecular reactions (i.e. those involving two reactant molecules/radicals – see Wayne [22], for a discussion of atmospheric reaction kinetics) will decrease, since a smaller number of reactant molecules will have sufficient energy to overcome the activation energy of the reaction. However, the rates of intermolecular reactions (i.e. those involving three reactant molecules/radicals) will generally increase, because the reactants will have less excess energy to get rid of in order to form stable products. The behaviour of heterogeneous reactions depends on the temperature dependence of the various steps outlined in Section 7.3.3. In general, the rates of heterogeneous reactions that affect stratospheric ozone will increase. Since many of the most important heterogeneous reactions for ozone depletion currently occur in localised regions where temperatures are low, for example the wintertime polar vortices, synoptic-scale temperature anomalies above anticyclonic weather systems and gravity wave crests; an increase in the area of low temperatures may be more important than the mean temperature reduction. There is already some indication that the area of Arctic stratosphere susceptible to ozone destruction may be changing, particularly in those years that are generally cold [90]. At the edge of the Antarctic vortex there is a broad region which is warmer than the vortex interior and so currently contains few PSCs. As increased greenhouse gases cool the stratosphere, there may be more ozone loss in this edge region despite the reduced chlorine from CFCs [91].

One can conceive of other climate effects on stratospheric ozone: changes in the Brewer–Dobson circulation due to changes in the energy or distribution of the tropospheric waves driving it; changes in stratospheric chemistry due to changes in the hydrological cycle and, hence, in HO*x* abundances; etc. These are active areas of research in the chemistry-climate modelling community [88] but the community is still some way off providing a quantitative understanding, often because of limitations in the treatment of aerosol processes. Including detailed microphysics and heterogeneous chemistry, to describe aerosol physics and ozone chemistry, is difficult within climate models that are already computationally expensive.

In this chapter, I have tried to give a general sense of the behaviour of the stratospheric aerosol layer, its impact on environmental chemistry and, hence, its impact wider impact on the earth system. In the half century since its discovery, the Junge layer has become very well characterised. There are, however, some surprisingly fundamental questions remaining: how much new particle formation occurs in the stratosphere, and, if any, where does the formation take place? What are the most important mechanisms for freezing of aerosol or PSC particles? How will the aerosol and PSC distribution respond to climate change, and will this response be driven by temperature alone, or by changes in circulation and changes in composition (especially of water vapour)? And how will changes in aerosol and cloud feedback onto climate? In the untangling of these questions, observations, detailed models, parameterisations and conceptual arguments will all play a part and will, in any event, build on the solid observational and theoretical understanding developed since the 1960s.

Acknowledgements

I am grateful to my Lancaster colleagues Doug Lowe and Chuansen Ren for polar stratospheric cloud model simulations and for help with figure drawing, respectively. I am grateful to Christiane Voigt and Hans Schlager (both of DLR, Germany) and to Francesco Cairo (ISAC-CNR, Italy) for permission to reproduce their measurements in Figure 7.3, which results from our collaboration in the EC EuPLEx project (EVK2-CT-2001-00119).

This review was carried out under the auspices of the SCOUT-O3 Integrated Project of the sixth framework of the European Commission (GOCE-CT-2004-505390).

References

1 Junge, C.E., Chagnon, C.W. and Manson, J.E. Stratospheric aerosols. *Journal of Meteorology* 1961; **18**(1): 81–108.

2 Mills, M.J. Volcanic aerosol and global atmospheric effects. In: *Encyclopedia of Volcanoes*. Academic Press, New York: 2000; pp. 931–943.

3 Hoinka, K.P. The tropopause: discovery, definition and demarcation. *Meteorologische Zeitschrift* 1997; **6**(6): 281–303.

4 Ohring, G. A most surprising discovery. *Bulletin of the American Meteorological Society* 1964; **45**(1): 12–14.

5 Davies, W.M. *Elementary Meteorology*. Ginn and Co., Boston: 1899.

6 Junge, C.E. *Air Chemistry and Radioactivity*. Academic Press, New York: 1963.

7 Dessler, A.E., Burrage, M.D., Grooss, J.U., Holton, J.R., Lean, J.L., Massie, S.T., Schoeberl, M.R., Douglass, A.R. and Jackman, C.H. Selected science highlights from the first 5 years of the Upper Atmosphere Research Satellite (UARS) program. *Reviews of Geophysics* 1998; **36**(2): 183–210.

8 McIntyre, M. Atmospheric dynamics: some fundamentals, with observational implications. In: J. Gille and G. Visconti, eds. *The Use of EOS for Studies of Atmospheric Physics*, CXV Course edn. North-Holland: 1992; pp. 313–386.

9 Shepherd, T.G. The middle atmosphere. *Journal of Atmospheric and Solar-Terrestrial Physics* 2000; **62**: 1587–1601.

10 Jones, R.L. and MacKenzie, A.R. Observational studies of the role of polar regions in mid-latitude ozone loss. *Geophysical Research Letters* 1995; **22**(24): 3485–3488.

11 Rex, M., Harris, N.R.P., vonderGathen, P., Lehmann, R., Braathen, G.O., Reimer, E., Beck, A., Chipperfield, M.P., Alfier, R., Allaart, M., O'Connor, F., Dier, H., Dorokhov, V., Fast, H., Gil, M., Kyrö, E., Litynska, Z., Mikkelsen, I.S., Molyneux, M.G., Nakane, H., Notholt, J., Rummukainen, M., Viatte, P. and Wenger, J. Prolonged stratospheric ozone loss in the 1995–1996 Arctic winter. *Nature* 1997; **389**(6653): 835–838.

12 Rex, M., Salawitch, R.J., Toon, G.C., Sen, B., Margitan, J.J., Osterman, G.B., Blavier, J.-F., Gao, R.S., Donnelly, S., Keim, E.,Neuman, J., Fahey, D.W., Webster, C.R., Scott, D.C., Herman, R.L., May, R.D., Moyer, E.J., Gunson, M.R., Irion, F.W., Chang, A.Y., Rinsland, C.P. and Bui, T.P. Subsidence, mixing, and denitrification of Arctic polar vortex air measured during POLARIS. *Journal of Geophysical Research – Atmospheres* 1999; **104**(D21): 26611–26623.

13 Holton, J.R., Haynes, P.H., McIntyre, M.E., Douglass, A.R., Rood, R.B. and Pfister, L. Stratosphere-troposphere exchange. *Reviews of Geophysics* 1995; **33**(4): 403–439.

14 Murphy, D.M., Thomson, D.S. and Mahoney, T.M.J. In situ measurements of organics, meteoritic material, mercury, and other elements in aerosols at 5 to 19 kilometers. *Science* 1998; **282**(5394): 1664–1669.

15 Thomason, L. and Peter, T., eds. *Assessment of Stratospheric Aerosol Properties (ASAP): Stratospheric Processes and Their Role in Climate.* SPARC Report No. 4, WCRP-124, WMO/TD-No. 1295, 2006.
16 Steele, H.M. and Hamill, P. Effects of temperature and humidity on the growth and optical-properties of sulfuric acid-water droplets in the stratosphere. *Journal of Aerosol Science* 1981; **12**(6): 517–528.
17 Carslaw, K.S., Peter, T. and Clegg, S.L. Modeling the composition of liquid stratospheric aerosols. *Reviews of Geophysics* 1997; **35**(2): 125–154.
18 Curtius, J., Weigel, R., Vossing, H.J., Wernli, H., Werner, A., Volk, C.M., Konopka, P., Krebsbach, M., Schiller, C., Roiger, A., Schlager, H., Dreiling, V. and Borrmann, S. Observations of meteoric material and implications for aerosol nucleation in the winter Arctic lower stratosphere derived from in situ particle measurements. *Atmospheric Chemistry and Physics* 2005; **5**: 3053–3069.
19 Brock, C.A., Hamill, P., Wilson, J.C., Jonsson, H.H. and Chan, K.R. Particle formation in the upper tropical troposphere: a source of nuclei for the stratospheric aerosol. *Science* 1995; **270**: 1650–1653.
20 Brasseur, G. and Solomon, S. *Aeronomy of the Middle Atmosphere*, 3rd edn. Springer, Dordrecht, The Netherlands: 2005.
21 Weisenstein, D., Bekki, S., Pitari, G., Timmreck, C. and MIlls, M. Modeling of stratospheric aerosols. In: L. Thomason and T. Peter, eds. *Assessment of Stratospheric Aerosol Properties (ASAP).* WCRP-124, WMO/TD No. 1295, SPARC Report No. 4, Toronto, Canada: 2006.
22 Wayne, R.P. *Chemistry of Atmospheres*, 3rd edn. Oxford University Press, Oxford, UK: 2000.
23 MacKenzie, A.R. Stratospheric chemistry and transport. In: C.N. Hewitt and A. Jackson, eds. *Handbook of Atmospheric Science.* Blackwell, Oxford, UK: 2003; pp. 188–210.
24 Roscoe, H.K., Jones, A.E. and Lee, A.M. Midwinter start to Antarctic ozone depletion: evidence from observations and models. *Science* 1997; **278**(5335): 93–96.
25 Farman, J.C., Gardiner, B.G. and Shanklin, J.D. Large losses of total ozone in Antarctica reveal seasonal Clox/Nox interaction. *Nature* 1985; **315**(6016): 207–210.
26 Cox, R.A., MacKenzie, A.R., Muller, R.H., Peter, T. and Crutzen, P.J. Activation of stratospheric chlorine by reactions in liquid sulfuric-acid. *Geophysical Research Letters* 1994; **21**(13): 1439–1442.
27 Hanson, D.R., Ravishankara, A.R. and Solomon, S. Heterogeneous reactions in sulfuric-acid aerosols – a framework for model-calculations. *Journal of Geophysical Research – Atmospheres* 1994; **99**(D2): 3615–3629.
28 MacKenzie, A.R. and Haynes, P.H. The influence of surface kinetics on the growth of stratospheric ice crystals. *Journal of Geophysical Research – Atmospheres* 1992; **97**(D8): 8057–8064.
29 Pruppacher, H.R. and Klett, J.D. *Microphysics of Clouds and Precipitation*, 2nd edn. Kluwer, Dordrecht, The Netherlands: 1997.
30 Hobbs, P.V. *Ice Physics.* Oxford University Press, Oxford, UK: 1974.
31 Bogdan, A., Kulmala, M., MacKenzie, A.R. and Laaksonen, A. Study of finely divided aqueous systems as an aid to understanding the surface chemistry of polar stratospheric clouds: case of HCl/H_2O and $HNO_3/HCl/H_2O$ systems. *Journal of Geophysical Research – Atmospheres* 2003; **108**(D10): 4303, doi: 10.1029/2002JD002606.
32 Pretor-Pinney, G. *The Cloudspotter's Guide*, 1st edn. Hodder and Stoughton, London: 2006.
33 Lowe, D., MacKenzie, A.R., Schlager, H., Voigt, C., Dörnbrack, A., Mahoney, M.J. and Cairo, F. Liquid particle composition and heterogeneous reactions in a mountain-wave polar stratospheric cloud. *Atmospheric Chemistry and Physics Discussions* 2005; **5**: 9547–9580.
34 Dye, J.E., Baumgardner, D., Gandrud, B.W., Kawa, S.R., Kelly, K.K., Loewenstein, M., Ferry, G.V., Chan, K.R. and Gary, B.L. Particle-size distributions in Arctic polar stratospheric clouds, growth and freezing of sulfuric-acid droplets, and implications for cloud formation. *Journal of Geophysical Research – Atmospheres* 1992; **97**(D8): 8015–8034.
35 Poole, L.R. and McCormick, M.P. Airborne lidar observations of Arctic polar stratospheric clouds – indications of 2 distinct growth-stages. *Geophysical Research Letters* 1988; **15**(1): 21–23.

36 Browell, E.V., Butler, C.F., Ismail, S., Robinette, P.A., Carter, A.F., Higdon, N.S., Toon, O.B., Schoeberl, M.R. and Tuck, A.F. Airborne lidar observations in the wintertime Arctic stratosphere – polar stratospheric clouds. *Geophysical Research Letters* 1990; **17**(4): 385–388.
37 Peter, T. Microphysics and heterogeneous chemistry of polar stratospheric clouds. *Annual Review of Physical Chemistry* 1997; **48**: 785–822.
38 Tsias, A., Wirth, M., Carslaw, K.S., Biele, J., Mehrtens, H., Reichardt, J., Wedekind, C., Weiss, V., Renger, W., Neuber, R., von Zahn, U., Stein, B., Santacesaria, V., Stefanutti, L., Fierli, F., Bacmeister, J. and Peter, T. Aircraft lidar observations of an enhanced type Ia polar stratospheric clouds during APE-POLECAT. *Journal of Geophysical Research – Atmospheres* 1999; **104**(D19): 23961–23969.
39 Tabazadeh, A. and Toon, O.B. The presence of metastable HNO_3/H_2O solid phases in the stratosphere inferred from ER 2 data. *Journal of Geophysical Research – Atmospheres* 1996; **101**(D4): 9071–9078.
40 Shibata, T., Iwasaka, Y., Fujiwara, M., Hayashi, M., Nagatani, M., Shiraishi, K., Adachi, H., Sakai, T., Susumu, K. and Nakura, N. Polar stratospheric clouds observed by lidar over Spitsbergen in the winter of 1994/1995: liquid particles and vertical 'sandwich' structure. *Journal of Geophysical Research – Atmospheres* 1997; **102**(D9): 10829–10840.
41 Santacesaria, V., MacKenzie, A.R. and Stefanutti, L. A climatological study of polar stratospheric clouds (1989–1997) from LIDAR measurements over Dumont d'Urville (Antarctica). *Tellus Series B – Chemical and Physical Meteorology* 2001; **53**(3): 306–321.
42 Steele, H.M., Hamill, P., McCormick, M.P. and Swissler, T.J. The formation of polar stratospheric clouds. *Journal of the Atmospheric Sciences* 1983; **40**(8): 2055–2067.
43 Carslaw, K.S., Luo, B.P., Clegg, S.L., Peter, T., Brimblecombe, P. and Crutzen, P.J. Stratospheric aerosol growth and HNO_3 gas-phase depletion from coupled HNO_3 and water-uptake by liquid particles. *Geophysical Research Letters* 1994; **21**(23): 2479–2482.
44 Tabazadeh, A., Turco, R.P. and Jacobson, M.Z. A model for studying the composition and chemical effects of stratospheric aerosols. *Journal of Geophysical Research – Atmospheres* 1994; **99**(D6): 12897–12914.
45 Drdla, K., Tabazadeh, A., Turco, R.P., Jacobson, M.Z., Dye, J.E., Twohy, C., and Baumgardner, D. Analysis of the physical state of one Arctic polar stratospheric cloud based on observations. *Geophysical Research Letters* 1994; **21**(23): 2475–2478.
46 Toon, O.B. and Tolbert, M.A. Spectroscopic evidence against nitric-acid trihydrate in polar stratospheric clouds. *Nature* 1995; **375**(6528): 218–221.
47 Sassen, K., Peter, T., Luo, B.P. and Crutzen, P.J. Volcanic bishop ring – evidence for a sulfuric-acid tetrahydrate particle aureole. *Applied Optics* 1994; **33**(21): 4602–4606.
48 Koop, T. and Carslaw, K.S. Melting of $H_2SO_4 \cdot 4H_2O$ particles upon cooling: implications for polar stratospheric clouds. *Science* 1996; **272**(5268): 1638–1641.
49 Koop, T., Luo, B.P., Tsias, A. and Peter, T. Water activity as the determinant for homogeneous ice nucleation in aqueous solutions. *Nature* 2000; **406**(6796): 611–614.
50 MacKenzie, A.R., Laaksonen, A., Batris, E. and Kulmala, M. The Turnbull correlation and the freezing of stratospheric aerosol droplets. *Journal of Geophysical Research – Atmospheres* 1998; **103**(D9): 10875–10884.
51 Carslaw, K.S., Wirth, M., Tsias, A., Luo, B.P., Dornbrack, A., Leutbecher, M., Volkert, H., Renger, W., Bacmeister, J.T. and Peter, T. Particle microphysics and chemistry in remotely observed mountain polar stratospheric clouds. *Journal of Geophysical Research – Atmospheres* 1998; **103**(D5): 5785–5796.
52 Tsias, A., Prenni, A.J., Carslaw, K.S., Onasch, T.P., Luo, B.P., Tolbert, M.A. and Peter, T. Freezing of polar stratospheric clouds in orographically induced strong warming events. *Geophysical Research Letters* 1997; **24**(18): 2303–2306.
53 Carslaw, K.S., Peter, T., Bacmeister, J.T. and Eckermann, S.D. Widespread solid particle formation by mountain waves in the Arctic stratosphere. *Journal of Geophysical Research – Atmospheres* 1999; **104**(D1): 1827–1836.

54 Waibel, A.E., Peter, T., Carslaw, K.S., Oelhaf, H., Wetzel, G., Crutzen, P.J., Pöschl, U., Tsias, A., Reimer, E. and Fischer, H. Arctic ozone loss due to denitrification. *Science* 1999; **283**(5410): 2064–2069.

55 Voigt, C., Schlager, H., Luo, B.P., Dornbrack, A.D., Roiger, A., Stock, P., Curtius, J., Vossing, H., Borrmann, S., Davies, S., Konopka, P., Schiller, C., Shur, G. and Peter, T. Nitric acid trihydrate (NAT) formation at low NAT supersaturation in polar stratospheric clouds (PSCs). *Atmospheric Chemistry and Physics* 2005; **5**: 1371–1380.

56 Tabazadeh, A., Toon, O.B., Gary, B.L., Bacmeister, J.T. and Schoeberl, M.R. Observational constraints on the formation of Type Ia polar stratospheric clouds. *Geophysical Research Letters* 1996; **23**(16): 2109–2112.

57 Larsen, N., Knudsen, B.M., Rosen, J.M., Kjome, N.T., Neuber, R. and Kyro, E. Temperature histories in liquid and solid polar stratospheric cloud formation. *Journal of Geophysical Research – Atmospheres* 1997; **102**(D19): 23505–23517.

58 Toon, O.B., Turco, R.P., Jordan, J., Goodman, J. and Ferry, G. Physical processes in polar stratospheric ice clouds. *Journal of Geophysical Research – Atmospheres* 1989; **94**(D9): 11359–11380.

59 Engel, A., Strunk, M., Muller, M., Haase, H.P., Poss, C., Levin, I., Schmidt, U. Temporal development of total chlorine in the high-latitude stratosphere based on reference distributions of mean age derived from CO_2 and SF_6. *Journal of Geophysical Research – Atmospheres* 2002; **107**(D12): 4136, doi: 10.1029/2001JD000584.

60 Weatherhead, E.C. and Andersen, S.B. The search for signs of recovery of the ozone layer. *Nature* 2006; **441**: 39–45.

61 Austin, J., Shindell, D., Beagley, S.R., Bruhl, C., Dameris, M., Manzini, E., Nagashima, T., Newman, P., Pawson, S., Pitari, G., Rozanov, E., Schnadt, C. and Shepherd, T.G. Uncertainties and assessments of chemistry-climate models of the stratosphere. *Atmospheric Chemistry and Physics* 2003; **3**: 1–27.

62 Taesler, I., Delaplane, R.G. and Olovsson, I. Hydrogen bond studies XCIV, diaquooxonium ion in nitric acid trihydrate. *Acta Crystallographica* 1975; Section B31, Structural Science: 1489–1492.

63 Hanson, D. and Mauersberger, K. Solubility and equilibrium vapor-pressures of HCl dissolved in polar stratospheric cloud materials – ice and the trihydrate of nitric-acid. *Geophysical Research Letters* 1988; **15**(13): 1507–1510.

64 Smith, R.H. Formation of nitric-acid hydrates – a chemical-equilibrium approach. *Geophysical Research Letters* 1990; **17**(9): 1291–1294.

65 Tolbert, M.A. and Toon, O.B. Solving the PSC mystery. *Science* 2001; **292**: 61–63.

66 Hanson, D. and Mauersberger, K. Laboratory studies of the nitric-acid trihydrate – implications for the south polar stratosphere. *Geophysical Research Letters* 1988; **15**(8): 855–858.

67 Murphy, D.M. and Koop, T. Review of the vapour pressures of ice and supercooled water for atmospheric applications. *Quarterly Journal of the Royal Meteorological Society* 2005; **131**(608): 1539–1565.

68 Oxtoby, D.W. Nucleation of crystals from the melt. In: D. Henderson, ed. *Fundamentals of Inhomogeneous Fluids.* Marcel Dekker, New York: 1992; pp. 407–442.

69 MacKenzie, A.R., Kulmala, M., Laaksonen, A. and Vesala, T. On the theories of type-1 polar stratospheric cloud formation. *Journal of Geophysical Research – Atmospheres* 1995; **100**(D6): 11275–11288.

70 McCormick, M.P., Thomason, L.W. and Trepte, C.R. Atmospheric effects of the Mt Pinatubo eruption. *Nature* 1995; **373**: 399–404.

71 Wild, M., Gilgen, H., Roesch, A., Ohmura, A., Long, C.N., Dutton, E.G., Forgan, B., Kallis, A., Russak, V., Tsvetkov, A. From dimming to brightening: decadal changes in solar radiation at Earth's surface. *Science* 2005; **308**(5723): 847–850.

72 Horseman, A.M., MacKenzie, A.R. and Timmis, R. A method of extracting a global dimming/brightening signal from sunshine records. Manuscript submitted for publication, 2006.

73 Thomason, L.W., Kent, G.S., Trepte, C.R. and Poole, L.R. A comparison of the stratospheric aerosol background periods of 1979 and 1989–1991. *Journal of Geophysical Research – Atmospheres* 1997; **102**(D3): 3611–3616.

74 Rampino, M.R. and Self, S. Volcanic winter and accelerated glaciation following the toba super-eruption. *Nature* 1992; **359**(6390): 50–52.
75 Gilbert, J.S., Lane, S.J., Sparks, R.S.J. and Koyaguchi, T. Charge measurements on particle fallout from a volcanic plume. *Nature* 1991; **349**(6310): 598–600.
76 Bursik, M.I., Sparks, R.S.J., Gilbert, J.S. and Carey, S.N. Sedimentation of tephra by volcanic plumes. 1. Theory and its comparison with a study of the fogo-a plinian deposit, Sao-Miguel (Azores). *Bulletin of Volcanology* 1992; **54**(4): 329–344.
77 Solomon, S., Portmann, R.W., Garcia, R.R., Thomason, L.W., Poole, L.R. and McCormick, M.P. The role of aerosol variations in anthropogenic ozone depletion at northern midlatitudes. *Journal of Geophysical Research – Atmospheres* 1996; **101**(D3): 6713–6727.
78 Solomon, S., Portmann, R.W., Garcia, R.R., Randel, W., Wu, F., Nagatani, R., Gleason, J., Thomason, L., Poole, L.R., and McCormick, M.P. Ozone depletion at mid-latitudes: coupling of volcanic aerosols and temperature variability to anthropogenic chlorine. *Geophysical Research Letters* 1998; **25**(11): 1871–1874.
79 Ceplecha, Z., Borovicka, J., Elford, W.G., Revelle, D.O., Hawkes, R.L., Porubcan, V. and Šimek, M. Meteor phenomena and bodies. *Space Science Review* 1998; **84**: 327–471.
80 Hunten, D.M., Turco, R.P. and Toon, O.B. Smoke and dust particles of meteoric origin in the mesosphere and stratosphere. *Journal of the Atmospheric Sciences* 1980; **37**(6): 1342–1357.
81 Klekociuk, A.R., Brown, P.G., Pack, D.W., ReVelle, D.O., Edwards, W.N., Spalding, R.E., Tagliaferri, E., Yoo, B.B. and Zagari, J. Meteoritic dust from the atmospheric disintegration of a large meteoroid. *Nature* 2005; **436**(7054): 1132–1135.
82 Bogdan, A., Molina, M.J., Kulmala, M., MacKenzie, A.R. and Laaksonen, A. Study of finely divided aqueous systems as an aid to understanding the formation mechanism of polar stratospheric clouds: case of HNO_3/H_2O and H_2SO_4/H_2O systems. *Journal of Geophysical Research – Atmospheres* 2003; **108**(D10): 4302, doi: 10.1029/2002JD002605.
83 Biermann, U.M., Presper, T., Koop, T., Mossinger, J., Crutzen, P.J. and Peter, T. The unsuitability of meteoritic and other nuclei for polar stratospheric cloud freezing. *Geophysical Research Letters* 1996; **23**(13): 1693–1696.
84 Fahey, D.W., Gao, R.S., Carslaw, K.S., Kettleborough, J., Popp, P.J., Northway, M.J., Holecek, J.C., Ciciora, S.C., McLaughlin, R.J., Thompson, T.L., Winkler, R.H., Baumgardner, D.G., Gandrud, B., Wennberg, P.O., Dhaniyala, S., McKinney, K., Peter, Th., Salawitch, R.J., Bui, T.P., Elkins, J.W., Webster, C.R., Atlas, E.L., Jost, H., Wilson, J.C., Herman, R.L., Kleinböhl, A. and von König, M. The detection of large HNO_3-containing particles in the winter Arctic stratosphere. *Science* 2001; **291**(5506): 1026–1031.
85 Hester, R.E. and Harrison, R.M. *Causes and Environmental Implications of Increased UV-B Radiation.* Royal Society of Chemistry, Cambridge, UK: 2000.
86 IPCC. Summary for policymakers. In: S. Solomon, D. Qin, M. Manning, Z. Chen, M. Marquis, K.B. Averyt, M. Tignor and H.L. Miller eds. *Climate Change 2007: The Physical Science Basis. Contribution of Working Group I to the Fourth Assessment Report of the Intergovernmental Panel on Climate Change.* Cambridge University Press, Cambridge, UK and New York, NY, USA: 2007.
87 Pyle, J.A., Shepherd, T.E., Bodeker, G., Canziani, P., Dameris, M., Forster, P., Gruzelev, A., Müller, R., Muthama, N.J., Pitari, G. and Randel, W. *IPCC/TEAP Special Report: Safeguarding the Ozone Layer and the Global Climate System*, 2005. Available from: http://arch.rivm.nl/env/int/ipcc/pages_media/SROC-final/SpecialReportSROC.html, cited 2006.
88 Eyring, V., Harris, N.R.P., Rex, M., Shepherd, T.G., Fahey, D.W., Amanatidis, G.T., Austin, J., Chipperfield, M.P., Dameris, M., De F. Forster, P.M., Gettelman, A., Graf, H.F., Nagashima, T., Newman, P.A., Pawson, S., Prather, M.J., Pyle, J.A., Salawitch, R.J., Santer, B.D. and Waugh, D.W. A strategy for process-oriented validation of coupled chemistry-climate models. *Bulletin of the American Meteorological Society* 2005; **86**(8): 1117–1133.

89 Ramaswamy, V., Chanin, M.L., Angell, J., Barnett, J., Gaffen, D., Gelman, M., Keckhut, P., Koshelkov, Y., Labitzke, K., Lin, J.-J.R., O'Neill, A., Nash, J., Randel, W., Rood, R., Shine, K., Shiotani, M. and Swinbank, R. Stratospheric temperature trends: observations and model simulations. *Reviews of Geophysics* 2001; **39**(1): 71–122.
90 Rex, M., Salawitch, R.J., von der Gathen, P., Harris, N.R.P., Chipperfield, M.P. and Naujokat, B. Arctic ozone loss and climate change. *Geophysical Research Letters* 2004; **31**(4): L04116, doi: 10.1029/2003GL018844.
91 Roscoe, H.K. and Lee, A.M. Increased stratospheric greenhouse gases could delay recovery of the ozone hole and of ozone loss at southern mid-latitudes. *Advances in Space Research* 2001; **28**: 965–970.

Chapter 8
Aerosol Chemistry in Remote Locations

Urs Baltensperger and Markus Furger

8.1 Introduction

8.1.1 Definition of 'remote locations'

A remote location in a geographical sense means 'far removed in space, distant, desolate, secluded' [1, 2] and indicates in connection with aerosols a large distance from the source of the aerosol. Immediately, images of solitary cabins in isolated mountain areas or the icy loneliness of polar areas come into one's mind. Such images, however, are misleading, and a closer look into the life cycle and behaviour of aerosols is necessary for an appropriate definition of remote locations.

To begin with, a distinction between anthropogenic and natural aerosol sources should be made. Anthropogenic sources of particulate matter relate to all areas directly affected by human management and all the combustion processes directly induced by human activity, like fuel combustion for transportation or heating, biomass burning for agricultural practices and industrial processes, to name a few. In contrast, natural sources of aerosols are any sources emitting particulate matter or gaseous precursors without intervention of human beings. This could basically be any part of the earth's surface not managed by humans, like ocean surfaces, deserts, ice-covered terrain, high mountain ranges, etc. Sea spray, volcanic eruptions and natural forest fires are just three processes emitting huge amounts of particulate matter into the atmosphere. Apart from the primary aerosols emitted directly into the atmosphere, airborne particles may be produced literally on the fly from gas-phase emissions and subsequent chemical reactions resulting in less volatile products which partition to the aerosol phase. Such secondary aerosols may form from both natural and anthropogenic precursor substances. They are usually formed at some distance from the source of the precursors when transport processes play a role. Aerosols may be transported by the wind, and the transport distance depends on the lifetime of the aerosol. The lifetime itself depends on the size of the particle and the local environment where the aerosol is transported. In general, aerosol lifetimes are rather short, that is of the order of a week near the surface, to months in the upper troposphere, but particles may remain in the stratosphere for years. Ageing processes take place during this residence time in the atmosphere, and the particle characteristics differ between aerosols freshly emitted (i.e. near the source) and aerosols transported over long distances. Vertical transport may have a direct effect on the lifetime of aerosols and on the transport distance (as seen, e.g., in the transport of Saharan dust [3]), and particle plumes may be detected over distances of thousands of kilometres. Therefore, distance alone is not a suitable criterion for remoteness, and its definition should be refined by including population density and human activity. Any human (industrial and agricultural) activity will increase the amount of airborne particles in the atmosphere. Because natural emission sources like oceans and deserts cover large areas of the globe and contribute the major part of the global primary emissions, we will restrict our definition of remote areas to places where the anthropogenic influence is minimal, and where we can expect an atmospheric composition close to the expected natural background composition. In that sense, oceans, deserts and forests may be considered 'remote' despite their at times enhanced aerosol loadings due to natural processes.

After having looked at oceans and deserts, we are left with the ice-covered ares at the poles and at high altitudes to define 'remote'. Of the polar areas, Antarctica may be considered far away from aerosol sources (apart from local emissions by the research stations), while the Arctic often suffers from long-range transport of polluted air masses mainly from the Eurasian continent. High-altitude stations may be within free tropospheric and thus less polluted air, even though this can be disturbed during strong convection, when boundary layer air is mixed up to higher levels [4]. Particles are usually emitted in the lowest layers of the atmosphere, where they may remain for their lifetime. Therefore, high-altitude mountain top stations are located in air that carries significantly less particulate matter than the air in the atmospheric boundary layer (ABL), and hence they can be regarded as remote stations in non-convective weather conditions. Thus we end up with a rather small number of stations performing measurements of particulate matter that characterise the remote atmosphere.

An alternative approach would be using the concept of 'natural background' of aerosols. The distinction between anthropogenic and natural sources of aerosols is based on the chemical characteristics of the source material (e.g. crustal material, sea salt and industrial emissions). For the remote atmosphere, the anthropogenic contribution should be minimal, not to say absent, and the natural background could be determined from the minima of aerosol concentration. It appears evident that these low-pollution episodes are representing the background concentration at that location [5]. With this approach, oceanic and desert areas can also be considered for the characterization of the natural aerosol background. A concept often used to characterise aerosol composition is the *enrichment factor*, defined as

$$\mathrm{EF(X)} = \frac{[C(\mathrm{X})/C(\mathrm{Y})]_{\mathrm{aerosol}}}{[C(\mathrm{X})/C(\mathrm{Y})]_{\mathrm{reference}}} \tag{8.1}$$

with $C(\mathrm{X})$ being the concentration of element X in the aerosol or the reference material, and $C(\mathrm{Y})$ being the reference element [6]. For crustal enrichment factors, Al is usually used as reference element Y. An enrichment factor of 1 means no enrichment compared to the reference material, while large deviations from this value indicate substantial enrichment or depletion and thus a strong deviation from the expected background concentration. A similar tracer for distinguishing marine aerosol from continental aerosol is the soluble Mg^{2+}/Na^{+} molar ratio, which should be close to 0.12 for sea salt [7]. The closer an actual measurement is to this ratio, the lesser is its contamination with continental air masses.

To summarise, natural background aerosols are characterised by their low abundance (low concentration) in anthropogenic constituents. Such low abundances originate from a number of interacting processes occurring during the transport from the emission source to the remote site. It is obvious that long transport distances both horizontally and vertically are a key factor to characterise remote area aerosol composition, as long travel times mean long times for interaction and chemical or physical processing, which is also called ageing (see Section 8.1.3). Remote area aerosol can thus be defined as mainly natural background aerosol at long distances from anthropogenic sources. This working definition will be used throughout this chapter.

Several types of aerosol background environments can thus be defined applying the criteria discussed above: polar ice covered, remote marine, tropical and boreal forest, high altitude and free troposphere (FT). These environments are expected to show little anthropogenic influence on the composition and characteristics of particulate matter, at least at favourable times and weather situations. The distinction of these compartments will serve as a good starting point for a proper description of chemical particle properties. However, under the aspect of global transport distances and observed lifetimes of particles, the different compartments are in continuous exchange among each other, and relevant reactions or structural changes have been observed to occur at the 'interfaces', making the concept of compartmentalisation to some degree artificial [8]. Anthropogenic aerosols have been found everywhere in the atmosphere, although in low concentrations under background conditions.

8.1.2 Aerosol measurement and monitoring

8.1.2.1 Instrumentation

Measurements in remote locations are technically challenging due to the rather low particle number concentrations encountered in clean environments, and due to the often harsh conditions at remote locations. Low particle number concentrations usually require prolonged sampling times to collect sufficient material for analysis, thus decreasing time resolution necessary to detect short-term variations. Chemical analysis also often requires a well-equipped laboratory that may be far away from the sampling site. Many of the chemical instruments require continuous surveillance during operation. All these complications together make chemical measurements difficult in remote locations.

Conventional instrumentation at remote sites comprise optical particle counters and particle sizers that characterise physical characteristics of the particles. Aethalometers measure a light-absorption coefficient from which an equivalent black-carbon content can be deduced. Impactors and filters collect particles for detailed elemental analysis off-line. Many of the latter measurements suffer from sampling artefacts, such as evaporation of volatile species. Sample contamination in the laboratories or during sample handling in the field is always an important issue.

Developments since the late nineties in chemical measurement technology have made it possible to equip numerous stations with instruments capable of characterising the aerosols chemically with reasonable time resolution for process studies. The sensitivity of the instruments has also increased, allowing for lower detection limits, higher time resolution and better particle size segregation. The latest instrumentation, such as the aerosol mass spectrometer (AMS [9]) or aerosol time-of-flight mass spectrometer (ATOFMS [10]), is capable of online analysis of chemical components of fine particles in situ in real time, with resolutions of a few seconds. Due to their high cost and required attendance at this time, these instruments are not yet used regularly for continuous monitoring, but are rather deployed to remote locations for intensive field campaigns. Nevertheless, they have brought huge advances in our knowledge and understanding of aerosol composition, especially for organic aerosols.

8.1.2.2 Stations, networks and databases

The quest for a better understanding of the global environment in view of global change has spawned the establishment of numerous monitoring stations and, more recently, networks of stations (Table 8.1) with similar or identical equipment and analysis procedures. Many networks have not yet collected prolonged time series of measurements extending for several years, and thus, their data may not yet be robust enough for a generalisation. In contrast, a number of monitoring stations have collected data for a decade or more. Such long-term series are a necessity to allow for a sound statistical analysis of chemical aerosol measurements, including trends. However, their sampling procedures may have evolved over time, or they may be different from other stations, and this has to be taken into account when comparing their measurements. The use of comparable sampling and analysis methods is however a prerequisite for a meaningful comparison of PM measurements. All networks given in Table 8.1 have therefore initiated the development of their own standard operating procedures (see, e.g., WMO, 2003 [11] for an overview of methods recommended by the Scientific Advisory Group for Aerosols of the Global Atmosphere Watch Program within the World Meteorological Organization). The great challenge remaining relates to harmonisation of procedures among the different networks. Thus, data presented here may have been obtained with different procedures and are not necessarily comparable with each other.

Measurement data, especially from monitoring stations, are usually archived in a database for further study. An important advantage of such databases is that the data are quality controlled and standardised, such that a comparison of different stations is possible. Archives will also contain prolonged time series of observations that are prerequisite for the study of trends. The URLs given in Table 8.1 provide a link to the corresponding databases.

This review will focus on a number of selected background stations scattered around the globe, representative for different environments, and various measurements taken during ship cruises and aircraft flights. The monitoring stations are given in Table 8.2. Note that upper tropospheric and stratospheric measurements are beyond the scope of this text.

8.1.3 Processes and mechanisms

The background aerosol observed in remote locations consists, in general, of a mixture of primary and secondary particles. While the majority of primary particles at remote sites can be considered as of nearby and (by our definition) natural origin, secondary particles have been formed during their journey from the emission source to the observation site. A multitude of possible reactions and transformations can occur during transport, and due to exchange between different atmospheric compartments, the chemical characteristics of the emitted material will be altered. The ensemble of possible alterations of the aerosol composition and structure is called *ageing* and shall be introduced in the following sections. It is emphasised that ageing does not only affect individual particles, but also the whole aerosol air mass.

Table 8.1 List of networks for aerosol chemistry monitoring

Network	Name	URL	Data and References
CMDL	Climate Monitoring and Diagnostics Laboratory (now: Earth System Research Laboratory, ESRL)	http://www.esrl.noaa.gov/gmd/	http://www.esrl.noaa.gov/gmd/publications/
EMEP	Cooperative Programme for Monitoring and Evaluation of the Long-Range Transmission of Air Pollutants in Europe	http://www.emep.int/	http://www.nilu.no/projects/ccc/emepdata.html
GAW	Global Atmosphere Watch	http://www.wmo.int/pages/prog/arep/gaw/gaw_home_en.html	http://wdca.jrc.it/
IMPROVE	Interagency Monitoring of Protected Visual Environments	http://vista.cira.colostate.edu/IMPROVE/	
NARSTO	North American Research Strategy for Tropospheric Ozone	http://www.narsto.org/; http://cdiac.esd.ornl.gov/programs/NARSTO/narsto.html	
CAPMoN	The Canadian Air and Precipitation Monitoring Network	http://text.msc.ec.gc.ca/capmon/index_e.cfm	
NAtChem	The Canadian National Atmospheric Chemistry	http://www.msc.ec.gc.ca/natchem/	

Table 8.2 List of stations for background chemical characterization of airborne particulate matter used in the text. Aircraft missions and ship cruises are not listed

Country	Station	Coordinates	Elevation (m asl)	Details
Ireland	Mace Head	53.33°N 9.90°W	5	Marine background
Germany	Neumayer	70.65°S 8.25°W	42	Antarctica
United States	South Pole	90.00°S 24.80°W	2841	Antarctica
Canada	Alert	82.45°S N 62.52°W	210	Arctic
Norway	Ny Ålesund	78.91°N 11.88°E	0	Arctic
Norway	Zeppelin Mountain (Ny Ålesund)	78.91°N 11.88°E	474	Arctic
United States	Point Barrow	71.32°N 156.6°W	8	Arctic
Brasil	Jarú, Rondônia	10.08°S 61.93°W	110	Tropical forest
Finland	Hyytiälä	61.85°N 24.28°E	181	Boreal forest
Finland	Sevettijarvi	69.35°N 28.50°E	130	Boreal forest
Norway	Birkenes	58.38°N 8.25°E	190	Boreal forest
Norway	Skreadalen	58.85°N 6.72°E	465	Boreal forest
Sweden	Asprveten	58.80°N 17.40°E	20	Boreal forest
Austria	Sonnblick	47.05°N 12.95°E	3106	High Alpine, free troposphere
France	Puy de Dôme	45.77°N 2.96°E	1465	High Alpine, free troposphere
Germany	Zugspitze	47.42°N 10.98°E	2950	High Alpine, free troposphere
Spain	Izaña (Tenerife)	28.30°N 16.50°W	2367	High Alpine, free troposphere
Switzerland	Jungfraujoch	46.55°N 7.99°E	3580	High Alpine, free troposphere
United States	Mauna Loa	19.54°N 155.6°W	3397	High Alpine, free troposphere

8.1.3.1 Sources

Aerosol particles can originate from two sources. Primary particles are directly emitted from the source. Natural sources are the ocean surface, the continents, volcanoes, biogenic processes in plants, biomass fires and meteoritic dust. Secondary particles are produced within the atmosphere from gases by condensation after chemical reactions. Sea salt is emitted when bubbles at the water surface burst. This process is more effective at higher wind speeds, when more bubbles are generated at the wave crests. Volcanic gases and particles are directly emitted into the atmosphere, and the gases may eventually transform into secondary aerosol. Dust is also lifted from the soil by wind and, if small enough, may become airborne and transported over long distances. Biological processes emit primary aerosols in the form of plant debris, humic matter and microbial particles (pollen, fungi, spores, bacteria etc.), as well as substances in the gas phase that contribute to secondary aerosol formation. An illustrative example is dimethylsulphide (DMS) emitted by plankton. Lightning produces NO_x, which subsequently reacts to form HNO_3 which partitions to the aerosol phase.

8.1.3.2 Transport

Good examples of the vast dimensions of transport processes are illustrated with satellite imagery of Saharan dust events (see Plate 8.1). Vertical and horizontal transport of atmospheric constituents by convection and advection is the most basic, yet very important process in aerosol alteration. It may occur with or without changing the chemical composition of an air mass. A physical separation between different particle sizes will occur, with heavier (larger) particles being deposited earlier than lighter (smaller) ones. Hence, long transport distances and times will tend to produce fine aerosol. In addition, particles may be scavenged by wet deposition. The result is in both cases a reduction in aerosol mass, and thus a cleansing of the advected air mass with increasing transport distance (or travel time). This is why remote locations are typically clean air sites. Horizontal transport occurs in layers, which enables mineral dust, smoke plumes and other material to be observed over long distances [3, 12–14].

Deep convection is another important process for vertical transport and, moreover, this path may alter the chemical characteristics of the aerosol, as particles that emerge from evaporating cloud droplets might have reacted with other species within the droplet. It is assumed that a separation between soluble and insoluble gas species takes place within clouds, which makes vertical motion an important process for aerosol chemistry. This is consistent with the observed amount of organic species in the FT in relation to sulphate [8]. Clouds further change the FT aerosol budget by scavenging and deposition of particles in precipitation.

Atmospheric thermal stratification may enhance or suppress convection. In the first case, aerosol and precursor substances may be transported into elevated layers of the atmosphere, where the lifetime of the particles is increased. In the second case, vertical mixing is reduced or omitted, but aerosol already transported to elevated layers may be horizontally advected over long distances, even globally. A stable ABL may also prevent efficient exchange with the FT, as is observed in Antarctica (see Section 8.2.2). Long-range transport is nicely demonstrated by dust plumes (see Plate 8.1), in which Saharan dust is transported over the Atlantic. Forest fire plumes have also been observed to cross continents and oceans [13, 15]. The radioactive plume of the Chernobyl accident has spread around the globe [16].

8.1.3.3 Fractionation and transformation: secondary aerosol formation and cloud processing

Long-range transport affects the aerosol composition in two ways. First, gas-phase species are oxidised in the atmosphere which process forms less volatile compounds. These compounds may either form new particles (as is often the case for sulphuric acid) or condense on the pre-existing surface of the particles. Alternatively, compounds may also react on the aerosol surface to form particulate matter. Second, if the aerosol particle gets activated to form a cloud droplet, heterogeneous reactions within the droplets occur, which after evaporation of the droplet also result in a change of the aerosol chemical composition. As larger particles are preferentially incorporated into cloud droplets [17], which then acquire more aerosol material by this cloud processing, this process favours the formation of a bimodal size distribution in the accumulation mode [18]. Both processes are most efficient for particles below 1 μm in diameter, and they do not produce particles larger than 1 μm in diameter. The involved reactions depend strongly on ambient conditions, that is radiation, temperature, humidity and available chemical species. For example, many organic gas-phase hydrocarbons become water soluble only after (photochemical) oxidation in the atmosphere [19].

These processes are called *secondary aerosol formation* and have been described by Graedel and Weschler [20] for inorganic material (e.g. nitrate and sulphate) and in a review by Blando and Turpin [19] for organic material. It was found that SO_2 oxidation in clouds and fogs accounts for up to 60% of the aerosol sulphate mass in certain areas. Secondary organic aerosol (SOA) formation is the cause of substantially increased atmospheric organic aerosol concentrations. Oligomerisation reactions may take place in the aerosol, thus reducing the volatility of these compounds, which further enhances the partitioning of these compounds to the aerosol [21].

All above processes result in a change in the physical and chemical properties of the aerosol particles, affecting their lifetimes, radiative properties and – ultimately – their effect on earth's climate. The end product is an aerosol that shows influences from the whole transport path.

8.1.3.4 Sinks

A number of removal processes end the life cycle of atmospheric particles. Dry deposition (sedimentation) is more efficient for larger particles. (Therefore, the size distribution of a Saharan dust aerosol is shifted to smaller sizes with increasing transport distance.) Wet deposition (by rain and snow) is the most efficient sink process for particles smaller than 1 μm. Here, in-cloud and below-cloud scavenging are distinguished. Black carbon (BC) contains little water-soluble material at the point of emission and is therefore less efficiently eliminated by in-cloud scavenging processes. However, the processes described in Section 8.1.3.3 transform these particles into an internal mixture with water-soluble compounds within about 1 day, which rapidly enhances their scavenging efficiency [22]. Particles are also lost due to coagulation to other particles, thereby reducing the particle number concentration, but not the total particle mass. The sinks are the cause for a reduction of particle number concentrations with time. Aged air masses arriving at remote locations usually have lost much of their original aerosol content.

8.2 Observations

In the following sections the characteristics of aerosol found in remote regions are described. Starting with the ocean as the largest area on the globe, we will continue via the poles (Antarctica being the most pristine environment) towards more and more particle-laden regions, to end up with tropical and boreal forests. In search for the natural background aerosol, we deliberately exclude measurements that contained episodes of anthropogenic pollution.

Continental sites are generally more influenced by anthropogenic activities than marine or polar sites. Background aerosols at remote continental locations mainly reflect the characteristics of the surrounding surface and vegetation (mineral dust and natural biogenic particles like fungal spores, pollen, micro-organisms etc.), but continent-wide long-range transport of mineral dust and sea salt has been documented in many places. In Sections 8.2.4 and 8.2.5 we will focus on tropical and boreal forests, which are regional-scale, homogeneous, continental areas having a major influence on the global aerosol budget. The regions are exemplified with the Amazon Basin and the Northern European boreal forests. These regions are well documented with respect to their aerosol characteristics and composition. They have been identified as sources of long-range transport of natural, continental aerosols in various studies, and they shall illuminate the nature of such aerosols for other, similar areas worldwide. Plate 8.2 gives examples of sampling sites in these six different environments.

8.2.1 Oceans and marine locations

Oceans are remote areas in a sense that they are far from populated areas and anthropogenic emissions unless near the coast. A concentration gradient from the continental coasts to remote ocean areas can be found, with lower concentrations of particulates farther away from the coasts of the continents. For this reason seas surrounded by landmasses such as the Mediterranean or the Baltic Sea are not treated in this overview.

Roughly 70% of the earth's surface is covered by oceans. The atmospheric layer mainly influenced by the sea–air interface, the marine boundary layer (MBL), has been considered as a location of clean air that characterises the natural atmosphere hardly influenced by anthropogenic activities, apart from punctual ship and aircraft emissions. While still approximately true, this picture may ask for revision with accumulating evidence based on recently collected data that indicate that even at remote sites the background chemistry of the MBL is perturbed [23]. The global marine atmosphere is not really pristine, as it is polluted everywhere by anthropogenic sulphur [24]. Aerosol measurements in these vast areas are sparse and mainly concentrated along ship tracks, flight tracks or on remote islands. Coastal stations may be considered when a distinction is made between onshore and offshore airflow, as done for example for Mace Head [25]. Heintzenberg *et al.* [24] compiled the available aerosol data at the time to sketch a global map. This map shows large areas with reasonable coverage (e.g. the Atlantic Ocean) besides areas with no data coverage at all (e.g. the Southern Pacific Ocean). One reason for the large gaps was that the available data were difficult to homogenise due to different sampling and calibration methods.

Recent field experiments with chemical measurements were ACE-1 [26], ACE-2 [27], Aerosols99 [7], NAMBLEX [23], ACE-ASIA [28], INDOEX [29] and PARFORCE [30] (see Sander *et al.* [31] for a comprehensive list of measurement campaigns). These experiments comprise ship cruises or episodic measurements with extended equipment at coastal stations that are complemented by occasional aircraft measurements. ACE-1 focused on remote marine aerosols, while ACE-2 studied the European continental outflow into the Atlantic atmosphere. Coastal stations are more prone to measuring a perturbed state of the MBL compared to measurements onboard ships or on remote islands. Enhanced turbulence in the surf zone near the coast increases the sea-salt concentration relative to other species compared to offshore locations [31].

8.2.1.1 Sources

Oceans are, due to their large surface area, the source of vast quantities of aerosols and aerosol precursors. They emit an estimated total of 3300 Tg per year of sea salt, mainly in the coarse mode [32].

Marine life is the source of biogenic substances like dimethylsulphide (DMS), which is produced by phytoplankton by enzymatic cleavage of dimethylsulphonium propionate [33]; DMS is emitted in gaseous form. After oxidation to dimethylsulphoxide (DMSO) in the presence of OH, it can further react with OH to methane sulphonic acid (MSA, $(CH_3)HSO_3$) in the gas phase, which can then partition to the particle phase [34]. In the aerosol phase, this compound is usually called methanesulphonate. Diurnal variations due to photochemical oxidation by OH show a maximum in the early morning and a minimum in the afternoon [7]. In the polluted marine atmosphere, however, the diurnal variation of DMS is different, with a concentration minimum around noon [35]. The oxidation with OH, NO_3 and halogenides (Cl atoms and BrO) produces particulate MSA and sulphate (SO_4^{2-}). There is an annual cycle in the emissions responding to the biological activity in the different oceanic regions, for example, the spring blooms. The DMS oxidation process is the basic process to produce non-sea-salt (nss-) sulphate ($nssSO_4^{2-}$ = the total measured sulphate minus that originating from sea salt) aerosol over the oceans.

Besides the autochthonous (i.e. formed in their present position) natural aerosol sources, significant amounts of continental aerosols and precursors are transported by the winds to the adjacent oceans. Crustal material is blown from the deserts to the oceans and may even reach other continents. This has been observed for the Atlantic, where Saharan dust was transported to Central and South America [36], as well as for the Pacific [37].

8.2.1.2 Meteorological aspects

Winds are a key factor for the exchange of aerosols and precursors between the water surface and the atmosphere. Many of the precursor gas emissions are well correlated with wind speed. Wind produces waves, and higher wind speeds lead to bubbles on the waves that can burst and thus emit substances like sea salt, or DMS from which secondary particles can form, to the atmosphere. Sea-salt emissions show the annual variations of the global wind field, whereas biogenic emissions show in addition the seasonal variation of biological activity.

Temperature plays an important role in the partitioning between gas phase and aerosol phase of substances. For example, MSA is found mainly in the particulate phase in the polar areas, whereas it occurs mostly in the gas phase in lower latitudes.

8.2.1.3 Composition

Sea salt

It is not surprising that sea salt is the predominant constituent of marine aerosol. It composes 80% of the ionic mass for particles <1.0 μm, and 99% of the particles between 1 and 10 μm [38]. Raes *et al.* [8] found the global mass emission of sea-salt particles to be more than 20 times the combined emissions of organics, black carbon, sulphate, nitrate and ammonium into the atmosphere. Sea salt showed mass concentration values in a range from less than 1000 to 15 000 ng m^{-3}, with the large values found in the mid-latitudes of both hemispheres [24]. Sea salt is mainly composed of sodium chloride. During ACE-Asia, sea-salt concentrations of 310 ng m^{-3} were reported in the Northwestern Pacific [39].

A more detailed analysis using ATOFMS allowed for a chemical classification of sea salt into three different classes [40]. Pure, mixed and aged sea salts were distinguished according to the extent of displacement of chloride by nitrate. ATOFMS spectra of pure sea salt revealed the presence of Na, K and Cl [40, 41]. In aged sea salt, nitrate contributes to the composition in addition to the sodium chloride. Sulphate is another component of aged sea-salt aerosol.

Sulphate

Sulphate makes up the lion's share of MBL aerosols smaller than about 250 nm. DMS concentrations of up to 806 ng m^{-3} were observed during Aerosols'99 [7]. ACE-Asia reported 720 ng m^{-3} of $nssSO_4^{2-}$ [39]. Natural MBL sulphate may also be entrained from the FT where cloud processing of marine emissions produces aerosols [42]. In the northern hemispheric oceans sulphate originates predominantly from anthropogenic and continental (e.g. volcanic) sources.

Ammonia as a part of the biological nitrogen cycle shows a net flux to the atmosphere over remote oceans and quickly reacts with acidic aerosols [7]. During Aerosols'99 only submicron aerosol showed measurable quantities of ammonia. The $NH_4^+/nssSO_4^{2-}$ molar ratio dropped from about 1 to 0.2 in the region where DMS concentrations began to increase. During ACE 1, this ratio showed average values of 0.9 for particles with geometric diameters less than 0.25 μm, and 0.3 for particle diameters between 0.25 and 0.68 μm [38]. ACE-Asia reported NH_4^+ concentrations of 170 ng m^{-3} [39].

Nitrate

Aerosol nitrate is predominantly anthropogenic (oxidation of NO_x), but can also be produced by lightning. Nitrate can be formed by the reaction of sea-salt aerosol with HNO_3 [43]. During ACE-Asia, NO_3^- concentrations of 10 ng m^{-3} were reported [39].

Carbon

Carbonaceous particles are usually distinguished into elemental carbon (EC), organic carbon (OC) and inorganic carbon (IC, mainly carbonates). EC is 'pure C' and is determined with thermal methods, while OC contains molecules with other elements such as O, H, N, etc. besides C. EC is often equated with BC, though BC is determined with optical methods, that is deduced from a light-absorption coefficient. In order to transform the latter, the mass-absorption efficiency must be known. This can vary substantially [44], which makes an independent calibration, for example, by a thermal method necessary. In the absence of such a calibration, BC should be called equivalent black carbon (EBC) rather than BC [45]. Carbonaceous aerosol is produced by emissions from vegetation, in combustion processes and biomass burning either by direct (primary) emission or by transformation of emitted gases.

EC has been measured recently during the ACE-Asia ship cruise. Quinn *et al.* [46] report average values of 30 ng m^{-3} for submicron particles in the remote Western Pacific. Supermicron particle concentrations were a factor of 10 smaller or even below detection limit.

Fresh sea-salt aerosol is substantially enriched in organic constituents [47]. A recent study at Mace Head, Ireland, has provided evidence for a biogenic organic contribution to marine aerosol [48]. They found a seasonal variation in the chemical properties of aerosol that followed the biological activity in the Northern Atlantic. During periods of algal blooming, 63% of the submicrometre aerosol mass was organic, while in winter organics comprised only 15%. In absolute values, the high biological activity aerosol mass concentration was 932 ng m^{-3}, and the low biological activity mass concentration 415 ng m^{-3}. The most likely origin of the organic aerosols are surface-active organic matter, such as lipidic and proteinaceous material and humic substances.

Dust

Dust particles, though evidently not produced by the oceans, can be found in large quantities in or above the MBL in the lee of continents. Jickells *et al.* estimated the total amount of dust deposited on the world's oceans to 438 Tg a^{-1} [49]. Crustal material is lifted from the earth's surface and transported over the ocean. The composition reflects the chemical fingerprint of the source region. Typical elements are, among others, Fe, Al, Ca, Ti, Li and Si [40, 50]. Fe has been shown to act as a fertilizer to marine biota.

Trace elements

A recent study by Witt *et al.* [51] shows trace metal concentrations in aerosols measured during ship cruises over the Atlantic and Indian Ocean. They found maximum values of a few ng m^{-3} for Cu, Ni, Ba, Zn and Pb, while the maximum Cd concentration was less than 0.03 ng m^{-3}. Apart from Ba, the trace metals were enriched in comparison to crustal sources except in the Saharan dust plume, indicating an anthropogenic origin. Fe, Mn and Ba had a major crustal source. The trace metals were mainly found on fine-mode aerosols (<1 μm).

Halogens

The depletion of chlorine relative to sodium is a tracer for the perturbation of the background composition of sea-salt aerosols. Background aerosol shows no deficit in Cl, whereas polluted or aged aerosol reveals substantial deficits in Cl compared to seawater values [31]. Dechlorination is mainly driven by acid displacement.

Particulate bromine has been measured in most regions of the MBL, and the total annual emission flux from the surface ocean has been estimated to be 6.2 Tg a^{-1} [31]. In this comprehensive review of inorganic bromine in the MBL, Sander *et al.* [31] present evidence that Br is depleted in supermicrometre sea-salt particles, while enriched in submicrometre particles. The reason for this is not yet clear and requires further study. However, such enrichments of submicrometre particles are practically absent in the remote MBL. No general latitudinal dependence of Br enrichment has been found outside the Arctic. Seasonal trends of Br enrichment with a winter maximum and a summer minimum were observed in unpolluted regions of the southern hemisphere. Bromine depletion is strongly correlated with MSA, and in high southern latitudes the loss of Br to the gas phase is a natural process [52].

Iodine in its oxidised form (IO) has been observed in coastal environments to nucleate, leading to new ultrafine particles [53]. Such iodine-containing vapours are the photolysis products of reactants emitted by marine algae. The reaction is therefore more important during daylight hours. The role of iodine in new particle production has been discussed in detail by Saiz-Lopez *et al.* [54]

Radionuclides

Radionuclides such as ^{222}Rn provide excellent tracers for air masses of continental origin. Activity concentrations of this nuclide are of the order of a couple of 100 mBq m^{-3} for background marine air, while continental air masses show values of up to 8000 mBq m^{-3} [27].

8.2.2 Antarctica

Antarctica is the most pristine site on the planet for aerosol studies and fulfils the requirement of a remote location best. It is no surprise that concentrations are extremely low, often at the detection limit of the instruments. Nonetheless, particulate matter has been found everywhere and at all times, but with some intrinsic variations. Aerosol measurements have started in Antarctica in the 1960s. After a series of prospective studies, measurement techniques have been further developed, and reliable, quantitative measurements have become available since

Plate 8.1 A Saharan dust storm off the west coast of Africa on 6 June 2006. The dust plumes are the thickest near their points of origin in the east. Although the dust dissipates somewhat as it moves westwards, it still remains thick over the Atlantic.

Plate 8.2 Selected stations. Left column from top: Mace Head (Ireland), Aspvreten (Sweden), Zeppelin (Svalbard, Norway). Right column from top: Cape Grim (Tasmania) (Reproduced with permission. Copyright CSIRO.), Santarém (Brazil), Jungfraujoch (Switzerland).

the late 1970s [55]. NOAA's Climate Monitoring and Diagnostics Laboratory (now merged into the Earth System Research Laboratory, ESRL) has run a continuous aerosol program at South Pole since 1982, with regular publication of the results. Newer studies have basically confirmed previous ones, but have also added new and refined knowledge of Antarctic aerosol chemistry. Of the latest field studies were ACE 1 [56], SCATE [57], ISCAT 1998 [58] and ISCAT 2000 [59]. Early longer term observations showed a seasonal variation of aerosol particle (condensation nuclei, CN) number concentrations at the South Pole [60], with a maximum in austral summer (November), and a minimum in austral winter. It was also observed that the chemistry differs significantly from season to season [61]. Coastal stations generally show higher concentrations of particulate matter than inland stations as a result of the transport mechanisms.

A key problem to solve beforehand was self-contamination of the samples, since Antarctic air was recognised to be notoriously clean when compared to a rural background situation. Local pollution, though still an order of magnitude smaller than at rural sites elsewhere, has been recognised in recent times near the research stations that may house up to 1000 scientists and supporting personnel during the summer season [62]. However, pollution events can usually be identified and eliminated from the data series to detect the background aerosol signal.

8.2.2.1 Sources

Any particulate matter found on the polar ice cap originates from a very limited number of possible local sources [63], or was brought to Antarctica by long-range transport (with or without chemical conversion). Local sources are the earth's crust, the surface of the surrounding seas, meteorites and volcanic eruptions (Mt Erebus) [61]. Long-range transport, on the other hand, can originate from any place on earth. A characteristic of aerosols observed at the South Pole is that it represents an aged or mature stage of particulate matter, since there are no local sources on the ice surface. Transport and secondary aerosol formation are fundamental to the composition of Antarctic aerosols. Particles with diameters of 0.01–1 μm predominate in the size spectrum, as larger particles are scavenged or deposited before they reach the interior of the continent.

Most of the sulphate aerosols can be traced back to biogenic processes at the surface of the ocean surrounding the continent, where also most of the sea-salt aerosols originate. Crustal material stems from weathering of the earth's crust, for example in the dry valleys of Antarctica or from the deserts of the neighbouring continents. Mt Erebus is an important source of sulphate and has been estimated to contribute about 3–10% to the sulphate budget of the continent [64, 65]. Meteoritic material mainly contains Fe and Co, but contributes only about 1% of the total aerosol mass of the continent. Volatile elements, though observed only in minor quantities, are believed to be due in part to volcanism [61]. Long-range transport plays a key role in carrying carbonaceous aerosols to Antarctica, for example from biomass burning in the southern hemisphere or from industrial processes.

8.2.2.2 Meteorological aspects

As already mentioned, the Antarctic continent is far away from any industrial emissions, and is shielded to a large degree from air pollution by its closed atmospheric circulation (the south polar vortex). Clouds and storm systems of the southern mid-latitudes provide another cleansing mechanism that prevents pollution from reaching the South Pole. Transport to the

Antarctic ice sheet happens mainly in the middle troposphere, but with transport distances of (tens of) thousands of kilometres. The strong surface inversion usually prevents efficient downmixing of aerosols towards the surface, such that aerosol measurements taken close to the surface may be quite different from measurements taken a couple of hundred metres above ground [66]. During cloudy episodes the inversion may weaken, and vertical transport towards the ground may increase considerably. Such events are known as sulphate or sodium storms. In austral spring, transport from coastal areas towards the pole may yield events of high sodium concentrations [67]. Such events may last several days before the atmospheric surface layer stabilises again and decouples from the overlying FT, where the transport takes place.

8.2.2.3 Composition

Sulphate

The Antarctic aerosol mass is dominated by sulphate (approximately 95% in summer and 60% in winter) [61]. Highly abundant are also sodium and chlorine, which both originate from sea salt emitted into the fine particle mode [61]. Na and Cl are carried by fine particles, and crustal elements like Al and Fe are carried by coarse particles [68].

In particulate form sulphur mainly occurs as sulphuric acid, H_2SO_4, in the form of droplets, or in the form of ammonium and sodium sulphate, $(NH_4)_2SO_4$ and Na_2SO_4 (see references in [60]). One finding of the ISCAT studies was that DMS and SO_2 are not evident at the South Pole [69], and hence marine air parcels containing DMS are thoroughly oxidised before reaching the South Pole [70]. Park *et al.* have shown that conditions at the South Pole are not favourable for local new particle formation and growth [71], since supersaturation is generally low. Rather, particle nucleation takes place in the surrounding ocean, and growth may take place during and after transport to the pole. During the ISCAT 2000 field campaign, all observed local particle formation events were related to local pollution caused by aircraft or the power generator.

Non-sea-salt sulphate aerosols make up the bulk of sulphate aerosols at the South Pole, where Arimoto and co-workers measured only 5% of the sulphate mass to originate from sea salt [58]. Interestingly, their study also indicated little or no evidence for a continental influence on non-biogenic $nssSO_4^{2-}$, which could amount to 35% of the total $nssSO_4^{2-}$ according to a rough estimate. At Dome C, $nssSO_4^{2-}$ represented 99.5 wt % of the fine fraction and 92.3% of the coarse fraction of sulphate during austral summer 2002–2003 [72].

Nitrate

Measurements of particulate nitrate have been performed during both ISCAT 1998 and ISCAT 2000, with the second field campaign yielding concentrations about four times as high (150 $ng\,m^{-3}$) as the first one (39 $ng\,m^{-3}$) [73]. These values suggest a stronger continental influence in 2000 than in 1998. The origin of aerosol NO_3^- cannot be easily identified. Possible sources are lightning, NO_x inputs from the lower stratosphere and long-range transport of peroxyacetylnitrate (PAN) within the southern hemisphere [73].

At coastal stations nitrogen aerosol in the form of ammonium originates, apart from biogenic oceanic emissions, from the ornithogenic (i.e. guano-enriched) soils as a consequence of the local penguin population [74]. Ornithogenic soils and the bacterial decomposition of uric acid are a source of ammonium, oxalate and potassium and calcium aerosol.

Carbon

Carbonaceous aerosol is a good tracer for long-range transport to Antarctica, as the continent does virtually not have any carbon sources. Multiyear time series of BC have been measured with aethalometers since the 1980s. At South Pole station, annual average concentrations for BC varied between 0.5 and 1 ng m^{-3}, while the values at the coast were roughly twice as high as 0.3–2 ng m^{-3} at Halley station [75]. Maximum concentrations are observed in austral summer, typically peaking in October, while the minimum is observed in April. The main source of BC in Antarctica is tropical biomass burning [75]. Local pollution at the research stations (heating, power generation and traffic) may be a nuisance to measurements, but does not (yet) affect the overall aerosol budget of the continent [62].

Recent work by Legrand *et al.* [76] at the coastal Antarctic site Dumont d'Urville has shown that formate and acetate aerosols occur at levels between <1 ng m^{-3} in winter and 8 ng m^{-3} in summer. The compounds are about 100 times more abundant in the gas phase and originate mainly from biogenic emissions of the Antarctic oceans. At the Japanese Antarctic research station Syowa, Kawamura *et al.* [77] investigated water-soluble organic compounds in aerosol. They found a homologous series of α,ω-dicarboxylic acids (C_2–C_{11}), ω-oxocarboxylic acids (C_2–C_9) and α-dicarbonyls (C_2–C_3), as well as pyruvic acid and aromatic (phthalic) diacid. Succinic (C_4) or oxalic (C_2) acid were the dominant diacid species, followed by azelaic (C_9), adipic (C_6) or malonic (C_3) acid. Concentrations averaged to 29 ng m^{-3}, with highest concentrations in summer. The predominant constituent was succinic acid, which accounted for about 3.5% of the total aerosol carbon at Syowa.

Trace elements

Abundances of trace elements like Al, V and Sc are similar to average crustal material from both Antarctica and the surrounding continents. Elements like Na, Cl and, to a lesser degree, Mg, K and Ca are typical tracers for marine sources (sea salt). Other, more volatile elements, like Zn, Cu, Sb, Se, Pb, Br, In, W, Au, As, Ag, Cd and I were orders of magnitude greater than expected if their sources were crustal dust [55]. Meteorites and volcanic emissions have been identified as the two main sources of the highly enriched elements [61], meteorites accounting for about 1% of total aerosol mass. Fe, Mn and Co correspond quite well with meteoritic abundances, while the more volatile elements Se and As follow the chemical composition patterns of terrestrial volcanic emissions. Interestingly, Mt Erebus on Ross Island seems to be a minor player as a volcanic source for volatile elements at the South Pole.

Studies in the 1980s and early 1990s did not reveal clear trends in trace element abundances. A recent study, however, detected a decrease in Pb at the South Pole [78]. A comparison with values measured in the 1970s showed a 20-fold decline in Pb concentrations at the South Pole by the year 2000 [73]. It is unclear whether this decrease is due to air pollution emissions control or due to contamination of earlier samples.

Halogens

The halogens Cl, Br and I were found in abundances that agreed well with expectations from marine sources. Typical Br concentrations are of the order of 1 ng m^{-3} over inland Antarctica [79], which is in striking contrast to the Arctic where Br was found in concentrations similar to mid-latitude suburban areas [80].

Radionuclides

^{7}Be is an important radioisotope produced by cosmic rays in the upper atmosphere. It is therefore a marker of stratosphere–troposphere exchange, but also shows some low-frequency variability near the surface. Besides the pronounced annual variability, ^{7}Be also exhibits an anti-correlation with the sunspot cycle, with an active sun preventing cosmic ray particles from entering the earth's atmosphere [73]. Other radionuclides observed have been ^{210}Pb and ^{210}Po, being tracers for continental and/or volcanic air.

During ISCAT, ^{7}Be activities of 7 mBq m^{-3} were measured [73]. This isotope shows long-term variations near the earth's surface, with a minimum around 1990 corresponding to the sunspot maximum. ^{210}Pb activity varied from 0.089 to 0.20 mBq m^{-3} between the two ISCAT campaigns [73].

8.2.3 Arctic

'The Antarctic is clean and pure, pristine and noble. (...) By contrast, the Arctic is indeterminate, broken up and dirty' [81]. There may be a grain of truth in this statement, which indicates substantial differences between the two polar regions. While the Antarctic is a continent well isolated from its neighbouring landmasses by the sea and a pronounced atmospheric vortex circulation, the Arctic is mainly an ice-covered ocean adjacent to the huge landmasses of Eurasia and North America. In winter, the north polar vortex often extends southwards beyond industrialised areas, thus enabling mixing of anthropogenic pollutants into the polar air. In summer, this is not the case. However, during that time of the year local anthropogenic emissions from biomass burning or from oil wells occur. As is typical for polar sites, most aerosols show characteristics of long-range transport and ageing. The bulk of the aerosol mass is within the diameter range of 0.1–1 μm (accumulation mode). There is, however, a significant amount of smaller particles with diameters <30 nm, indicating that active nucleation occurs even during the polar night [82]. Distinct features of the Arctic that influence the atmospheric chemistry are its extremely low temperatures, the long Arctic night, varying snow cover and sea-ice extent, biological activity in the ocean and injections of polluted air in winter [80].

No permanent observing station exists at the North Pole, but monitoring stations are scattered along the continental coasts or on arctic islands. Episodic measurements have been obtained from aircraft missions and ship cruises (icebreakers) up to the pole. Systematic air samples were not taken before 1976 [81], but are routinely collected nowadays. Still, there is a lack of observations of Arctic aerosol characteristics, especially over longer times [83].

Field campaigns started in 1983 with the AGASP (Arctic Gas and Aerosol Sampling Program) experiment [84]. Subsequent experiments were ABLE 3A [89] in 1988, IAOE-91 [86] and ASTAR 2000 [83], to name a few. Besides these episodic measurement campaigns, a couple of permanent field stations are operated by different nations (Table 8.2).

Arctic haze is an important anthropogenic pollution phenomenon that has attracted attention since the 1980s. Arctic haze seems to be a post-World War II phenomenon originating in the industrialised areas of – mainly – Eurasia. Its occurrence is linked to the combined effects of low temperatures, lack of sunlight and a stabilising of the lowermost atmosphere in winter due to near-surface cooling, which boost the lifetimes of aerosol particles to the order of 7–15 days [85].

8.2.3.1 Sources

In a detailed study performed during a summer ship cruise, Maenhaut *et al.* [86] determined possible sources of arctic particulate matter in the coarse (2.5–10 μm) and the fine modes (<2.5 μm). They listed three likely processes: (i) mechanical generation of coarse-mode aerosol by bubble bursting at the sea surface; (ii) long-range transport of fine-mode biogenic materials from the open waters and areas further south; (iii) generation due to gas-to-particle reactions, in particular during the cycling of air through clouds. Sedimentation and wet deposition were considered as the main aerosol-removal processes. To their surprise, typical anthropogenic metals seemed to originate from areas influenced by continental river run-off from the Siberian coast. It has been observed that the ice near the Siberian river deltas is covered with clay spots. Material from such spots can be lifted and transported by the wind.

Polissar *et al.* [87] used three-way positive matrix factorisation and potential source contribution function analysis to identify source regions for aerosols measured at Barrow, Alaska. They mention several mechanisms for the observed aerosol variability, namely seasonal variability in the long-range transport, in the blocking phenomenon, in the oxidation rate of SO_2, in pollutant-removal processes and in the thickness of the surface temperature inversion. Despite some interannual variability, strong pollution sources were identified in central Siberia in winter and in eastern North America in spring.

8.2.3.2 Meteorological aspects

Seasonal differences in the weather are a key factor in determining the composition of the Arctic aerosol [88]. Long-range transport routes change sensibly due to changes in the circulation patterns, and backward trajectories then originate in completely different source regions. Arctic haze events are typically coupled to special transport conditions. The annual movement of the polar front is another important determinant of the origin and composition of Arctic aerosols. As the polar front moves to the north it provides some isolation from the area south of the front, where many emission sources are located. At the Arctic margin during the summer months, high temperature, increased turbulence and clouds and wet deposition reduce the lifetime of aerosol particles to 3–5 days [89], decreasing the dominance of long-range transport to the Arctic.

Strong winds during the summer may significantly increase the number and area of open leads in the pack ice, from which coarse-mode sea salt and precursors for NSS aerosols may escape. The high Arctic boundary layer during summer, on the other hand, is only very little influenced by natural and anthropogenic aerosols and aerosol precursors [86].

A pollution phenomenon unique to the Arctic is arctic haze [90]. It mainly consists of aerosol sulphate and, to a lesser extent, of soot and sea salt, but anthropogenic tracers like Cu, Pb and Ni have been found as well during ASTAR 2000 by Yamanouchi *et al.* [83]. They also found increased aerosol loadings in the mid-troposphere. During an arctic haze event, soot particles were predominantly externally mixed, while during background conditions, soot was mainly internally mixed with sulphate particles.

There is a distinct maximum of occurrence of arctic haze in late winter/early spring (March–May) during anti-cyclonic conditions. This is the time of greatest strength and largest extent of the arctic front, which at these times encompasses several Eurasian industrial areas. These areas are also considered to be the origin of the particles forming the arctic haze. Therefore, arctic haze is not a background phenomenon, but rather an example of anthropogenic pollution. Furthermore, the atmospheric winter/spring conditions reduce chemical reactions due to low temperatures and reduced turbulence, and decrease deposition due to stable stratification of the ABL. These mechanisms significantly increase the lifetimes of the aerosol particles. Therefore, arctic haze is the result of decreased removal of particulate matter rather than increased inflow of pollutants into the arctic reservoir. The anthropogenic origin of arctic haze is further corroborated by the fact that it has not been observed before the 1950s [91]. Numerous aircraft observations have testified the patchy structure of the haze. Antarctica is free from such haze because it is more distant and more isolated from anthropogenic sources by the Antarctic vortex, which does not reach to industrialised areas.

8.2.3.3 Composition

Sulphate

Sulphur levels were found to be lower in the high Arctic than at the South Pole [86]. Non-sea-salt S was found to be predominantly associated with the fine size fraction, with a median concentration over the pack ice of 25 ng m^{-3}. NSS sulphur originates from biogenic processes in the Arctic sea, and from anthropogenic sources further south. Their predominance in the fine size fraction indicates a formation from gaseous precursors by gas-to-particle conversion. The fine fraction concentrations of S and sea-salt elements in the open ocean were about five to ten times larger than for the high Arctic samples.

Staebler and co-workers [92] observed fine-mode inorganic aerosol compounds in Ny Ålesund to be composed of slightly ammoniated sulphuric acid, which is equivalent to an equal mixture of H_2SO_4 and NH_4HSO_4, when the source area was Eurasia. When the air arrived from central northern Russia, NH_4^+ was virtually absent. NSS sulphate was present with 2000 ng m^{-3} at this station during the AGASP experiment in spring 1989 [93].

Nitrate

Winter measurements in Ny Ålesund on Spitsbergen [94] and at Alert in the Canadian Arctic revealed that most of the inorganic nitrate (>85%) was present as particulate NO_3^- in the coarse mode. Formation of particulate inorganic nitrate is considered to be a heterogeneous process where reactive nitrogen oxides like N_2O_5 and NO_3 and HNO_3 react with wet sea-salt particles. Particulate mass concentrations at Ny Ålesund were less than 370 ng m^{-3}, with typical values around 100 ng m^{-3}. At Alert [95], concentrations were about 14 ng m^{-3}, with peak concentrations of 55 ng m^{-3}.

Carbon

Like sulphur, EC is associated with the fine size fraction and is thus also linked to long-range transport, mainly from northern Europe and Russia [86]. Even more distant sources can be considered in cases when air subsides from the FT. During IAOE-91 [86] values from the pack ice and from the open ocean were practically identical, with about 1 ng m^{-3}. Sharma and co-workers [45] found decreasing trends of BC from 1990 to about 2000, and slightly increasing trends afterwards for Barrow and Alert. Their mean BC concentrations were 30 ng m^{-3} at Alert and 40 ng m^{-3} at Barrow in 2004, but the annual variations around the mean values spread over two orders of magnitude, with high winter concentrations and low summer concentrations. It should be noted that Sharma *et al.* [45] did not calibrate their light-absorption measurements with an independent method and consequently called their data equivalent BC rather than BC.

Particulate organic matter (POM) can mainly be attributed to anthropogenic (pollution) sources [88]. Particulate organic matter (POM) and polycyclic aromatic hydrocarbons (PAH) show a distinct seasonal dependence that can be attributed to aged aerosol and long-range transport from anthropogenic pollution sources. POM in winter originates mainly from coal and oil combustion processes. In contrast, summer organic aerosol contains mainly pollen and sea salt. PAH concentrations at Barrow are of the order of 1 $ng\,m^{-3}$ in spring and about an order of magnitude lower in August. PAH originate from combustion processes like fossil-fuel burning, forest fires and agricultural biomass burning.

Staebler *et al.* [92] found 6–30% in mass attributable to non-carbonate carbon in Ny Ålesund in spring. They hypothesised that most of their residual aerosol mass besides sea salt, nss-sulphate and other inorganic ions are carbonaceous compounds.

Trace elements

A large number of metallic elements have been analysed during IAOE-91 [86]. The anthropogenic metals and metalloids Zn, As and Sb were mainly associated with the coarse fraction (>2 μm), thus originating from a more local source, which was hypothesised to be the pack ice influenced by the run-off of the Siberian rivers. Mineral elements show a similar behaviour. Earlier studies had already revealed the presence of elements like Fe, Si and K in coarse particles, while that of Pb, Zn and Cu was found in fine particles [96]. Vanadium, a tracer for oil burning, was observed at concentrations of 0.5 $ng\,m^{-3}$ at Barrow, Alaska, in winter [88]. A recent study by Shotyk *et al.* [97] showed that 95–99% of atmospheric Pb input into Arctic snow and ice is anthropogenic, despite the practically complete elimination of leaded fuel in the US and in Europe in the 1970s and 1980s. Isotope ratios $^{206}Pb/^{207}Pb$ can be used to attribute source areas to the contaminants. They also noted that in winter, the Canadian Arctic is mainly influenced by haze from Eurasia, while in summer the transport originates in North America.

Halogens

Halogens (chlorine, bromine and iodine) have attracted considerable attention especially during arctic winter and early spring conditions, as they play an important role in ozone depletion [80, 98]. Particulate Cl^- is well correlated with sea salt, peaking in winter with values of 1–3 $\mu g\,m^{-3}$. Summer Cl^- concentrations are two orders of magnitude smaller.

Particulate Br^- shows an excess concentration with respect to sea salt. The enrichment of Br^- increases from winter to early spring, and Br^- continuously accumulates in the Arctic boundary layer. Br concentrations can reach 100 $ng\,m^{-3}$ in March–May (spring pulse), from a winter concentration of 5–25 $ng\,m^{-3}$ throughout the Arctic [80].

Iodine is enriched with respect to seawater by chemical fractionation at the sea surface [99]. Enrichment factors range between 1000 and 10 000. Arctic concentrations were found to be 1 $ng\,m^{-3}$ or less [80].

8.2.4 Tropical forests

From all the tropical rain forest areas on earth, the Amazon Basin is the largest. It is an intense source of biogenic gases and aerosols. Field campaigns with aerosol chemistry measurements in the Amazonian Basin started with the Amazon Boundary-Layer Experiment (ABLE) [100] in the 1980s. Recent campaigns were conducted under the enveloping Large-Scale Biosphere-Atmosphere Experiment in Amazonia (LBA). Subprojects were the Cooperative LBA Airborne Regional Experiment (CLAIRE) 2001 [6], the European Studies on Trace Gases and Atmospheric Chemistry (LBA-EUSTACH) [101] and the Smoke Aerosols, Clouds, Rainfall and Climate (LBA-SMOCC) Experiment [102].

8.2.4.1 Sources

The Amazon Basin is a huge source of particulate and gaseous emissions, from which secondary aerosols can be produced. Apart from the natural biogenic emissions, anthropogenic activities like pasture clearing fires and deforestation emit large quantities of smoke and dust into the atmosphere. Therefore large differences in aerosol concentrations and composition between the wet and dry season were observed [103]. On the other hand, the basin receives seasonally variable amounts of sea-salt aerosols and Saharan dust due to long-range transport [104].

8.2.4.2 Meteorological aspects

The basin is influenced by the annual displacement of the Inter Tropical Convergence Zone (ITCZ), which divides the year into a dry (May–October) and a wet (November–April) season. During the dry season, biomass burning is widespread and pyrogenic aerosols are dominant. During the wet season, the aerosol characteristics of the mostly biogenic emissions resemble the natural background. The basin is also a region of intense convection [100]. Rapid mixing of biogenic gases and aerosols to high altitudes will support a reservoir from where substances can be transported and dispersed globally.

Graham *et al.* [6] found concentration differences between day and night that were attributable partly to emission behaviour of the vegetation [103], and partly to the diurnal development of the ABL and photochemistry. Convective downward mixing during the day led to enhanced OC, sea salt and soil-dust aerosol concentration close to the ground. The latter two were attributed to long-range transport from the Atlantic Ocean by backward trajectory analyses and to Saharan dust by elemental analysis.

8.2.4.3 Composition

Sulphate

Sulphate is mainly confined to the fine mode [6], with concentrations being about six times higher than those in the coarse mode (324 and 51 ng m^{-3}, respectively). This can be interpreted as the result of gas-to-particle production of sulphate from sulphur-containing gases (DMS, H_2S and CS_2). It has been suggested that the main background source is oxidation of marine DMS from the Atlantic Ocean [105].

Ammonium mainly occurs in the fine mode with concentrations of 50 ng m^{-3}, probably in the form of ammonium sulphate or bisulphate [6].

Nitrate

Nitrate is prevalent in the coarse mode. Concentrations show values around 113 ng m^{-3} in the coarse mode and 23 ng m^{-3} in the fine mode [6]. The latter is probably present as ammonium nitrate formed from gaseous NO and NH_3 emitted from the rainforest.

Carbon

Organic matter was found during CLAIRE 2001 in both coarse and fine particles, making up about 70–80% by mass [6]. OC concentrations in the fine mode were 1.13 ng m^{-3} and in the coarse mode 2.26 ng m^{-3}. EC values were 0.05 ng m^{-3} in the coarse mode and 0.02 ng m^{-3} in the fine mode. Biogenic element concentrations are elevated during nighttime in the coarse mode in the shallow nocturnal boundary layer, while in the fine mode daytime concentrations of OC are higher due to convective downward mixing from the FT. The latter has been attributed to SOA formation. Giant primary biogenic particles such as pollen, fern spores, fungal spores and even leaf fragments dominate the coarse mode during the day at the beginning of the dry season. Pyrogenic particles contain more BC due to combustion, but otherwise their elemental composition resembles that of biogenic particles [103]. Oxalate was measured in significant amounts, and is also suspected to originate from SOA formation due to the oxidation of volatile organic compounds (VOCs) [6].

Trace elements

As described by Graham *et al.* [6], soil-dust elements (Al, Si, Ca, Ti, Mn and Fe) ranged from 0.2 ng m^{-3} (Mn) to 25 ng m^{-3} (Si). They were found in various wet season field campaigns in both the fine and coarse aerosol fractions in comparable amounts, which hint towards significant long-range transport. Crustal enrichment factors close to unity indicate soil dust as the main source. Wind erosion is minimal in the Amazon Basin due to the dense vegetation, and hence, concentrations of these species are very low. A major fraction of soil dust in the basin originates from the Sahara. Coarse-mode P, S, K, Cu and Zn enrichment factors are larger than unity and thus typical for primary biogenic emissions at tropical forest sites. Na and Cl derive mainly from sea salt from the Atlantic, although Cl is also emitted from plants. Ni, Cr, Pb and V were observed in very low concentrations, often below the detection limit.

Halogens

I and Br over the Amazon Basin originate from marine, biogenic and pyrogenic emissions [6]. These elements were observed with concentrations of 0.27 ng m^{-3} for I and 1.7 ng m^{-3} for Br in the fine size fraction, which are comparable to the coarse size fraction.

8.2.4.4 Biomass burning

A special aspect of Amazonia is biomass burning in the dry season (July–September). In forest areas, trees are typically slashed 20–60 days prior to burning, while at Savanna-type sites (called 'cerrados'), fire is put to standing vegetation [106]. Artaxo *et al.* [107] in an earlier study found a fine mass concentration of 2.1 μg m^{-3} and a coarse mass concentration of 6.1 μg m^{-3}. In a subsequent study by Yamasoe *et al.* [106] the total aerosol mass was further divided into water-soluble ion components, organic matter and 19 trace elements. They further distinguished flaming and smouldering phases of the biomass fires. They found that organic matter made up to 92% of the fine particulate matter in emission plumes over forest (assuming that organic matter was the residual between fine particulate matter and the sum of the measured components).

The global relevance of biomass burning is that it is an important source of heavy metals and BC. Tropical and savanna forest biomass burning could emit 2, 3 and 12% of the global budget of Cu, Zn and BC [106].

8.2.5 Boreal forests

Boreal coniferous forests cover about 8% of the earth's surface [108]. They are a relevant source of biogenic gases that act as precursors to aerosol formation, especially in clean marine polar or arctic air masses. The newly formed particles may grow into cloud condensation nuclei (CCN) and thus exert an influence on the global climate. Yet, many aspects of this process chain are still debated, especially for organic CCN [109].

Particle formation has been observed to occur in *nucleation bursts* [110]. Polar air mass nucleation events are distinct from those with advection of polluted continental subtropical air masses, which do not lead to particle nucleation. During the non-events (days without particle formation), the pre-existing aerosol acts as an effective condensational sink depleting the gaseous precursors for new particle formation [111]. The co-operative action of sources and sinks produces a steady-state aerosol particle population [112].

A number of experiments have been conducted to study aerosols over boreal forests, among them are BIOFOR [113] (Biological Aerosol Formation in the Boreal Forest) and QUEST (Quantification of Aerosol Nucleation in the European Boundary Layer) [111].

8.2.5.1 Sources

Boreal forests are a source of terpenes that are oxidised to pinonic and pinic acids. These organic acids may condense onto freshly nucleated particles. Photo-oxidation products of α-pinene are probably the most likely species that contribute to the growth of nanometre-sized particles at the Finnish site of Hyytiälä [111, 114]. Precursors for particle formation are emitted by trees as well as by soils and leaf litter [115].

8.2.5.2 Meteorological aspects

Nucleation events typically occur in spring and autumn during daytime, but are rarely observed in summer and winter, in good agreement with synoptic-scale air mass characteristics [116]. During the QUEST campaign, Cavalli *et al.* [111] have found that new particle formation was favoured at this site in northerly flow, when clean, marine (polar or arctic) air masses were advected. In contrast, southerly advection (usually subtropical air masses which travelled over populated areas) resulted in a complete suppression of particle formation. Even at days with arctic or polar air masses, aerosol formation was prevented when global radiation was low (reduced photochemistry) or when precipitation led to wet scavenging of precursor gases or the newly formed aerosol. In addition, low temperatures favour gases to partition into the aerosol phase. Furthermore, cold air masses are linked to low mixing heights, reducing the volume available for dispersion of precursors [116].

8.2.5.3 Composition

Sulphate

Northern air masses during nucleation events consisted mainly of sulphate in the accumulation mode [117]. Sulphate was not fully neutralised by ammonium, and therefore the particles were mainly acidic in nature. During an ideal (i.e. maritime arctic/polar air) nucleation event at

Hyytiälä, Finland, a submicron nss-sulphate mass concentration of 270 ng m^{-3} was measured [111]. This is in good agreement with background marine concentrations [118].

Nitrate

Nitrate was found in the supermicron mode with a concentration of 180 ng m^{-3} [111].

Carbon

The majority of the material involved in particle formation is organic in nature during most of their growth stages, although organic species are not necessarily involved in the initial nucleation itself [117]. Carbonaceous compounds play an important role over boreal forests, as they are emitted from biogenic processes. Key precursor gas species are terpenes, whose condensable photo-oxidation products (mostly organic acids, such as pinonic, pinic, maleic and fumaric acid) favour particle growth by condensation [108]. SOA formation from α-pinene was observed, and biogenic compounds may account for a significant portion of aerosol water-soluble organic carbon (WSOC). Cavalli and co-workers [111] found submicron WSOC concentrations of 308 ng m^{-3} during a nucleation event. They also made detailed analyses of WSOC and found neutral compounds, mono- and dicarboxylic acids and polycarboxylic acids in the aerosol. Water-insoluble organic carbon (WINSOC, BC, soil-derived insoluble material) was not determined in their study, but can account for the 'unaccounted' category that makes up 32% of the mass. Allan *et al.* [117] found that much of the water-insoluble organic mass in nucleation events were alkanes. They also found hints of oxidation products of monoterpenes and sesquiterpenes, which originate from plant waxes. However, Janson *et al.* [119] found that these products take part in particle growth only, not in new particle formation. Kourtchev *et al.* [120] also found evidence for SOA formation from isoprene over boreal forests. They even speculate that isoprene photo-oxidation products might be important in the nucleation process.

8.2.6 High elevation and free troposphere

From the general rule that aerosol characteristics eventually attain background values far away from their sources, the FT may be considered as remote when it is sufficiently distant from air traffic routes, active volcanoes or deep convection. Sampling of aerosol can only be performed with airborne instruments or at suitably located mountain stations. The latter are prone to regional emissions and to transport from the ABL over the adjacent terrain [4, 121], which requires a careful distinction of background air from ABL-influenced air. This is relatively easily achieved in winter when atmospheric stability usually prevents ABL air from reaching the tops of high mountains. Mountain stations are also often within clouds. In such situations, remarkable changes to the composition and chemistry of particles take place. Real background situations are best achieved above the cloud layer.

8.2.6.1 Sources

Any aerosol source is able to contribute to the FT aerosol composition, provided the transport velocity is fast enough to encompass the characteristic lifetime of the particle. During transport, secondary aerosol is formed by condensation of oxidised compounds. Through the latter process the aerosol becomes internally mixed. Nucleation in the FT has been observed and modelled in the vicinity of convective clouds, especially in areas where cloud droplets evaporate and free H_2SO_4 in the gas phase becomes available. Together with a high actinic flux that favours the production of OH, H_2SO_4 starts nucleation in such regions [122].

It has been reported already in 1980 that in the FT the differences between the marine and the continental FT in aerosol concentrations are small except for ammonium [123]. In the ABL, however, a clear distinction can be made between these two areas.

8.2.6.2 Meteorological aspects

Perhaps the most important influence on FT composition of aerosol is atmospheric stability that inhibits or enables vertical exchange of ABL air with higher altitudes. This is best seen in the diurnal and annual variations of aerosol concentrations observed at high mountain stations such as the Jungfraujoch [4]. Typically, winter concentrations are quite low and represent the FT background, while summer concentrations are much higher, especially in the afternoon, and show characteristics of (polluted) ABL air that is transported by either thermal circulation or convection to elevated layers. The capping inversion at the top of the ABL is especially strong over the oceans and separates the FT from the MBL. Huebert and Lazrus [123] even speak of a dichotomy of the troposphere. Over land, the capping inversion is generally not that strong and is more variable due to terrain inhomogeneities that lead to a more variable depth of the ABL, a situation that is even more complicated above mountainous terrain [124, 125].

Modelling studies indicate that deep convection transports ABL aerosol to the middle and upper FT. For example, SO_2 in the northern hemispheric upper troposphere stems mainly from anthropogenic sources, while in the southern hemisphere marine emission of DMS and its subsequent oxidation provide the main contribution [8].

8.2.6.3 Composition

Sulphate

At Mount Sonnblick in Austria, winter sulphate concentrations of 346 ng m^{-3} were reported [126]. Summer concentrations were about sixfold higher. Similar concentrations and summer/winter ratios were observed at Jungfraujoch [127], where coarse-mode sea-salt sulphate comprised only 3% of total suspended matter sulphate. At this high alpine site, 82% of the sulphate originates from oxidation of SO_2 and can be assumed to originate from anthropogenic sources. Non-sea-salt sulphate has been found to be the second most important contribution to the submicron FT aerosol mix after OC [8]. An overview of sulphate concentrations at various places is given by Kasper and Puxbaum [126].

Nitrate

Relatively few measurements of particulate nitrate in the remote FT have been made until recently, and little data have been published so far. Values for different places are listed in Kasper and Puxbaum [126]. Values vary between 33 ng m^{-3} in winter and 397 ng m^{-3} in summer at Sonnblick Observatory. They also quote values for Mauna Loa [128], Hawaii, which are comparable in winter, but only about two-thirds the Sonnblick values in summer. Over the North American continent, FT values vary between 30 and 120 ng m^{-3} [129]. Zellweger *et al.* [130] give seasonal values of particulate nitrate for the Jungfraujoch obtained during the

2 years from April 1997 to March 1999. They found, for undisturbed FT conditions, mean concentration values of 44 ng m^{-3} in spring, 213 ng m^{-3} in summer, 20 ng m^{-3} in autumn and 35 ng m^{-3} in winter. In their study, particulate nitrate made up some 3% of the total reactive nitrogen mixture for undisturbed conditions. The nitrate/sulphate ratio is much lower than that in the ABL, which is explained by the strong vertical profile of NH_3: available NH_3 will first neutralise sulphuric acid, such that only NH_3 in excess of sulphuric acid is available for neutralisation of nitric acid [127]. As a result of this, the nitrate found in the coarse mode as a result of neutralisation reactions of nitric acid with carbonates becomes relatively more important: Henning *et al.* (2003) report that about 50% of nitrate was found in the coarse mode [127].

Carbon

An extensive review of OC and BC concentrations is given by Chung and Seinfeld [22]. Carbonaceous material is the main component of FT particulate matter. Background FT values of up to 940 ng m^{-3} OC and 610 ng m^{-3} BC (or EC) have been reported for high alpine sites in previous studies [131]. Recent AMS measurements at the Jungfraujoch have shown that more than 50% of fine particles (<1 μm) in summer are composed of organic matter [132]. Putaud *et al.* [133] found a relative amount between 16 and 29%, depending on particle size, for European background sites. Raes *et al.* [8] give a range from 31 to 56% for the submicron OC contribution in the background FT over the North Atlantic. A separation of total carbon into BC, WSOC and WINSOC fractions revealed that BC amounted to 22%, WSOC to 47% and WINSOC to 31% during a summer period at the Jungfraujoch [131]. Carboxylic acids were found in mean concentrations of 49 ng m^{-3} (oxalic acid), 15 ng m^{-3}(malonic acid), 11 ng m^{-3}(succinic acid) and 9 ng m^{-3}(formic acid), contributing 6.5% to the total WSOS mass.

Heald *et al.* [134] report a free tropospheric background of 1000–3000 ng m^{-3} of OC over the Pacific Ocean, observed by aircraft during ACE-Asia [28]. The high OC values observed are best explained as the result of SOA formation due to VOC oxidation and condensation favoured through the low temperatures. In their study they show that the observed OC concentrations are a factor of 10–100 higher than calculated with current models. They conclude that there is a large sustained source of SOA in the FT currently not reflected in today's models, and that SOA is the dominant component of aerosol mass in the FT. SOA formation is also a characteristic of ageing air masses, typical for the FT.

Trace elements

Long-term measurements of trace elements have been performed at Mauna Loa observatory on Hawaii. Holmes *et al.* [135] report on data obtained between 1979 and 1991. Weekly average crustal concentrations were 130 ng m^{-3} for the non-dust (background) season. Asian dust events in spring are the major contributor to trace metal concentrations at Mauna Loa, and crustal material makes up about half the total mass of the mid-tropospheric aerosol in the North Pacific. Marine material contributed about 40% to the total mass, and anthropogenic and biogenic sources made up only 5%.

Radionuclides

Gaseous ^{222}Rn as well as their daughter nuclides are excellent tracers for air masses that have been in contact with landmasses during their travel, as this isotope originates only from landmasses. They allow for information on how long and where or whether an air mass has been in contact

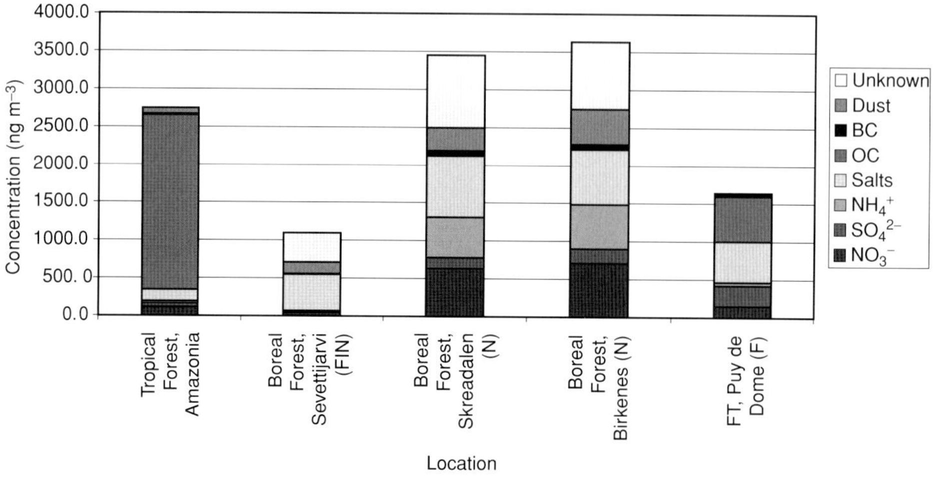

Figure 8.1 Background coarse aerosol concentrations and composition for selected remote locations. Coarse aerosol is defined as PM_{10}–$PM_{2.5}$ except for Puy de Dome where it is PM_{10}. Organic acids have been added to the OC fraction in the Amazon Basin and at Puy de Dome. They were not separately determined at the other places. In Sevettijarvi PM_{10}–$PM_{2.5}$ was <0 and thus set to 0.

with the solid earth's surface [136]. ^{222}Rn decays into ^{214}Pb, which becomes attached to aerosol particles. At the Jungfraujoch, ^{214}Pb activities are 150 mBq m^{-3} in winter and between 600 and 1200 mBq m^{-3} in summer, with the maximum activity in the afternoon due to convective transport of boundary layer air [137].

^{7}Be is a tracer for upper tropospheric air. It was found that ^{7}Be correlated with the fine particle number concentration [138]. ^{7}Be activities in non-dusty situations at Izaña, Canary Islands, were on the order of 10 mBq m^{-3}.

Dust

Long-range transport of mineral dust affects many FT sites, where due to the low overall aerosol load its contribution can be substantial. At the Jungfraujoch, Saharan dust comprises about 24% of the total aerosol on average [3]. Asian dust was the most dominant natural source of trace metals in the FT at Mauna Loa. During the non-dust season, a weekly average concentration of 130 ng m^{-3} was found [135]. Elements with a crustal component were Al, Mg, K, Ca, Sc, Ti, V, Mn, Fe, Ga, Cs, La, Ce, Sm, Eu , Hf, W and Th. These elements were highly enriched during the dust season. Elements with less enrichment were Na, Cl, Co, As, Br, Sb, Hg, Ba, Nd, Tb, Yb, Ta and U. They also have sources other than dust. Cr, Zn, Se, I and Lu showed no significant concentration differences between dust and non-dust seasons. Abundances of all these elements were in the range of <80 ng m^{-3} to a few pg m^{-3}.

8.3 Summary

Figures 8.1 and 8.2 summarise the chemical composition from a variety of sites. Only examples can be given here, and a multitude of further data is found in the data centres of the networks

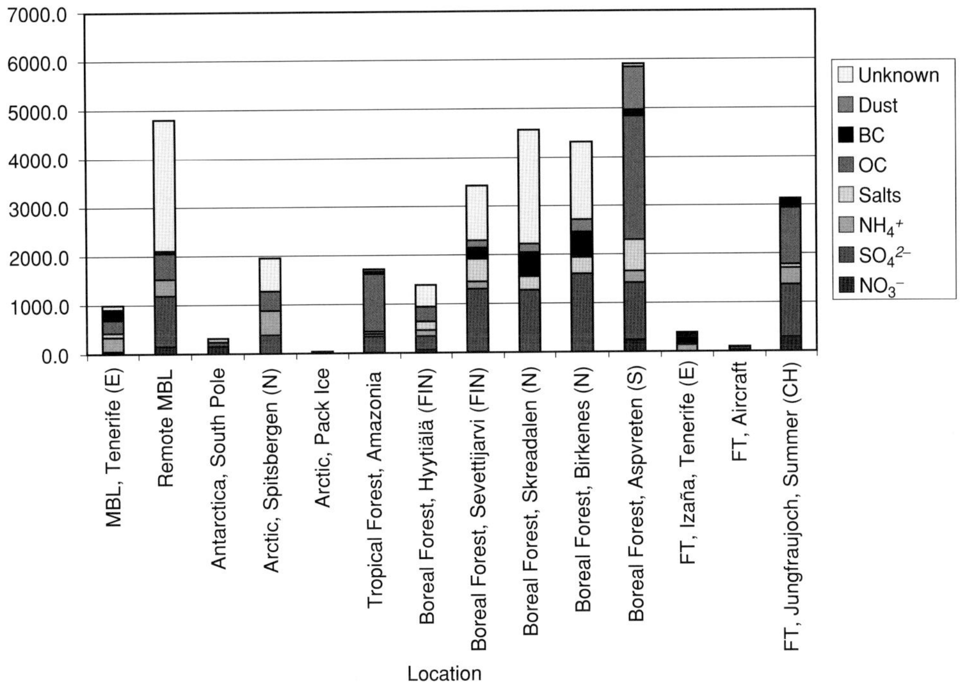

Figure 8.2 Background fine aerosol concentrations and composition for selected remote locations. Fine aerosol is defined as PM_1 for MBL and pack-ice measurements, and in Hyytiälä and Izaña, as PM_2 at Aspvreten, as PM_5 in the Arctic and as $PM_{2.5}$ at all other stations. Organic acids have been added to the OC fraction in the Amazon Basin, at Izaña and for the MBL measurements. They were not separately determined at the other places.

given in Table 8.1. It should furthermore be noted that the data were obtained with different procedures, which may be the cause for some differences found between the stations. Clearly, more emphasis on harmonised procedures worldwide is needed. In addition, concentrations can vary by up to three orders of magnitude such that the situation may be quite different for a specific episode. These huge variations make it especially difficult to calculate trends, such that often a time series of more than a decade is needed before a reliable trend can be calculated. The mere chemical measurements may not be sufficient to discriminate between different sources, which is best exemplified for OC. Here, new instrumentation brings along new possibilities for a better source apportionment of the various sources contributing to the aerosol load of a specific site.

Acknowledgements

This work was supported by the European commission project ACCENT, The European Network of Excellence on Atmospheric Composition Change.

References

1 Wikipedia. Remote. http://en.wikipedia.org/wiki/Remote. Accessed 13 March 2006.

2 Merriam-Webster. Remote. http://www.m-w.com/dictionary/remote. Accessed 13 March 2006.

3 Collaud Coen, M., Weingartner, E., Schaub, D., Hueglin, C., Corrigan, C., Henning, S., Schwikowski, M. and Baltensperger, U. Saharan dust events at the Jungfraujoch: detection by wavelength dependence of the single scattering albedo and first climatology analysis. *Atmospheric Chemistry and Physics* 2004; **4**: 5547–5594.

4 Lugauer, M., Baltensperger, U., Furger, M., Gaggeler, H.W., Jost, D.T., Schwikowski, M. and Wanner, H. Aerosol transport to the high Alpine sites Jungfraujoch (3454 m asl) and Colle Gnifetti (4452 m asl). *Tellus Series B –Chemical and Physical Meteorology* 1998; **50**(1): 76–92.

5 Putaud, J.P., Van Dingenen, R., Mangoni, M., Virkkula, A., Raes, F., Maring, H., Prospero, J.M., Swietlicki, E., Berg, O.H., Hillamo, R. and Makela, T. Chemical mass closure and assessment of the origin of the submicron aerosol in the marine boundary layer and the free troposphere at Tenerife during ACE 2. *Tellus B* 2000; **52**(2): 141–168.

6 Graham, B., Guyon, P., Maenhaut, W., Taylor, P.E., Ebert, M., Matthias-Maser, S., Mayol-Bracero, O.L., Godoi, R.H.M., Artaxo, P., Meixner, F.X., Moura, M.A.L., Rocha, C., Van Grieken, R., Glovsky, M.M., Flagan, R.C. and Andreae, M.O. Composition and diurnal variability of the natural Amazonian aerosol. *Journal of Geophysical Research – Atmospheres* 2003; **108**(D24): 4765, doi: 10.1029/2003JD004049.

7 Bates, T.S., Quinn, P.K., Coffman, D.J., Johnson, J.E., Miller, T.L., Covert, D.S., Wiedensohler, A., Leinert, S., Nowak, A. and Neusüss, C. Regional physical and chemical properties of the marine boundary layer aerosol across the Atlantic during Aerosols99: an overview. *Journal of Geophysical Research* 2001; **106**(D18): 20767–20782.

8 Raes, F., Van Dingenen, R., Vignati, E., Wilson, J., Putaud, J.P., Seinfeld, J.H. and Adams, P. Formation and cycling of aerosols in the global troposphere. *Atmospheric Environment* 2000; **34**(25): 4215–4240.

9 Jayne, J.T., Leard, D.C., Zhang, X.F., Davidovits, P., Smith, K.A., Kolb, C.E. and Worsnop, D.R. Development of an aerosol mass spectrometer for size and composition analysis of submicron particles. *Aerosol Science and Technology* 2000; **33**(1–2): 49–70.

10 Gard, E., Mayer, J.E., Morrical, B.D., Dienes, T., Fergenson, D.P. and Prather, K.A. Real-time analysis of individual atmospheric aerosol particles: design and performance of a portable ATOFMS. *Analytical Chemistry* 1997; **69**(20): 4083–4091.

11 WMO. *WMO/GAW Aerosol Measurement Procedures, Guidelines and Recommendations*, no. 153. WMO TD no. 1178. World Meteorological Organization, Geneva: 2003; p. 67. Available from http://www.wmo.ch/pages/prog/arep/gaw/gaw-reports.html. Accessed 5 September 2007.

12 Riemer, N., Doherty, O.M. and Hameed, S. On the variability of African dust transport across the Atlantic. *Geophysical Research Letters* 2006; **33**(13): L13814, doi: 10.1029/2006GL026163.

13 Forster, C., Wandinger, U., Wotawa, G., James, P., Mattis, I., Althausen, D., Simmonds, P., O'Doherty, S., Jennings, S.G., Kleefeld, C., Schneider, J., Trickl, T., Kreipl, S., Jager, H. and Stohl, A. Transport of boreal forest fire emissions from Canada to Europe. *Journal of Geophysical Research – Atmospheres* 2001; **106**(D19): 22887–22906.

14 Lowenthal, D.H., Borys, R.D., Chow, J.C., Rogers, F. and Shaw, G.E. Evidence for long-range transport of aerosol from the Kuwaiti oil fires to Hawaii. *Journal of Geophysical Research – Atmospheres* 1992; **97**(D13): 14573–14580.

15 Damoah, R., Spichtinger, N., Forster, C., James, P., Mattis, I., Wandinger, U., Beirle, S., Wagner, T. and Stohl, A. Around the world in 17 days – hemispheric-scale transport of forest fire smoke from Russia in May 2003. *Atmospheric Chemistry and Physics* 2004; **4**: 1311–1321.

16 Anspaugh, L.R., Catlin, R.J. and Goldman, M. The global impact of the Chernobyl reactor accident. *Science* 1988; **242**(4885): 1513–1519.

17 Dusek, U., Frank, G.P., Hildebrandt, L., Curtius, J., Schneider, J., Walter, S., Chand, D., Drewnick, F., Hings, S., Jung, D., Borrmann, S. and Andreae, M.O. Size matters more than chemistry for cloud-nucleating ability of aerosol particles. *Science* 2006; **312**(5778): 1375–1378.
18 Weingartner, E., Nyeki, S. and Baltensperger, U. Seasonal and diurnal variation of aerosol size distributions (10 < D 750 nm) at a high-alpine site (Jungfraujoch 3580 m asl). *Journal of Geophysical Research – Atmospheres* 1999; **104**(D21): 26809–26820.
19 Blando, J.D. and Turpin, B.J. Secondary organic aerosol formation in cloud and fog droplets: a literature evaluation of plausibility. *Atmospheric Environment* 2000; **34**(10): 1623–1632.
20 Graedel, T.E. and Weschler, C.J. Chemistry within aqueous atmospheric aerosols and raindrops. *Reviews of Geophysics* 1981; **19**(4): 505–539.
21 Kalberer, M., Paulsen, D., Sax, M., Steinbacher, M., Dommen, J., Prevot, A.S.H., Fisseha, R., Weingartner, E., Frankevich, V., Zenobi, R. and Baltensperger, U. Identification of polymers as major components of atmospheric organic aerosols. *Science* 2004; **303**(5664): 1659–1662.
22 Chung, S.H. and Seinfeld, J.H. Global distribution and climate forcing of carbonaceous aerosols. *Journal of Geophysical Research – Atmospheres* 2002; **107**(D19): 4407, doi: 10.1029/2001JD001397.
23 Heard, D.E., Read, K.A., Methven, J., Al-Haider, S., Bloss, W.J., Johnson, G.P., Pilling, M.J., Seakins, P.W., Smith, S.C., Sommariva, R., Stanton, J.C., Still, T.J., Ingham, T., Brooks, B., De Leeuw, G., Jackson, A.V., McQuaid, J.B., Morgan, R., Smith, M.H., Carpenter, L.J., Carslaw, N., Hamilton, J., Hopkins, J.R., Lee, J.D., Lewis, A.C., Purvis, R.M., Wevill, D.J., Brough, N., Green, T., Mills, G., Penkett, S.A., Plane, J.M.C., Saiz-Lopez, A., Worton, D., Monks, P.S., Fleming, Z., Rickard, A.R., Alfarra, M.R., Allan, J.D., Bower, K., Coe, H., Cubison, M., Flynn, M., McFiggans, G., Gallagher, M., Norton, E.G., O'Dowd, C.D., Shillito, J., Topping, D., Vaughan, G., Williams, P., Bitter, M., Ball, S.M., Jones, R.L., Povey, I.M., O'Doherty, S., Simmonds, P.G., Allen, A., Kinnersley, R.P., Beddows, D.C.S., Dall'Osto, M., Harrison, R.M., Donovan, R.J., Heal, M.R., Jennings, S.G., Noone, C. and Spain, G. The North Atlantic Marine Boundary Layer Experiment (NAMBLEX). Overview of the campaign held at Mace Head, Ireland, in summer 2002. *Atmospheric Chemistry and Physics* 2006; **6**: 2241–2272.
24 Heintzenberg, J., Covert, D.C. and Van Dingenen, R. Size distribution and chemical composition of marine aerosols: a compilation and review. *Tellus Series B – Chemical and Physical Meteorology* 2000; **52**(4): 1104–1122.
25 Junker, C., Jennings, S.G. and Cachier, H. Aerosol light absorption in the North Atlantic: trends and seasonal characteristics during the period 1989 to 2003. *Atmospheric Chemistry and Physics* 2006; **6**: 1913–1925.
26 Bates, T.S., Huebert, B.J., Gras, J.L., Griffiths, F.B. and Durkee, P.A. International Global Atmospheric Chemistry (IGAC) project's first aerosol characterization experiment (ACE 1): overview. *Journal of Geophysical Research – Atmospheres* 1998; **103**(D13): 16297–16318.
27 Quinn, P.K., Bates, T.S., Coffman, D.J., Miller, T.L., Johnson, J.E., Covert, D.S., Putaud, J.P., Neususs, C. and Novakov, T. A comparison of aerosol chemical and optical properties from the 1st and 2nd aerosol characterization experiments. *Tellus Series B – Chemical and Physical Meteorology* 2000; **52**(2): 239–257.
28 Huebert, B.J., Bates, T., Russell, P.B., Shi, G.Y., Kim, Y.J., Kawamura, K., Carmichael, G. and Nakajima, T. An overview of ACE-Asia: strategies for quantifying the relationships between Asian aerosols and their climatic impacts. *Journal of Geophysical Research – Atmospheres* 2003; **108**(D23): 8633, doi: 10.1029/2003JD003550.
29 Ramanathan, V., Crutzen, P.J., Lelieveld, J., Mitra, A.P., Althausen, D., Anderson, J., Andreae, M.O., Cantrell, W., Cass, G.R., Chung, C.E., Clarke, A.D., Coakley, J.A., Collins, W.D., Conant, W.C., Dulac, F., Heintzenberg, J., Heymsfield, A.J., Holben, B., Howell, S., Hudson, J., Jayaraman, A., Kiehl, J.T., Krishnamurti, T.N., Lubin, D., McFarquhar, G., Novakov, T., Ogren, J.A., Podgorny, I.A., Prather, K., Priestley, K., Prospero, J.M., Quinn, P.K., Rajeev, K., Rasch, P., Rupert, S., Sadourny, R., Satheesh, S.K., Shaw, G.E., Sheridan, P. and Valero, F.P.J. Indian ocean experiment: an integrated analysis of the

climate forcing and effects of the great Indo-Asian haze. *Journal of Geophysical Research – Atmospheres* 2001; **106**(D22): 28371–28398.
30 O'Dowd, C.D., Hameri, K., Makela, J.M., Pirjola, L., Kulmala, M., Jennings, S.G., Berresheim, H., Hansson, H.C., de Leeuw, G., Kunz, G.J., Allen, A.G., Hewitt, C.N., Jackson, A., Viisanen, Y. and Hoffmann, T. A dedicated study of New Particle Formation and Fate in the Coastal Environment (PARFORCE): overview of objectives and achievements. *Journal of Geophysical Research – Atmospheres* 2002; **107**(D19): 8108, doi: 10.1029/2001JD000555.
31 Sander, R., Keene, W.C., Pszenny, A.A.P., Arimoto, R., Ayers, G.P., Baboukas, E., Cainey, J.M., Crutzen, P.J., Duce, R.A., Honninger, G., Huebert, B.J., Maenhaut, W., Mihalopoulos, N., Turekian, V.C. and Van Dingenen, R. Inorganic bromine in the marine boundary layer: a critical review. *Atmospheric Chemistry and Physics* 2003; **3**: 1301–1336.
32 IPCC. *Third Assessment Report – Climate Change 2001*. Cambridge University Press, Cambridge: 2001.
33 Bates, T.S., Kiene, R.P., Wolfe, G.V., Matrai, P.A., Chavez, F.P., Buck, K.R., Blomquist, B.W. and Cuhel, R.L. The cycling of sulfur in surface seawater of the Northeast Pacific. *Journal of Geophysical Research – Oceans* 1994; **99**(C4): 7835–7843.
34 Legrand, M., Sciare, J., Jourdain, B. and Genthon, C. Subdaily variations of atmospheric dimethylsulfide, dimethylsulfoxide, methanesulfonate, and non-sea-salt sulfate aerosols in the atmospheric boundary layer at Dumont d'Urville (coastal Antarctica) during summer. *Journal of Geophysical Research – Atmospheres* 2001; **106**(D13): 14409–14422.
35 Bardouki, H., Berresheim, H., Vrekoussis, M., Sciare, J., Kouvarakis, G., Oikonomou, K., Schneider, J. and Mihalopoulos, N. Gaseous (DMS, MSA, SO_2, H_2SO_4 and DMSO) and particulate (sulfate and methanesulfonate) sulfur species over the northeastern coast of Crete. *Atmospheric Chemistry and Physics* 2003; **3**: 1871–1886.
36 Talbot, R.W., Andreae, M.O., Berresheim, H., Artaxo, P., Garstang, M., Harriss, R.C. and Beecher, K.M. Aerosol chemistry during the wet season in Central Amazonia: the influence of long-range transport. *Journal of Geophysical Research* 1990; **95**(D10): 16955–16969.
37 Jaffe, D., McKendry, I., Anderson, T. and Price, H. Six 'new' episodes of trans-Pacific transport of air pollutants. *Atmospheric Environment* 2003; **37**(3): 391–404.
38 Quinn, P.K., Coffman, D.J., Kapustin, V.N., Bates, T.S. and Covert, D.S. Aerosol optical properties in the marine boundary layer during the first aerosol characterization experiment (ACE 1) and the underlying chemical and physical aerosol properties. *Journal of Geophysical Research – Atmospheres* 1998; **103**(D13): 16547–16563.
39 Bates, T.S., Quinn, P.K., Coffman, D.J., Covert, D.S., Miller, T.L., Johnson, J.E., Carmichael, G.R., Uno, I., Guazzotti, S.A., Sodeman, D.A., Prather, K.A., Rivera, M., Russell, L.M. and Merrill, J.T. Marine boundary layer dust and pollutant transport associated with the passage of a frontal system over eastern Asia. *Journal of Geophysical Research – Atmospheres* 2004; **109**: D19S19, doi: 10.1029/2003JD004094.
40 Dall'Osto, M., Beddows, D.C.S., Kinnersley, R.P., Harrison, R.M., Donovan, R.J. and Heal, M.R. Characterization of individual airborne particles by using aerosol time-of-flight mass spectrometry at Mace Head, Ireland. *Journal of Geophysical Research – Atmospheres* 2004; **109**: D21302, doi: 10.1029/2004JD004747.
41 Liu, D.Y., Wenzel, R.J. and Prather, K.A. Aerosol time-of-flight mass spectrometry during the Atlanta Supersite Experiment: 1. Measurements. *Journal of Geophysical Research – Atmospheres* 2003; **108**(D7): 8427, doi: 10.1029/2001JD001563.
42 Raes, F. Entrainment of free tropospheric aerosols as a regulating mechanism for cloud condensation nuclei in the remote marine boundary-layer. *Journal of Geophysical Research – Atmospheres* 1995; **100**(D2): 2893–2903.
43 Pakkanen, T.A. Study of formation of coarse particle nitrate aerosol. *Atmospheric Environment* 1996; **30**(14): 2475–2482.

44 Andreae, M.O. and Gelencser, A. Black carbon or brown carbon? The nature of light-absorbing carbonaceous aerosols. *Atmospheric Chemistry and Physics* 2006; **6**: 3131–3148.
45 Sharma, S., Andrews, E., Barrie, L.A., Ogren, J.A. and Lavoué, D. Variations and sources of the equivalent black carbon in the high Arctic revealed by long-term observations at Alert and Barrow: 1989–2003. *Journal of Geophysical Research* 2006; **111**(D14): D14208, doi: 10.1029/2005JD006581.
46 Quinn, P.K., Coffman, D.J., Bates, T.S., Welton, E.J., Covert, D.S., Miller, T.L., Johnson, J.E., Maria, S., Russell, L., Arimoto, R., Carrico, C.M., Rood, M.J. and Anderson, J. Aerosol optical properties measured on board the Ronald H. Brown during ACE-Asia as a function of aerosol chemical composition and source region. *Journal of Geophysical Research* 2004; **109**(D19): 1–28.
47 Turekian, V.C., Macko, S.A. and Keene, W.C. Concentrations, isotopic compositions, and sources of size-resolved, particulate organic carbon and oxalate in near-surface marine air at Bermuda during spring. *Journal of Geophysical Research – Atmospheres* 2003; **108**(D5): 4157, doi: 10.1029/2002JD002053.
48 O'Dowd, C.D., Facchini, M.C., Cavalli, F., Ceburnis, D., Mircea, M., Decesari, S., Fuzzi, S., Yoon, Y.J. and Putaud, J.P. Biogenically driven organic contribution to marine aerosol. *Nature* 2004; **431**(7009): 676–680.
49 Jickells, T.D., An, Z.S., Andersen, K.K., Baker, A.R., Bergametti, G., Brooks, N., Cao, J.J., Boyd, P.W., Duce, R.A., Hunter, K.A., Kawahata, H., Kubilay, N., laRoche, J., Liss, P.S., Mahowald, N., Prospero, J.M., Ridgwell, A.J., Tegen, I. and Torres, R. Global iron connections between desert dust, ocean biogeochemistry, and climate. *Science* 2005; **308**(5718): 67–71.
50 Niemi, J.V., Tervahattu, H., Virkkula, A., Hillamo, R., Teinila, K., Koponen, I.K. and Kulmala, M. Continental impact on marine boundary layer coarse particles over the Atlantic Ocean between Europe and Antarctica. *Atmospheric Research* 2005; **75**(4): 301–321.
51 Witt, M., Baker, A.R. and Jickells, T.D. Atmospheric trace metals over the Atlantic and South Indian oceans: investigation of metal concentrations and lead isotope ratios in coastal and remote marine aerosols. *Atmospheric Environment* 2006; **40**(28): 5435–5451.
52 Ayers, G.P., Gillett, R.W., Cainey, J.M. and Dick, A.L. Chloride and bromide loss from sea-salt particles in southern ocean air. *Journal of Atmospheric Chemistry* 1999; **33**(3): 299–319.
53 O'Dowd, C.D., Jimenez, J.L., Bahreini, R., Flagan, R.C., Seinfeld, J.H., Hameri, K., Pirjola, L., Kulmala, M., Jennings, S.G. and Hoffmann, T. Marine aerosol formation from biogenic iodine emissions. *Nature* 2002; **417**(6889): 632–636.
54 Saiz-Lopez, A., Shillito, J.A., Coe, H. and Plane, J.M.C. Measurements and modelling of I2, IO, OIO, BrO and NO3 in the mid-latitude marine boundary layer. *Atmospheric Chemistry and Physics* 2006; **6**: 1513–1528.
55 Shaw, G.E. Antarctic aerosols – a review. *Reviews of Geophysics* 1988; **26**(1): 89–112.
56 Hainsworth, A.H.W., Dick, A.L. and Gras, J.L. Climatic context of the first aerosol characterization experiment (ACE 1): a meteorological and chemical overview. *Journal of Geophysical Research – Atmospheres* 1998; **103**(D13): 16319–16340.
57 Berresheim, H. and Eisele, F.L. Sulfur chemistry in the Antarctic troposphere experiment: an overview of project SCATE. *Journal of Geophysical Research – Atmospheres* 1998; **103**(D1): 1619–1627.
58 Arimoto, R., Nottingham, A.S., Webb, J., Schloesslin, C.A. and Davis, D.D. Non-sea salt sulfate and other aerosol constituents at the South Pole during ISCAT. *Geophysical Research Letters* 2001; **28**(19): 3645–3648.
59 Davis, D.D., Eisele, F., Chen, G., Crawford, J., Huey, G., Tanner, D., Slusher, D., Mauldin, L., Oncley, S. and Lenschow, D. An overview of ISCAT 2000. *Atmospheric Environment* 2004; **38**(32): 5363–5373.
60 Bodhaine, B.A., Deluisi, J.J., Harris, J.M., Houmere, P. and Bauman, S. Aerosol measurements at the South Pole. *Tellus* 1986; **38B**: 223–235.
61 Cunningham, W.C. and Zoller, W.H. The chemical composition of remote area aerosols. *Journal of Aerosol Science* 1981; **12**(4): 367–384.

62 Mazzera, D.M., Lowenthal, D.H., Chow, J.C. and Watson, J.G. Sources of PM_{10} and sulfate aerosol at McMurdo station, Antarctica. *Chemosphere* 2001; **45**(3): 347.
63 Tuncel, G., Aras, N.K. and Zoller, W.H. Temporal variations and sources of elements in the south-pole atmosphere. 1. Nonenriched and moderately enriched elements. *Journal of Geophysical Research – Atmospheres* 1989; **94**(D10): 13025–13038.
64 Rose, W.I., Kyle, P.R., Chuan, R.L. and Palais, J. Emission rates of sulfur dioxide and particulate material from Mt. Erebus, Ross Island. *Antarctic Journal of the United States* 1984; **19**: 195–196.
65 Zreda-Gostynska, G., Kyle, P.R., Finnegan, D. and Prestbo, K.M. Volcanic gas emissions from Mount Erebus and their impact on the Antarctic environment. *Journal of Geophysical Research* 1997; **102B**(B7): 15039–15056.
66 Rankin, A.M. and Wolff, E.W. Aerosol profiling using a tethered balloon in coastal Antarctica. *Journal of Atmospheric and Oceanic Technology* 2002; **19**(12): 1978–1985.
67 Bodhaine, B.A., Deluisi, J.J., Harris, J.M., Houmere, P. and Bauman, S. Pixe analysis of south-pole aerosol. *Nuclear Instruments & Methods in Physics Research Section B – Beam Interactions With Materials and Atoms* 1987; **22**(1–3): 241–247.
68 Parungo, F., Bodhaine, B. and Bortniak, J. Seasonal-variation in Antarctic aerosol. *Journal of Aerosol Science* 1981; **12**(6): 491–504.
69 Mauldin, R.L., Eisele, F.L., Tanner, D.J., Kosciuch, E., Shetter, R., Lefer, B., Hall, S.R., Nowak, J.B., Buhr, M., Chen, G., Wang, P. and Davis, D. Measurements of OH, H_2SO_4, and MSA at the South Pole during ISCAT. *Geophysical Research Letters* 2001; **28**(19): 3629–3632.
70 Huey, L.G., Tanner, D.J., Slusher, D.L., Dibb, J.E., Arimoto, R., Chen, G., Davis, D., Buhr, M.P. and Nowak, J.B. CIMS measurements of $HNO_3$and SO_2 at the South Pole during ISCAT 2000. *Atmospheric Environment* 2004; **38**(32): 5411–5421.
71 Park, J., Sakurai, H., Vollmers, K. and McMurry, P.H. Aerosol size distributions measured at the South Pole during ISCAT. *Atmospheric Environment* 2004; **38**(32): 5493–5500.
72 Fattori, I., Becagli, S., Bellandi, S., Castellano, E., Innocenti, M., Mannini, A., Severi, M., Vitale, V. and Udisti, R. Chemical composition and physical features of summer aerosol at Terra Nova Bay and Dome C, Antarctica. *Journal of Environmental Monitoring* 2005; **7**(12): 1265–1274.
73 Arimoto, R., Hogan, A., Grube, P., Davis, D., Webb, J., Schloesslin, C., Sage, S. and Raccah, F. Major ions and radionuclides in aerosol particles from the South Pole during ISCAT-2000. *Atmospheric Environment* 2004; **38**: 5473–5484.
74 Legrand, M., Ducroz, F., Wagenbach, D., Mulvaney, R. and Hall, J. Ammonium in coastal Antarctic aerosol and snow: role of polar ocean and penguin emissions. *Journal of Geophysical Research – Atmospheres* 1998; **103**(D9): 11043–11056.
75 Wolff, E.W. and Cachier, H. Concentrations and seasonal cycle of black carbon in aerosol at a coastal Antarctic station. *Journal of Geophysical Research – Atmospheres* 1998; **103**(D9): 11033–11041.
76 Legrand, M., Preunkert, S., Jourdain, B. and Aumont, B. Year-round records of gas and particulate formic and acetic acids in the boundary layer at Dumont d'Urville, coastal Antarctica. *Journal of Geophysical Research* 2004; **109**(D06313): 1–11, doi: 10.1029/2003JD003786.
77 Kawamura, K. Semere, R., Imai, Y., Fujii, Y. and Hayashi, M. Water soluble dicarboxylic acids and related compounds in Antarctic aerosols. *Journal of Geophysical Research – Atmospheres* 1996; **101**(D13): 18721–18728.
78 Arimoto, R., Schloesslin, C., Davis, D., Hogan, A., Grube, P., Fitzgerald, W. and Lamborg, C. Lead and mercury in aerosol particles collected over the South Pole during ISCAT-2000. *Atmospheric Environment* 2004; **38**(32): 5485–5491.
79 Duce, R.A., Zoller, W.H. and Moyers, J.L. Particulate and gaseous halogens in Antarctic atmosphere. *Journal of Geophysical Research* 1973; **78**(33): 7802–7811.
80 Sturges, W.T. and Barrie, L.A. Chlorine, bromine and iodine in arctic aerosols. *Atmospheric Environment* 1988; **22**(6): 1179–1194.

81 Rahn, K.A. Progress in Arctic air chemistry, 1980–1984. *Atmospheric Environment (1967)* 1985; **19**(12): 1987.
82 Shaw, G.E. On the aerosol-particle size distribution spectrum in Alaskan air-mass systems – Arctic haze and non-haze episodes. *Journal of the Atmospheric Sciences* 1983; **40**(5): 1313–1320.
83 Yamanouchi, T., Treffeisen, R., Herber, A., Shiobara, M., Yamagata, S., Hara, K., Sato, K., Yabuki, M., Tomikawa, Y., Rinke, A., Neuber, R., Schumachter, R., Kriews, M., Strom, J., Schrems, O. and Gernandt, H. Arctic study of tropospheric aerosol and radiation (ASTAR) 2000: Arctic haze case study. *Tellus Series B – Chemical and Physical Meteorology* 2005; **57**(2): 141–152.
84 Hileman, B. Arctic haze. *Environmental Science & Technology* 1983; **17**(6): A232–A236.
85 Shaw, G.E. Chemical air-mass systems in Alaska. *Atmospheric Environment* 1988; **22**(10): 2239–2248.
86 Maenhaut, W., Ducastel, G., Leck, C., Nilsson, E.D. and Heintzenberg, J. Multi-elemental composition and sources of the high Arctic atmospheric aerosol during summer and autumn. *Tellus Series B – Chemical and Physical Meteorology* 1996; **48**(2): 300–321.
87 Polissar, A.V., Hopke, P.K., Paatero, P., Kaufmann, Y.J., Hall, D.K., Bodhaine, B.A., Dutton, E.G. and Harris, J.M. The aerosol at Barrow, Alaska: long-term trends and source locations. *Atmospheric Environment* 1999; **33**(16): 2441–2458.
88 Daisey, J.M., McCaffrey, R.J. and Gallagher, R.A. Polycyclic aromatic hydrocarbons and total extractable particulate organic matter in the Arctic aerosol. *Atmospheric Environment (1967)* 1981; **15**(8): 1353.
89 Gregory, G.L., Anderson, B.E., Warren, L.S., Browell, E.V., Bagwell, D.R. and Hudgins, C.H. Tropospheric ozone and aerosol observations – the Alaskan Arctic. *Journal of Geophysical Research – Atmospheres* 1992; **97**(D15): 16451–16471.
90 Shaw, G.E. The arctic haze phenomenon. *Bulletin of the American Meteorological Society* 1995; **76**(12): 2403–2413.
91 Mitchell, J.M., Jr. Visual range in the polar regions with particular reference to the Alaskan Arctic. *Atmospheric and Terrestrial Physics* 1957; (Special Supplement): 195–211.
92 Staebler, R., Toom-Sauntry, D., Barrie, L., Langendorfer, U., Lehrer, E., Li, S.M. and Dryfhout-Clark, H. Physical and chemical characteristics of aerosols at Spitsbergen in spring of 1996. *Journal of Geophysical Research –Atmospheres* 1999; **104**(D5): 5515–5529.
93 Covert, D.S. and Heintzenberg, J. Size distributions and chemical-properties of aerosol at Ny Alesund, Svalbard. *Atmospheric Environment Part A – General Topics* 1993; **27**(17–18): 2989–2997.
94 Hara, K., Osada, K., Hayashi, M., Matsunaga, K., Shibata, T., Iwasaka, Y. and Furuya, K. Fractionation of inorganic nitrates in winter Arctic troposphere: coarse aerosol particles containing inorganic nitrates. *Journal of Geophysical Research – Atmospheres* 1999; **104**(D19): 23671–23679.
95 Barrie, L.A., Li, S.M., Toom, D.L., Landsberger, S. and Sturges, W. Lower tropospheric measurements of halogens, nitrates, and sulfur-oxides during polar sunrise experiment 1992. *Journal of Geophysical Research –Atmospheres* 1994; **99**(D12): 25453–25467.
96 Heintzenberg, J., Hansson, H.C. and Lannefors, H. The chemical-composition of Arctic haze at Ny-Alesund, Spitsbergen. *Tellus* 1981; **33**(2): 162–171.
97 Shotyk, W., Zheng, J.C., Krachler, M., Zdanowicz, C., Koerner, R. and Fisher, D. Predominance of industrial Pb in recent snow (1994–2004) and ice (1842–1996) from Devon Island, Arctic Canada. *Geophysical Research Letters* 2005; **32**(21): L21814, doi: 10.1029/2005GL023860.
98 Hara, K., Osada, K., Matsunaga, K., Iwasaka, Y., Shibata, T. and Furuya, K. Atmospheric inorganic chlorine and bromine species in Arctic boundary layer of the winter/spring. *Journal of Geophysical Research – Atmospheres* 2002; **107**(D18): 4361, doi: 10.1029/2001JD001008.
99 Whitehead, D.C. The distribution and transformations of iodine in the environment. *Environment International* 1984; **10**(4): 321–339.
100 Harriss, R.C., Wofsy, S.C., Garstang, M., Browell, E.V., Molion, L.C.B., McNeal, R.J., Hoell, J.M., Bendura, R.J., Beck, S.M., Navarro, R.L., Riley, J.T. and Snell, R.L. The Amazon boundary-layer

experiment (Able-2a) – Dry Season 1985. *Journal of Geophysical Research – Atmospheres* 1988; **93**(D2): 1351–1360.

101 Andreae, M.O., Artaxo, P., Brandao, C., Carswell, F.E., Ciccioli, P., da Costa, A.L., Culf, A.D., Esteves, J.L., Gash, J.H.C., Grace, J., Kabat, P., Lelieveld, J., Malhi, Y., Manzi, A.O., Meixner, F.X., Nobre, A.D., Nobre, C., Ruivo, M., Silva-Dias, M.A., Stefani, P., Valentini, R., von Jouanne, J. and Waterloo, M.J. Biogeochemical cycling of carbon, water, energy, trace gases, and aerosols in Amazonia: the LBA-EUSTACH experiments. *Journal of Geophysical Research – Atmospheres* 2002; **107**(D20): 8066, doi: 10.1029/2001JD000524.

102 Andreae, M.O., Rosenfeld, D., Artaxo, P., Costa, A.A., Frank, G.P., Longo, K.M. and Silva-Dias, M.A.F. Smoking rain clouds over the Amazon. *Science* 2004; **303**(5662): 1337–1342.

103 Guyon, P., Graham, B., Roberts, G.C., Mayol-Bracero, O.L., Maenhaut, W., Artaxo, P. and Andreae, M.O. In-canopy gradients, composition, sources, and optical properties of aerosol over the Amazon forest. *Journal of Geophysical Research* 2003; **108**(D18): 4591, doi:10.1029/2003JD003465.

104 Formenti, P., Andreae, M.O., Lange, L., Roberts, G., Cafmeyer, J., Rajta, I., Maenhaut, W., Holben, B.N., Artaxo, P. and Lelieveld, J. Saharan dust in Brazil and Suriname during the large-scale biosphere-atmosphere experiment in Amazonia (LBA) – cooperative LBA regional experiment (CLAIRE) in March 1998. *Journal of Geophysical Research – Atmospheres* 2001; **106**(D14): 14919–14934.

105 Andreae, M.O., Berresheim, H., Bingemer, H., Jacob, D.J., Lewis, B.L., Li, S.M. and Talbot, R.W. The atmospheric sulfur cycle over the Amazon Basin. 2. Wet season. *Journal of Geophysical Research – Atmospheres* 1990; **95**(D10): 16813–16824.

106 Yamasoe, M.A., Artaxo, P., Miguel, A.H. and Allen, A.G. Chemical composition of aerosol particles from direct emissions of vegetation fires in the Amazon Basin: water-soluble species and trace elements. *Atmospheric Environment* 2000; **34**(10): 1641–1653.

107 Artaxo, P., Maenhaut, W., Storms, H. and Vangrieken, R. Aerosol characteristics and sources for the Amazon basin during the wet season. *Journal of Geophysical Research – Atmospheres* 1990; **95**(D10): 16971–16985.

108 Kulmala, M., Hari, P., Laaksonen, A., Vesala, T. and Viisanen, Y. Research unit of physics, chemistry and biology of atmospheric composition and climate change: overview of recent results. *Boreal Environment Research* 2005; **10**(6): 459–477.

109 Sun, J. and Ariya, P.A. Atmospheric organic and bio-aerosols as cloud condensation nuclei (CCN): a review. *Atmospheric Environment* 2006; **40**(5): 795–820.

110 Kulmala, M., Vehkamaki, H., Petajda, T., Dal Maso, M., Lauri, A., Kerminen, V.M., Birmili, W. and McMurry, P.H. Formation and growth rates of ultrafine atmospheric particles: a review of observations. *Journal of Aerosol Science* 2004; **35**(2): 143–176.

111 Cavalli, F., Facchini, M.C., Decesari, S., Emblico, L., Mircea, M., Jensen, N.R. and Fuzzi, S. Size-segregated aerosol chemical composition at a boreal site in southern Finland, during the QUEST project. *Atmospheric Chemistry and Physics* 2006; **6**: 993–1002.

112 Lyubovtseva, Y.S., Sogacheva, L., Dal Maso, M., Bonn, B., Keronen, P. and Kulmala, M. Seasonal variations of trace gases, meteorological parameters, and formation of aerosols in boreal forests. *Boreal Environment Research* 2005; **10**(6): 493–510.

113 Kulmala, M., Hameri, K., Aalto, P.P., Makela, J.M., Pirjola, L., Nilsson, E.D., Buzorius, G., Rannik, U., Dal Maso, M., Seidl, W., Hoffman, T., Janson, R., Hansson, H.C., Viisanen, Y., Laaksonen, A. and O'Dowd, C.D. Overview of the international project on biogenic aerosol formation in the boreal forest (BIOFOR). *Tellus Series B –Chemical and Physical Meteorology* 2001; **53**(4): 324–343.

114 Tunved, P., Hansson, H.C., Kerminen, V.M., Strom, J., Dal Maso, M., Lihavainen, H., Viisanen, Y., Aalto, P.P., Komppula, M. and Kulmala, M. High natural aerosol loading over boreal forests. *Science* 2006; **312**(5771): 261–263.

115 Bigg, E.K. Gas emissions from soils and leaf litter as a source of new particle formation. *Atmospheric Research* 2004; **70**(1): 33–42.

116 Nilsson, E.D., Paatero, J. and Boy, M. Effects of air masses and synoptic weather on aerosol formation in the continental boundary layer. *Tellus Series B – Chemical and Physical Meteorology* 2001; **53**(4): 462–478.
117 Allan, J.D., Alfarra, M.R., Bower, K.N., Coe, H., Jayne, J.T., Worsnop, D.R., Aalto, P.P., Kulmala, M., Hyotylainen, T., Cavalli, F. and Laaksonen, A. Size and composition measurements of background aerosol and new particle growth in a Finnish forest during QUEST 2 using an Aerodyne aerosol mass spectrometer. *Atmospheric Chemistry and Physics* 2006; **6**: 315–327.
118 Cavalli, F., Facchini, M.C., Decesari, S., Mircea, M., Emblico, L., Fuzzi, S., Ceburnis, D., Yoon, Y.J., O'Dowd, C.D., Putaud, J.P. and Dell'Acqua, A. Advances in characterization of size-resolved organic matter in aerosol over the North Atlantic. *Journal of Geophysical Research – Atmospheres* 2004; **109**: D24215, doi: 10.1029/2004JD005137.
119 Janson, R., Rosman, K., Karlsson, A. and Hansson, H.C. Biogenic emissions and gaseous precursors to forest aerosols. *Tellus Series B – Chemical and Physical Meteorology* 2001; **53**(4): 423–440.
120 Kourtchev, I., Ruuskanen, T., Maenhaut, W., Kulmala, M. and Claeys, M. Observation of 2-methyltetrols and related photo-oxidation products of isoprene in boreal forest aerosols from Hyytiala, Finland. *Atmospheric Chemistry and Physics* 2005; **5**: 2761–2770.
121 Seibert, P., Kromp-Kolb, H., Kasper, A., Kalina, M., Puxbaum, H., Jost, D.T., Schwikowski, M. and Baltensperger, U. Transport of polluted boundary layer air from the Po Valley to high-alpine sites. *Atmospheric Environment* 1998; **32**(23): 3953–3965.
122 Clarke, A.D., Kapustin, V.N., Eisele, F.L., Weber, R.J. and McMurry, P.H. Particle production near marine clouds: sulfuric acid and predictions from classical binary nucleation. *Geophysical Research Letters* 1999; **26**(16): 2425–2428.
123 Huebert, B.J. and Lazrus, A.L. Bulk composition of aerosols in the remote troposphere. *Journal of Geophysical Research – Oceans and Atmospheres* 1980; **85**(NC12): 7337–7344.
124 Henne, S., Furger, M., Nyeki, S., Steinbacher, M., Neininger, B., de Wekker, S.F.J., Dommen, J., Spichtinger, N., Stohl, A. and Prévôt, A.S.H. Quantification of topographic venting of boundary layer air to the free troposphere. *Atmospheric Chemistry and Physics* 2004; **4**: 497–509.
125 Nyeki, S., Kalberer, M., Colbeck, I., de Wekker, S., Furger, M., Gäggeler, H.W., Kossmann, M., Lugauer, M., Steyn, D.G., Weingartner, E., Wirth, M. and Baltensperger, U. Convective boundary layer evolution to 4 km asl over high-alpine terrain: airborne lidar observation in the Alps. *Geophysical Research Letters* 2000; **27**(5): 689–692.
126 Kasper, A. and Puxbaum, H. Seasonal variation of SO_2, HNO_3, NH_3 and selected aerosol components at Sonnblick (3106 m asl). *Atmospheric Environment* 1998; **32**(23): 3925–3939.
127 Henning, S., Weingartner, E., Schwikowski, M., Gaggeler, H.W., Gehrig, R., Hinz, K.P., Trimborn, A., Spengler, B. and Baltensperger, U. Seasonal variation of water-soluble ions of the aerosol at the high-alpine site Jungfraujoch (3580 m asl). *Journal of Geophysical Research – Atmospheres* 2003; **108**(D1): 4030, doi: 10.1029/2002JD002439.
128 Galasyn, J.F., Tschudy, K.L. and Huebert, B.J. Seasonal and diurnal variability of nitric-acid vapor and ionic aerosol species in the remote free troposphere at Mauna-Loa, Hawaii. *Journal of Geophysical Research – Atmospheres* 1987; **92**(D3): 3105–3113.
129 Huebert, B.J. and Lazrus, A.L. Tropospheric gas-phase and particulate nitrate measurements. *Journal of Geophysical Research – Oceans and Atmospheres* 1980; **85**(NC12): 7322–7328.
130 Zellweger, C., Forrer, J., Hofer, P., Nyeki, S., Schwarzenbach, B., Weingartner, E., Ammann, M. and Baltensperger, U. Partitioning of reactive nitrogen (NO_y) and dependence on meteorological conditions in the lower free troposphere. *Atmospheric Chemistry and Physics* 2003; **3**: 779–796.
131 Krivacsy, Z., Hoffer, A., Sarvari, Z., Temesi, D., Baltensperger, U., Nyeki, S., Weingartner, E., Kleefeld, S. and Jennings, S.G. Role of organic and black carbon in the chemical composition of atmospheric aerosol at European background sites. *Atmospheric Environment* 2001; **35**(36): 6231.
132 Alfarra, M.R. *Insights into Atmospheric Organic Aerosols Using an Aerosol Mass Spectrometer*. University of Manchester, Manchester: 2004.

133 Putaud, J.-P., Raes, F., Van Dingenen, R., Bruggemann, E., Facchini, M.C., Decesari, S., Fuzzi, S., Gehrig, R., Huglin, C., Laj, P., Lorbeer, G., Maenhaut, W., Mihalopoulos, N., Müller, K., Querol, X., Rodriguez, S., Schneider, J., Spindler, G., ten Brink, H., Tørseth, K. and Wiedensohler, A. A European aerosol phenomenology – 2: chemical characteristics of particulate matter at kerbside, urban, rural and background sites in Europe. *Atmospheric Environment* 2004; **38**(16): 2579–2595.

134 Heald, C.L., Jacob, D.J., Park, R.J., Russell, L.M., Huebert, B.J., Seinfeld, J., Liao, H. and Weber, R.J. A large organic aerosol source in the free troposphere missing from current models. *Geophysical Research Letters* 2005; **32**: L18809, doi: 10.1029/2005GL023831.

135 Holmes, J., Samberg, T., McInnes, L., Zieman, J., Zoller, W. and Harris, J. Long-term aerosol and trace acidic gas collection at Mauna Loa Observatory 1979–1991. *Journal of Geophysical Research – Atmospheres* 1997; **102**(D15): 19007–19019.

136 Prospero, J.M. and Carlson, T.N. Radon-222 in North Atlantic trade winds: its relationship to dust transport from Africa. *Science* 1970; **167**(3920): 974–977.

137 Lugauer, M., Baltensperger, U., Furger, M., Gaggeler, H.W., Jost, D.T., Nyeki, S. and Schwikowski, M. Influences of vertical transport and scavenging on aerosol particle surface area and radon decay product concentrations at the Jungfraujoch (3454 m above sea level). *Journal of Geophysical Research – Atmospheres* 2000; **105**(D15): 19869–19879.

138 Maring, H., Savoie, D.L., Izaguirre, M.A., McCormick, C., Arimoto, R., Prospero, J.M. and Pilinis, C. Aerosol physical and optical properties and their relationship to aerosol composition in the free troposphere at Izana, Tenerife, Canary Islands, during July 1995. *Journal of Geophysical Research – Atmospheres* 2000; **105**(D11): 14677–14700.

Index